Mechanics Over Micro and Nano Scales

Suman Chakraborty
Editor

Mechanics Over Micro and Nano Scales

Springer

Editor
Suman Chakraborty
Department of Mechanical Engineering
Indian Institute of Technology (IIT)
Kharagpur, India
suman@mech.iitkgp.ernet.in

ISBN 978-1-4419-9600-8 e-ISBN 978-1-4419-9601-5
DOI 10.1007/978-1-4419-9601-5
Springer New York Dordrecht Heidelberg London

Library of Congress Control Number: 2011928701

Printed on acid-free paper

Springer is part of Springer Science+Business Media (www.springer.com)

Preface

The science and technology of micro-scale devices and applications have received unprecedented levels of attention from the research community, at large, especially in the last couple of decades, thanks largely due to the advancements in micro- and nano-fabrication techniques. These advancements have made it possible to translate mental ideas and theoretical predictions into physical reality. From another perspective, the attention afforded to this micro-realm is motivated by a slew of novel applications particularly in bioengineering, electronics, and electromechanical fields. This is, of course, over and above the primary scientific motivation of delving deeper into the microscopic realm with the aim of uncovering new phenomena coupled with a deep and fundamental understanding behind them. Certain areas of this broad micro-realm, primarily electronics, have already developed to a mature stage; others remain mostly confined to the abstruse domains in the higher echelons of research labs. Still others are at a crucial junction of graduation from adolescence to a mature and well-established technology. The most striking feature in the scientific bases of all these areas is that they require the application of fundamental knowledge from diverse backgrounds so that any comprehensive study requires a truly interdisciplinary approach. Finally, as is true for any scientific discipline, the ultimate onus of the development of this micro-scale realm lies with the pool of human talent devoted to it. As micro-scale research and development grows and threads out into niche areas, an ever-increasing number of well-trained researchers are required to handle the burgeoning volume of study areas and to explore their potential applications. As the subject grows in its applications, continuous challenges are faced towards exploring newer scientific facets and developing deeper fundamental understanding for transforming research dreams over miniaturized scales into a practical reality.

This book is written to address, primarily, this issue of expanding and nourishing the talent pool geared toward the development of micro- and nano-scale research. The seven chapters comprising this book are mostly based on the invited lectures presented by various renowned speakers at the Indo-US Short Term Course: *Mechanics Over Micro- and Nano-scales* which was organized jointly by Indian Institute of Technology Kharagpur, India, and Bengal Engineering and Science University, India, during December 21–22, 2009. In keeping with the spirit of these lecture sessions, the book chapters are aimed at the senior undergraduate and graduate student to first expose him/her to various exciting sub-areas and then to apprise him/her of the fundamental issues involved in micro- and nano-scale research. The term "mechanics" in the present context should not be viewed in a narrow sense as it pertains to the well-known standard undergraduate curriculum. Here, "mechanics" encompasses the broad mechanistic approach that is necessary to elucidate the fundamentals of the pertinent areas borrowing heavily, as and when necessary, from numerous physical and chemical concepts. In this respect, this book aims to augment the specialized training received by an individual student from his undergraduate training to prepare him well for more sophisticated work in these interdisciplinary areas. Indeed, the common thread coursing through all the chapters is the pedagogical stance adopted by each author. The editor and the team of authors have taken special care not to fall into the temptation of discussing, at length, their pet research areas and highlighting recent major advances. Rather, emphasis has been laid to highlight the essential fundamentals that are necessary to kindle deeper research thoughts. Nevertheless, references to ongoing research topics have been suitably placed in the text to highlight motivation for a particular kind of study and to keep the study material practically relevant.

The editor wholeheartedly thanks all the contributing authors who, in spite of their busy work schedules, managed to meet the deadlines within acceptable tolerances! The relentless efforts of Mr. Steven Elliot from Springer ensured that the intricate logistics of the publication procedure never got derailed from the initially planned time frame of completion. The editor also expresses his sincere gratitude to the 'Indo-US Centre for Research Excellence in Fabrionics' for financing the lecture series that acted as a prelude to this edited volume. A special note of thanks is due to Dr. Arabinda Mitra, executive director, Indo-US Science and Technology Forum, for his support in creating the center. The editor also acknowledges the immense help from all his research students in general, and Mr. Jeevanjyoti Chakraborty in particular, for working on various aspects of this book. The editor also acknowledges the continuous mental support that he has been receiving from his parents and his wife, without which this project could have never been materialized. Last but not the least, the editor wishes to dedicate this book to his son, who has first seen the light of the earth on November 20, 2010.

Kharagpur, India Suman Chakraborty

Contents

Contributors

K. Anantheshwara Department of Mechanical Engineering, Indian Institute of Science, Bangalore, India

M.S. Bobji Department of Mechanical Engineering, Indian Institute of Science, Bangalore, India, bobji@mecheng.iisc.ernet.in

Jeevanjyoti Chakraborty Advanced Technology Development Centre, Indian Institute of Technology (IIT), Kharagpur, India, jeevanjyoti4@gmail.com

Suman Chakraborty Department of Mechanical Engineering, Indian Institute of Technology (IIT), Kharagpur, India, suman@mech.iitkgp.ernet.in

Animangsu Ghatak Department of Chemical Engineering, Indian Institute of Technology (IIT), Kanpur, India, aghatak@iitk.ac.in

Amitabha Ghosh Indian National Science Academy, New Delhi, India; Indian Institute of Technology (IIT), Kanpur, India; Bengal Engineering and Science University, Howrah, India, amitabha@iitk.ac.in

U.B. Jayadeep Department of Mechanical Engineering, Indian Institute of Science, Bangalore, India

Qinjun Kang Los Alamos National Laboratory, Los Alamos, NM, USA

Partha P. Mukherjee Computer Science and Mathematics Division, Oak Ridge National Laboratory, Oak Ridge, TN, USA, mukherjeepp@ornl.gov

Rabibrata Mukherjee Department of Chemical Engineering, Indian Institute of Technology (IIT), Kharagpur, India, rabibrata@iitkgp.ac.in

Puneet K. Sinha Electrochemical Energy Research Laboratory, General Motors Global R&D, Honeoye Falls, NY, USA, puneet.sinha@gm.com

Chapter 1
Some Fundamental Aspects of Fluid Mechanics over Microscopic Length Scales

Jeevanjyoti Chakraborty and Suman Chakraborty

Abstract In this chapter we recapitulate some of the fundamental theories of fluid flow at the micro-scale and then introduce the theoretical models of various phenomena which enrich the mechanics of this fluid flow. Specifically, we concentrate on the rudiments of electrokinetics, surface tension, non-Newtonian fluid, and acoustofluidics and their applications in fluid flow at the micro-scale. We resort to analytically addressable mathematical treatments for convenience of an introductory reading.

1 Introduction

The uninitiated reader, with a healthy background in engineering science, may wonder that "flow physics at the micro-scale" is nothing but fluid mechanics applied at the scale of micrometers. In reality, though, this topic is much broader than is admitted by such a viewpoint. Indeed, the traditional way of manipulating fluids in conduits/channels has been through the application of mechanically applied pressure gradients or through exploiting the natural effects of gravity. These driving agents are on the total volume of the fluid. However, as dimensions scale down, the effects of the interfaces between the fluid and the confining surfaces become increasingly prominent, which is in keeping with the intuitive expectation that surface effects (as compared to volume effects) should become more important as physical dimensions decrease. This gives us a cue that with increasing levels of miniaturization, driving agents which exploit surface effects (in contrast to pressure gradients and/or gravity) might become increasingly important. In fact, that is so. Electrokinetic and surface tension effects are two extremely important surface effects. These are vast areas in themselves. This, together with the fact that there exist myriad ways to exploit these effects for flow manipulation, gives us an idea about the magnificent breadth of micro-scale flow physics.

It follows naturally from this preliminary discussion that flow physics at the micro-scale is truly an interdisciplinary subject. But, what, after all, is the motivation behind these studies and of bringing them all together? The answer lies in the modern technology of microfluidics, which is a broad term subsuming within it all the

S. Chakraborty (✉)
Department of Mechanical Engineering, Indian Institute of Technology (IIT), Kharagpur, India
e-mail: suman@mech.iitkgp.ernet.in

S. Chakraborty (ed.), *Mechanics Over Micro and Nano Scales*,
DOI 10.1007/978-1-4419-9601-5_1,

science and technology that is concerned with fluid flows over micron or sub-micron lengths. Specifically, it pertains to the precise manipulation and control of minute volumes of fluids through miniaturized conduits. The prime mover behind the emergence of microfluidics has been the advancements in micro- and nano-fabrication technologies. The application of microfluidic technology itself is diverse – ranging from the most popular biotechnological/biomedical engineering realm to inkjet printing and thermal management of electrokinetic devices. Many of these microfluidic applications are at the bleeding edge of research. The basics of these advanced applications are pertinently the fundamental principles of micro-scale flow physics.

We begin with a revision of the fundamentals of fluid mechanics with special emphasis on the applicability of the equations of fluid motion at the micro-scale.

2 Recapitulation of Fundamentals

Newton formulated his laws of mechanics in his book *Principia* (1687). However, it was not until 1752 that a mathematically clear description of fluid mechanics (albeit inviscid) was put forward by Euler. It was finally Cauchy who in 1822 introduced the concept of stress tensor and incorporated it into Euler's laws, thus presenting a very general theory for the motion of any continuous body [1]. These equations of Cauchy were not just limited to a fluid medium. It is important to note that the frame of reference used in these formulations is an Eulerian one. This is a frame of reference that focuses attention on a particular region in space as opposed to a Lagrangian frame of reference which focuses attention on a particular set of material points. Due to the inherent flowing nature of fluids it is the Eulerian frame that is the preferred one in fluid medium.

Instead of a presentation following the chronology of historical development, we will take up a more pedagogical approach with a uniform mathematical framework. The backbone of such a mathematical framework is the Reynolds transport theorem. This theorem relates the system approach with the control volume approach:

$$\left.\frac{DA}{Dt}\right|_{\text{sys}} = \frac{\partial}{\partial t}\int_{\text{CV}} \rho a dV + \int_{\text{CS}} \rho a \bar{v}_r \cdot \hat{n} dS \tag{1.1}$$

where A is any scalar or vector function denoting an extensive property, a is A per unit mass, $\bar{v}_r$ is the velocity relative to the control surface, dV is a differentially small element of the control volume, dS is an elemental arc on the control surface with a unit normal $\hat{n}$, and ρ is the density of the fluid.

2.1 Conservation of Mass

To use the Reynolds transport theorem, the relevant extensive property is the total mass. Thus, $A = M$ (the total mass) so that $a = 1$. Now, mass conservation, by definition, means

$$\frac{DM}{Dt} = 0$$

or,

$$\int_{\text{CV}} \frac{D(\rho\vec{u})}{Dt} dV = 0 \tag{1.2}$$

Therefore,

$$\frac{\partial}{\partial t}\int_{\text{CV}} \rho dV + \int_{\text{CS}} \rho\vec{u}\cdot\hat{n}dS = 0 \tag{1.3}$$

Using Gauss' divergence theorem, we get

$$\frac{\partial}{\partial t}\int_{\text{CV}} \rho dV + \int_{\text{CV}} \nabla\cdot(\rho\vec{u})\, dV = 0$$

or, for a non-deformable control volume

$$\int_{\text{CV}} \left\{\frac{\partial\rho}{\partial t} + \nabla\cdot(\rho\vec{u})\right\} dV = 0 \tag{1.4}$$

Now, this is true for any arbitrary non-deformable control volume of elementary volume dV implying the necessary condition

$$\frac{\partial\rho}{\partial t} + \nabla\cdot(\rho\vec{u}) = 0 \tag{1.5}$$

For a stationary control volume, $\vec{u}$ is the fluid flow velocity.

2.2 Conservation of Linear Momentum

The relevant extensive property, this time, is the linear momentum so that $A = M\vec{u}$ and $a = \vec{u}$. Now, using Reynolds' transport theorem for the motion of any continuous media, and assuming, as before, that the control volume is non-deformable, we have

$$\left.\frac{D(M\vec{u})}{Dt}\right|_{\text{sys}} = \int_{\text{CV}} \frac{\partial(\rho\vec{u})}{\partial t} dV + \int_{\text{CS}} \rho\vec{u}\vec{u}\cdot\hat{n}dS \tag{1.6}$$

Noting that

$$\frac{D(M\vec{u})}{Dt} = \int_{\text{CV}} \frac{D(\rho\vec{u})}{Dt} dV$$

we have from Cauchy's general theory of motion for any continuous medium,

$$\frac{D(\rho u)}{Dt} = \nabla \cdot [\sigma] + \vec{f}_B \tag{1.7}$$

Therefore, from (1.6) and (1.7), we get

$$\int_{\text{CV}} \frac{\partial (\rho \vec{u})}{\partial t} dV + \int_{\text{CS}} \rho \vec{u}\vec{u} \cdot \hat{n} dS = \int_{\text{CV}} \left\{ \nabla \cdot [\sigma] + \vec{f}_B \right\} dV \tag{1.8}$$

Now, using Gauss' divergence theorem as before, we obtain

$$\int_{\text{CV}} \frac{\partial (\rho \vec{u})}{\partial t} dV + \int_{\text{CV}} \nabla \cdot (\rho \vec{u}\vec{u}) \, dV = \int_{\text{CV}} \left\{ \nabla \cdot [\sigma] + \vec{f}_B \right\} dV \tag{1.9}$$

This must be true for any arbitrary control volume, implying the necessary condition

$$\frac{\partial (\rho \vec{u})}{\partial t} + \nabla \cdot (\rho \vec{u}\vec{u}) = \nabla \cdot [\sigma] + \vec{f}_B \tag{1.10}$$

We reiterate, here, that this is the general equation of motion of a fluid in an inertial frame of reference. It is valid irrespective of the nature of the fluid material.

Now, let us simplify the left-hand side of (1.10), i.e., $\partial (\rho \vec{u}) / \partial t + \nabla \cdot (\rho \vec{u}\vec{u})$. Let us first focus our attention on the $\nabla \cdot (\rho \vec{u}\vec{u})$ term. This is a vector, and thus, $\rho \vec{u}\vec{u}$ must be a tensor. It is a special product called the dyadic product and is explicitly denoted by the symbol $\otimes$. Thus, $\rho \vec{u}\vec{u} = \rho \vec{u} \otimes \vec{u}$

We expand this as follows:

$$\rho \vec{u} \otimes \vec{u} = \begin{bmatrix} \rho u \\ \rho v \\ \rho w \end{bmatrix} [u \; v \; w]$$

$$= \begin{bmatrix} \rho uu & \rho uv & \rho uw \\ \rho vu & \rho vv & \rho vw \\ \rho wu & \rho wv & \rho ww \end{bmatrix}$$

Now,

$$\nabla \cdot (\rho \vec{u} \otimes \vec{u}) = \left(\begin{bmatrix} \frac{\partial}{\partial x} & \frac{\partial}{\partial y} & \frac{\partial}{\partial z} \end{bmatrix} \begin{bmatrix} \rho uu & \rho uv & \rho uw \\ \rho vu & \rho vv & \rho vw \\ \rho wu & \rho wv & \rho ww \end{bmatrix} \right)^T$$

$$= \begin{bmatrix} \frac{\partial (\rho uu)}{\partial x} + \frac{\partial (\rho vu)}{\partial y} + \frac{\partial (\rho wu)}{\partial z} \\ \frac{\partial (\rho uv)}{\partial x} + \frac{\partial (\rho vv)}{\partial y} + \frac{\partial (\rho wv)}{\partial z} \\ \frac{\partial (\rho uw)}{\partial x} + \frac{\partial (\rho vw)}{\partial y} + \frac{\partial (\rho ww)}{\partial z} \end{bmatrix}$$

$$= \begin{bmatrix} \rho u \frac{\partial (u)}{\partial x} + u \frac{\partial (\rho u)}{\partial x} + \rho v \frac{\partial (u)}{\partial y} + u \frac{\partial (\rho v)}{\partial y} + \rho w \frac{\partial (u)}{\partial z} + u \frac{\partial (\rho w)}{\partial z} \\ \rho u \frac{\partial (v)}{\partial x} + v \frac{\partial (\rho u)}{\partial x} + \rho v \frac{\partial (v)}{\partial y} + v \frac{\partial (\rho v)}{\partial y} + \rho w \frac{\partial (v)}{\partial z} + v \frac{\partial (\rho w)}{\partial z} \\ \rho u \frac{\partial (w)}{\partial x} + w \frac{\partial (\rho u)}{\partial x} + \rho v \frac{\partial (w)}{\partial y} + w \frac{\partial (\rho v)}{\partial y} + \rho w \frac{\partial (w)}{\partial z} + w \frac{\partial (\rho w)}{\partial z} \end{bmatrix}$$

$$= \left[\rho \vec{u} \cdot \nabla u + u \nabla \cdot (\rho \vec{u}) \;\; \rho \vec{u} \cdot \nabla v + v \nabla \cdot (\rho \vec{u}) \;\; \rho \vec{u} \cdot \nabla w + w \nabla \cdot (\rho \vec{u}) \right]^T$$

$$= \rho \vec{u} \cdot \nabla \vec{u} + \vec{u} \nabla \cdot (\rho \vec{u})$$

Therefore,

$$\begin{aligned} \frac{\partial (\rho \vec{u})}{\partial t} + \nabla \cdot (\rho \vec{u} \vec{u}) &= \rho \frac{\partial \vec{u}}{\partial t} + \vec{u} \frac{\partial \rho}{\partial t} + \rho \vec{u} \cdot \nabla \vec{u} + \vec{u} \nabla \cdot (\rho \vec{u}) \\ &= \rho \frac{\partial \vec{u}}{\partial t} + \rho \vec{u} \cdot \nabla \vec{u} + \vec{u} \left(\frac{\partial \rho}{\partial t} + \nabla \cdot (\rho \vec{u}) \right) \\ &= \rho \frac{\partial \vec{u}}{\partial t} + \rho \vec{u} \cdot \nabla \vec{u} \quad \text{(using the continuity equation (1.5))} \end{aligned}$$

The general equation of motion of a fluid (of any material property) in an inertial Eulerian frame of reference is, thus,

$$\rho \frac{\partial \vec{u}}{\partial t} + \rho \vec{u} \cdot \nabla \vec{u} = \nabla \cdot [\sigma] + \vec{f}_B \tag{1.11}$$

2.3 The Navier–Stokes Equation

We now focus our attention on a special class of fluid, called Newtonian fluid, where the shear stress is proportional to the rate of shearing strain. The general constitutive law for a homogeneous, isotropic Newtonian fluid is, in indicial notation,

$$\sigma_{ij} = -p_d\delta_{ij} + \lambda\varepsilon_{kk}\delta_{ij} + 2\mu\varepsilon_{ij} \tag{1.12}$$

or, in tensorial notation,

$$[\sigma] = -p_d\,[I] + \lambda\,\mathrm{tr}\,([\varepsilon])\,[I] + 2\mu\,[\varepsilon] \tag{1.13}$$

where p_d is the thermodynamic pressure, δ_{ij} is the Kronecker delta symbol, $[\varepsilon]$ is the strain tensor and $\mathrm{tr}\,([\varepsilon])$ is the trace of the strain tensor, λ is the volume dilation coefficient, and μ is the viscosity coefficient.

From kinematics it follows that

$$[\varepsilon] = \frac{1}{2}\nabla \otimes \vec{u} + \frac{1}{2}(\nabla \otimes \vec{u})^T$$

$$= \frac{1}{2}\begin{bmatrix} 2\dfrac{\partial u}{\partial x} & \dfrac{\partial v}{\partial x} + \dfrac{\partial u}{\partial y} & \dfrac{\partial w}{\partial x} + \dfrac{\partial u}{\partial z} \\ \dfrac{\partial u}{\partial y} + \dfrac{\partial v}{\partial x} & 2\dfrac{\partial v}{\partial y} & \dfrac{\partial w}{\partial y} + \dfrac{\partial v}{\partial z} \\ \dfrac{\partial u}{\partial z} + \dfrac{\partial w}{\partial x} & \dfrac{\partial v}{\partial z} + \dfrac{\partial w}{\partial y} & 2\dfrac{\partial w}{\partial z} \end{bmatrix}$$

Therefore,

$$\mathrm{tr}\,([\varepsilon]) = \varepsilon_{kk} = \frac{\partial u}{\partial x} + \frac{\partial v}{\partial y} + \frac{\partial w}{\partial z} = \nabla \cdot \vec{u}$$

Using this expression in (1.13) we get

$$[\sigma] = -p_d\,[I] + \lambda\nabla \cdot \vec{u}\,[I] + 2\mu\,[\varepsilon] \tag{1.14}$$

Now we note from the general equation of motion (1.10) that we need the term $\nabla \cdot [\sigma]$. For a Newtonian fluid,

$$\begin{aligned} \nabla \cdot [\sigma] &= \nabla \cdot (-p_d\,[I] + \lambda\nabla \cdot \vec{u}\,[I] + 2\mu\,[\varepsilon]) \\ &= -\nabla p_d + \lambda\nabla\,(\nabla \cdot \vec{u}) + \mu\nabla\,(\nabla \cdot \vec{u}) + \mu\nabla^2\vec{u} \\ &= -\nabla p_d + \mu\nabla^2\vec{u} + (\lambda + \mu)\,\nabla\,(\nabla \cdot \vec{u}) \end{aligned}$$

The combination $\lambda + \frac{2}{3}\mu$ is referred to as the bulk viscosity coefficient and denoted by μ_v. Therefore,

$$\nabla \cdot [\sigma] = -\nabla p_d + \mu \nabla^2 \vec{u} + \left(\mu_v + \frac{1}{3}\mu\right) \nabla (\nabla \cdot \vec{u})$$

The general equation of motion, then, becomes

$$\rho \left[\frac{\partial \vec{u}}{\partial t} + \vec{u} \cdot (\nabla \vec{u})\right] = -\nabla p_d + \mu \nabla^2 \vec{u} + \left(\mu_v + \frac{1}{3}\mu\right) \nabla (\nabla \cdot \vec{u}) + \vec{f}_B \qquad (1.15)$$

2.3.1 Further Simplifications

Let us denote $\sigma_{xx} + \sigma_{yy} + \sigma_{zz} = -3p$, where p is called the mechanical pressure. Using (1.14) we write

$$\sigma_{xx} + \sigma_{yy} + \sigma_{zz} = -3p_d + 3\lambda \nabla \cdot \vec{u} + 2\mu \left(\frac{\partial u}{\partial x} + \frac{\partial v}{\partial y} + \frac{\partial w}{\partial z}\right)$$

or,

$$-3p = -3p_d + 3\lambda \nabla \cdot \vec{u} + 2\mu \left(\frac{\partial u}{\partial x} + \frac{\partial v}{\partial y} + \frac{\partial w}{\partial z}\right)$$

or,

$$-3p = -3p_d + 3\lambda \nabla \cdot \vec{u} + 2\mu \nabla \cdot \vec{u}$$

or,

$$3p_d - 3p = (3\lambda + 2\mu) \nabla \cdot \vec{u}$$

Now, there are two ways in which the thermodynamic pressure can become equal to the mechanical pressure:

(a) **Incompressible flow**: For an incompressible flow $\nabla \cdot \vec{u} = 0$, so that $3(p_d - p) = 0$ implying $p_d = p$. For an incompressible flow the bulk viscosity remains unspecified and the equation of motion becomes

$$\rho \left[\frac{\partial \vec{u}}{\partial t} + \vec{u} \cdot (\nabla \vec{u})\right] = -\nabla p + \mu \nabla^2 \vec{u} + \vec{f}_B \qquad (1.16)$$

(b) **Stokes' hypothesis**: This hypothesis states that for a wide class of fluid flow problems $3\lambda + 2\mu = 0$. It holds true when the characteristic timescales in the system are large compared to the molecular relaxation time. It is also true for monoatomic gases. Using $\lambda + \frac{2}{3}\mu = 0$, i.e., $\mu_v = 0$, we obtain the general equation of motion as

$$\rho \left[\frac{\partial \vec{u}}{\partial t} + \vec{u} \cdot (\nabla \vec{u})\right] = -\nabla p + \mu \nabla^2 \vec{u} + \frac{1}{3}\mu \nabla (\nabla \cdot \vec{u}) + \vec{f}_B \qquad (1.17)$$

2.4 Poiseuille Flow

The steady-state flow actuated solely by a pressure difference between the two ends of a channel is called Poiseuille flow. The mathematical modeling is simplified with a couple of considerations:

(a) The gradient of the flow velocity along the flow direction is zero.
(b) The only non-zero component of the flow velocity is the one along the flow direction, i.e., all components normal to the flow direction are zero.

With these considerations, for an incompressible flow, we have

$$0 = -\frac{\partial p}{\partial x} + \mu \nabla^2 u \tag{1.18}$$

where x and u are, respectively, the coordinate component and the velocity component along the flow direction. Here p is piezometric pressure which contains the effect of gravity as well. The zero velocity components perpendicular to this flow direction imply zero gradients of the pressure along those directions. Hence, $p \equiv p(x)$ and $\partial p / \partial x - \equiv dp / dx$. Now, $dp / dx = -\Delta p / L$ where Δp is the pressure difference between the two ends of the channel of length L. From (1.18) we have

$$\mu \left(\frac{\partial^2 u}{\partial y^2} + \frac{\partial^2 u}{\partial z^2} \right) = -\frac{\Delta p}{L} \quad \left(\frac{\partial u}{\partial x} = 0 \right)$$

This simple equation is surprisingly successful in correctly modeling fluid flow in a wide variety of microfluidic channel shapes.

The simplest geometrical shape that can be considered within this class of Poiseuille flow is that of the slit channel which is a two-dimensional idealization of a three-dimensional rectangular channel where the width is "infinite," meaning that the width is so large that it does not significantly affect the flow profile. In this case,

$$\mu \frac{\partial^2 u}{\partial y^2} = -\frac{\Delta p}{L} \tag{1.19}$$

where y is the coordinate along the channel height. This governing equation together with the no-slip boundary condition at the walls (u=0 at y=0 and H) gives us the velocity profile:

$$u = -\frac{\Delta p}{2\mu L} y (y - H) \tag{1.20}$$

It is informative and useful (from the point of view of the experiment) to find a relation between the total volumetric flow rate and the applied pressure difference.

$$
\begin{aligned}
Q &= \int_0^H u dy \\
&= -\int_0^H \frac{\Delta p}{2\mu L} y\,(y-H)\,dy \\
&= \frac{\Delta p H^3}{12\mu L}
\end{aligned}
$$

Thus, $\Delta p = \left(12\mu L/H^3\right) Q$. The factor $12\mu L/H^3$ is referred to as the *hydraulic resistance* for this particular channel. Similar expressions for the hydraulic resistance may be derived for other channel shapes.

It may be noted from (1.18) that the governing differential equation has been reduced from the general nonlinear form of the Navier–Stokes equation to a linear one. Such a reduction based on consideration (a) (mentioned just before (1.18)) is what is referred to as the fully developed flow condition. Interestingly, even when the flow is not fully developed but the Reynolds number is small, the inertia terms in the Navier–Stokes equation become negligible (though not identically zero), as compared to the viscous terms. Under such circumstances, the nonlinear terms in the Navier–Stokes equation become virtually inconsequential. The resultant simplified equation without the nonlinear terms is known as the Stokes equation, which is often the starting point of analyzing low Reynolds number flows.

2.5 Physical Justification of Linearization

Let us consider the Navier–Stokes equation for a steady, incompressible flow, without any body force term:

$$\rho\vec{u}\cdot\nabla\vec{u} = -\nabla p + \mu\nabla^2\vec{u} \tag{1.21}$$

A systematic way of investigating the order of the different terms present in an equation is to non-dimensionalize it using characteristic scaling parameters. Let the pertinent parameters in this equation be L and V for scaling length and velocity, respectively. Using these, we can write

$$\rho\vec{u}\cdot\nabla\vec{u} = \rho\,\underset{\sim}{\vec{u}}\cdot\underset{\sim}{\nabla}\,\underset{\sim}{\vec{u}}\left(\frac{V^2}{L}\right) \tag{1.22}$$

$$\mu\nabla^2\vec{u} = \mu\underset{\sim}{\nabla}^2\,\underset{\sim}{\vec{u}}\left(\frac{V}{L^2}\right) \tag{1.23}$$

where symbols with tilde (~) under them represent dimensionless terms.

Substituting these expressions, in (1.21) we get

$$\rho \underset{\sim}{\vec{u}} \cdot \underset{\sim}{\nabla} \underset{\sim}{\vec{u}} \left(\frac{V^2}{L}\right) = -\nabla p + \mu \underset{\sim}{\nabla}^2 \underset{\sim}{\vec{u}} \left(\frac{V}{L^2}\right)$$

or,

$$\left(\frac{\rho VL}{\mu}\right) \underset{\sim}{\vec{u}} \cdot \underset{\sim}{\nabla} \underset{\sim}{\vec{u}} = -\frac{L}{\mu V} \underset{\sim}{\nabla} p + \underset{\sim}{\nabla}^2 \underset{\sim}{\vec{u}} \tag{1.24}$$

The combination $\rho VL/\mu$ is non-dimensional and is known by the name Reynolds number *Re*. If $Re \ll 1$ then, clearly, the nonlinear term can be neglected. This fact is of utmost importance in microfluidics because in many situations the length scales and velocity regimes are such that the Reynolds number is *indeed* much less than 1. It is in appreciation of this fact that the study of microfluidics is often termed as low Reynolds number hydrodynamics.

The physical justification of the Navier–Stokes equation linearization (this linearized version is called the Stokes equation) for most microfluidic modeling purposes must be appreciated within certain caveats:

(a) **Unsteady case**:
A valid temporal scale when there is no external temporal perturbation imposed on the system is $T = L/V$. Then

$$\rho \frac{\partial \vec{u}}{\partial t} = \rho \frac{\partial \underset{\sim}{\vec{u}}}{\partial \underset{\sim}{t}} \left(\frac{V^2}{L}\right)$$

A comparison of this expression with (1.22) clearly shows that the unsteady term will also be present in the non-dimensional equation with *Re* as a coefficient. Since *Re* is small, the term with this temporal dependence can be neglected.

It is extremely important to understand the fact that such a scaling would not be valid if there is an externally imposed temporal variation in the system. If, for example, one of the boundaries is moved periodically with frequency ω, the correct temporal scale would be $1/\omega$ and not L/V which is inherent to the flow. In such a case, the velocity scale could be expressed in terms of the length scale and the timescale so that $V{=}L\omega$. Then,

$$\rho \frac{\partial \vec{u}}{\partial t} = \rho \frac{\partial \underset{\sim}{\vec{u}}}{\partial \underset{\sim}{t}} (\omega^2 L)$$

$$\rho \vec{u} \cdot \nabla \vec{u} = \rho \underset{\sim}{\vec{u}} \cdot \underset{\sim}{\nabla} \underset{\sim}{\vec{u}} (\omega^2 L)$$

and

$$\mu\nabla^2\vec{u} = \mu \underset{\sim}{\nabla}^2 \underset{\sim}{\vec{u}} \left(\frac{\omega}{L}\right)$$

The unsteady non-dimensional equation becomes

$$Re\left(\frac{\partial \underset{\sim}{\vec{u}}}{\partial \underset{\sim}{t}} + \underset{\sim}{\vec{u}} \cdot \underset{\sim}{\nabla} \underset{\sim}{\vec{u}}\right) = -\frac{1}{\mu\omega} \underset{\sim}{\nabla p} + \underset{\sim}{\nabla}^2 \underset{\sim}{\vec{u}} \tag{1.25}$$

where $Re = \rho\omega L^2/\mu$. Just as before, if $Re = \rho\omega L^2/\mu \ll 1$ then all the inertial terms (i.e., the left-hand side of the equation) can be neglected.

(b) **Different length scales**:
We have hitherto defined the Reynolds numbed based on a characteristic length scale of the system. However, if the system under consideration is characterized by more than one length scale (a common occurrence in many real systems) then how do we define the Reynolds number? The convention in such a scenario is to choose the smallest length scale for the Reynolds number definition. To illustrate this point, let us consider the example of steady flow of a thin film of viscous fluid between two rigid boundaries at z=0 and $z = h(x, y)$. The vertical length scale is h while the horizontal length scale is L. Here, $h \ll L$ so the Reynolds number (by convention) is $Re = \rho V h/\mu$.
Now,

$$\frac{\partial}{\partial z} \gg \frac{\partial}{\partial x}, \frac{\partial}{\partial y} \quad \left(\text{because } \frac{1}{h} \gg \frac{1}{L}\right)$$

Therefore,

$$\nabla^2\vec{u} \approx \frac{\partial^2 u}{\partial z^2}$$

Furthermore, the incompressibility condition $\nabla \cdot \vec{u} = 0$, i.e.,

$$\frac{\partial u}{\partial x} + \frac{\partial v}{\partial y} + \frac{\partial w}{\partial z} = 0 \text{ implies } w \sim \frac{h}{L}V$$

Therefore,

$$\rho\vec{u}\cdot\nabla\vec{u} \sim \rho\left(\frac{V^2}{L}, \frac{V^2}{L}, \frac{V^2 h}{L^2}\right) = \rho\frac{V^2}{L}\left(1, 1, \frac{h}{L}\right)$$

$$\mu\nabla^2\vec{u} \sim \mu\frac{\partial^2\vec{u}}{\partial z^2} = \mu\left(\frac{V}{h^2}, \frac{V}{h^2}, \frac{Vh}{h^2 L}\right) = \mu\frac{V}{h^2}\left(1, 1, \frac{h}{L}\right)$$

Then,

$$\frac{|\rho\vec{u}\cdot\nabla\vec{u}|}{|\mu\nabla^2\vec{u}|} = \frac{\rho V^2/L}{\mu V/h^2} = \frac{\rho V h^2}{\mu L} = \frac{\rho V h}{\mu}\left(\frac{h}{L}\right) = Re\left(\frac{h}{L}\right)$$

Thus, the nonlinear terms can be neglected if $Re\,(h/L) \ll 1$ even if Re is not $\ll 1$. We could say that the linearization can be done provided the effective Reynolds number $Re_{\mathrm{eff}} = Re\,(h/L) \ll 1$.

3 Electrokinetics

We had mentioned the importance of surface effects as it particularly pertains to micro-scales in Section 1. Of all the flow manipulation techniques which exploit surface effects, electrokinetic effects are, arguably, the most popular ones in microfluidics. Fundamental to the use of electrokinetic effects is an understanding of the electrical double layer (EDL). Generally, most solid surfaces tend to acquire a net surface charge (positive or negative) when brought into contact with an aqueous (polar) medium. There are various mechanisms behind this surface charging phenomenon such as ionization of covalently bound surface groups or ion adsorption. We will not delve deeper into these mechanisms; instead we direct the interested reader to comprehensive expositions of the topic in [4, 13].

Aqueous solutions invariably have dissolved ions (e.g., from dissolved salts or dissociated water groups) present in them (even distilled water is not perfectly devoid of ions). The ions which are charged oppositely to the charged surface are called counterions, and the ions which have the same charge polarity as the surface are called co-ions. The charged surface, naturally, attracts the counterions and repels the co-ions. If there were no thermal motion of the ions, the charged surface would be perfectly shielded by a layer of counterions stacked against the surface. However, ions have non-zero absolute temperature and the concomitant random thermal motion of the ions precludes such a physical picture. What happens, then, is that a balance is established between the electrostatic forces and the thermal interactions so that, at equilibrium, a certain charge distribution prevails adjacent to the surface (of course, with the predominant presence of the counterions in the vicinity of the surface). This charge distribution with the predominant distribution of counterions is called the electrical double layer (EDL).

There are various theories which attempt to give models for the physical picture of the EDL.[1] We present here the Gouy–Chapman model with the Stern modification (GCS model).

[1] For a historical development of the attempts to theoretically model the EDL, the interested reader may refer to [16].

A schematic depicting the charge and the potential distribution within an EDL is shown in Fig. 1.1. A layer of immobile counterions is present just next to the charged surface. This layer is known as the compact layer or the Stern layer or the Helmholtz layer. The thickness of this layer is about a few Angstroms and, hence, the potential distribution within it may be assumed to be linear. From this Stern layer to the electrically neutral bulk liquid, the ions are mobile. This layer of mobile ions beyond the Stern layer is called the Gouy–Chapman layer or the diffuse layer of the EDL. Besides this, there is a plane called the shear plane or surface which is considered to be the plane at which the mobile portion of the EDL can flow past the charged surface. The potential at the shear plane is called the zeta potential (ζ). The characteristic thickness of the EDL is known as the Debye length (λ) which is the

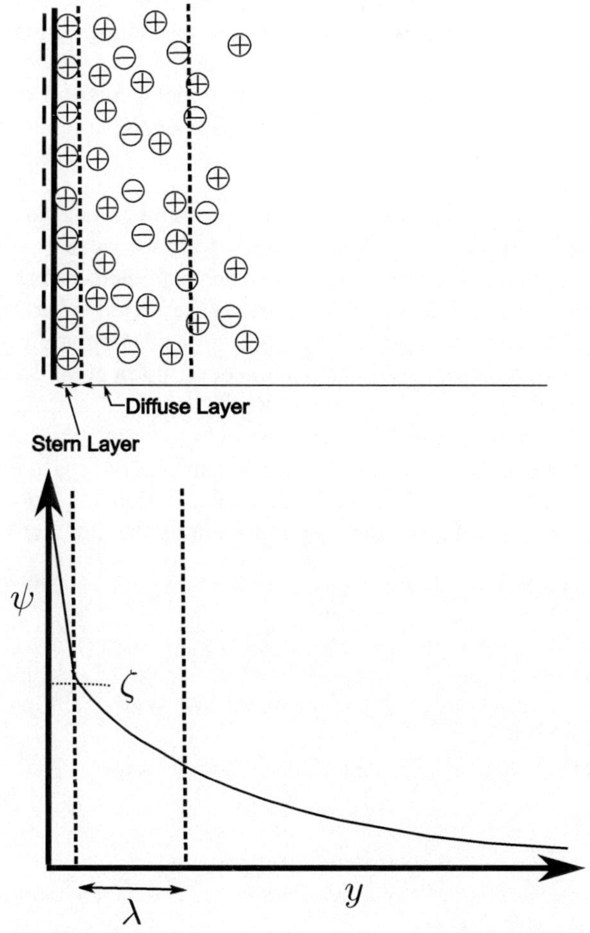

Fig. 1.1 Distribution of counterions and co-ions in an EDL (*top*) and potential profile screening the surface charge (*bottom*)

length from the shear plane over which the EDL potential reduces to $1/e$ of ζ (where e is the Euler number).

The four primary electrokinetic effects are as follows:

1. Electroosmosis: It refers to the relative movement of liquid over a stationary charged surface, with an external electric field acting as the actuator.
2. Streaming potential: It refers to the electric potential that is induced when a liquid, containing ions, is driven by a pressure gradient to flow along a stationary charged surface.
3. Electrophoresis: It refers to the movement of a charged surface (e.g., a charged particle) relative to a stationary liquid due to the application of an external electric field. The phenomena of electrophoresis and electroosmosis are closely related. It is also important to note that, strictly following the definition, these phenomena are not dependent on the establishment of the EDL. They just depend on the surface being charged irrespective of the mechanism bringing about the charge.
4. Sedimentation potential: It refers to the potential that is induced when a charged particle moves relative to a stationary liquid (for example, under the effect of gravity).

Besides these, closely related electrokinetic phenomena like dielectrophoresis and diffusioosmosis are also important. In what follows, we will concentrate on electroosmosis, streaming potential, electrophoresis, and dielectrophoresis because these have received the greatest attention (due to the surfeit of practical applications) in the microfluidics community. But, before we can delve into these topics individually, we need to develop a mathematical description of the EDL.

We begin the mathematical description of the EDL by considering a single plate or surface in an infinite liquid phase. For the system to be in equilibrium, the electrochemical potential of the ions needs to be constant everywhere to ensure that the system is in equilibrium. Thus, for a one-dimensional system of a solvent in contact with a planar surface with the y-coordinate representing the direction normal to the surface, we have

$$\frac{d\bar{\mu}_i}{dy} = 0 \tag{1.26}$$

where $\bar{\mu}_i$ is the electrochemical potential of ions of type i. The electrochemical potential is defined as

$$\bar{\mu}_i = \mu_i + z_i e\psi \tag{1.27}$$

where μ_i and z_i are, respectively, the chemical potential and valence of ions of type i and e is the protonic charge. Now, from thermodynamics, the chemical potential can be expressed as

$$\mu_i = \mu_i^0 + k_{\mathrm{B}} T \ln n_i \tag{1.28}$$

where μ_i^0 is a constant for ions of type i, k_B is the Boltzmann constant, T is the absolute temperature of the solution, and n_i is the ionic number concentration of type i. Differentiating (1.28) with respect to y

$$\frac{d\mu_i}{dy} = \frac{k_B T}{n_i}\frac{dn_i}{dy} \tag{1.29}$$

Therefore, from (1.27), we have

$$\frac{d\bar{\mu}_i}{dy} = \frac{d\mu_i}{dy} + z_i e \frac{d\psi}{dy}$$

or,

$$0 = \frac{k_B T}{n_i}\frac{dn_i}{dy} + z_i e \frac{d\psi}{dy} \quad \text{(using (1.29))}$$

or,

$$\frac{dn_i}{n_i} = -\frac{z_i e}{k_B T} d\psi \tag{1.30}$$

Now, (1.30) can be solved in conjunction with appropriate boundary conditions. A very common scenario is one where the potential far away from the surface is 0 and the ionic number concentration corresponds to a bulk value indicating the absence of any surface effect. In mathematical terms, this physical scenario translates to the following: at $y \to \infty$, $n_i = n_i^0$ and $\psi = 0$. The solution, with such a boundary condition, is

$$n_i = n_i^0 \exp\left(-\frac{z_i e \psi}{k_B T}\right) \tag{1.31}$$

This is the Boltzmann distribution of ions near a charged surface. The assumptions underlying this derivation are as follows:

- The system is in equilibrium with no macroscopic advection/diffusion of ions
- The solid surface is microscopically homogeneous
- The far-stream boundary condition is applicable, meaning that the charged surface is in contact with a large enough liquid medium such that at a distance far in bulk no surface effect is experienced

Surprisingly, however, the Boltzmann distribution is successfully employed in scenarios where fluid flow is taking place, i.e., a system which is certainly not in thermodynamic equilibrium. Although this blatant violation of an underlying assumption might seem like a paradox, it can be shown that for low velocities with $Re \ll 10$, the Boltzmann distribution is, indeed, a good approximation. Let us see how.

We use the ionic species conservation equation (for the steady case) which describes the flux of the ith species as

$$\nabla \cdot \vec{j}_i = 0 \tag{1.32}$$

Now, $\vec{j}_i$ can be described by

$$\vec{j}_i = n_i\vec{u} - D_i\nabla n_i - \left(\frac{z_ieD_in_i}{k_BT}\right)\nabla\psi \tag{1.33}$$

Using (1.33) in (1.32), we have the Nernst–Planck equation

$$\nabla \cdot (n_i\vec{u}) = \nabla \cdot (D_i\nabla n_i) + \nabla \cdot \left[\left(\frac{z_ieD_in_i}{k_BT}\right)\nabla\psi\right] \tag{1.34}$$

For low velocities, the advection term can be neglected, thus

$$D_i\nabla^2 n_i + D_i\nabla \cdot \left[\left(\frac{z_ien_i}{k_BT}\right)\nabla\psi\right] = 0 \text{ (for constant } D_i) \tag{1.35}$$

In the case of a one-dimensional situation (along y)

$$\frac{d^2n_i}{dy^2} + \frac{d}{dy}\left(\frac{z_ien_i}{k_BT}\frac{d\psi}{dy}\right) = 0 \tag{1.36}$$

Integrating (1.36) and imposing the condition $d\psi/dy = 0$, $dn_i/dy = 0$ as $y \to \infty$ gives

$$\frac{dn_i}{dy} + \left(\frac{z_ien_i}{k_BT}\right)\frac{d\psi}{dy} = 0 \tag{1.37}$$

This is, basically, the same as (1.30) derived previously. This shows that, indeed, for low velocities the Boltzmann distribution is valid even though the fluid flow violates the equilibrium condition.

Now, in order to get a full picture of the ionic distribution we need to know the potential distribution. With this aim we start off from Gauss' law:

$$\int_S \vec{E} \cdot \hat{n} ds = \frac{Q_{\text{enclosed}}}{\varepsilon_0}$$

or, using the Gauss' divergence theorem to transform the surface integral to a volume integral:

$$\int_V \left(\nabla \cdot \vec{E} - \frac{\rho_e}{\varepsilon_0}\right) dV = 0$$

where Q_{enclosed} is the total charge enclosed within the volume V bound by the surface S and ρ_e is the charge density.

This must be true for any arbitrary volume, so

$$\nabla \cdot \vec{E} = \frac{\rho_e}{\varepsilon_0} \tag{1.38}$$

It is important to note that since the fluid medium within which the electric field is calculated is a dielectric one, we have to account for the induced (or, bound) charge appearing as a result of polarization. This means, $\rho_e = \rho_{e,\text{free}} + \rho_{e,\text{bound}}$ where $\rho_{e,\text{free}}$ is volumetric charge density of the free ions in the dielectric medium while $\rho_{e,\text{bound}}$ is the volumetric density of the bound charges. It is known that $\rho_{e,\text{bound}} = -\nabla \cdot \vec{P}$ where $\vec{P}$ is the polarization vector. Assuming that the dielectric medium is linear and isotropic $\vec{P} = \varepsilon_0 \chi_e \vec{E}$ where χ_e is the electrical susceptibility. Therefore,

$$\varepsilon_0 \nabla \cdot \vec{E} = \rho_{e,\text{free}} - \nabla \cdot \left(\varepsilon_0 \chi_e \vec{E}\right)$$

or,

$$\nabla \cdot \left\{\vec{E}\left(\varepsilon_0 + \varepsilon_0 \chi_e\right)\right\} = \rho_{e,\text{free}}$$

or,

$$\nabla \cdot \left(\varepsilon \vec{E}\right) = \rho_{e,\text{free}} \tag{1.39}$$

where $\varepsilon = \varepsilon_0 (1 + \chi_e)$ is the permittivity of the dielectric medium. If ε is assumed to be invariant spatially then from (1.39) we have

$$\nabla \cdot \vec{E} = \frac{\rho_{e,\text{free}}}{\varepsilon} \tag{1.40}$$

Since the electric field $\vec{E}$ may be expressed in terms of a potential field ψ as $\vec{E} = -\nabla \psi$, (1.40) may be expressed as

$$\nabla^2 \psi = -\frac{\rho_{e,\text{free}}}{\varepsilon} = -\frac{e \sum z_i n_i}{\varepsilon} \tag{1.41}$$

where the summation is taken over all species of ions. For a symmetric electrolyte $z_i = z = z_+ = -z_- =$ constant and (1.41) reduces to

$$\nabla^2 \psi = -\frac{e\left(n_+ z_+ + n_- z_-\right)}{\varepsilon} = -\frac{ez\left(n_+ - n_-\right)}{\varepsilon} \tag{1.42}$$

It is important to note that the special case of a symmetric electrolyte is a useful consideration because it is possible to treat most electrolytes as though they were symmetric with valency z equal to the counterion valency.[2]

[2]For a full-fledged justification of this assertion refer to Section 2.3.3.3 of [9].

3.1 Electroosmosis

The presence of charges (usually in the form of ions) in the liquid can be usefully exploited to actuate bulk motion of the liquid by the application of an electric field. Basically, the electric field acts upon these charges. As they move due to the electrical force they drag the liquid along with them. This phenomenon is called *electroosmosis*; the fluid flow is referred to as electroosmotic flow (EOF).

In order to model this fluid actuation we need to incorporate an extra body force term in the momentum equation for a fluid, i.e., the Navier–Stokes equation. Close to the surface of a substrate, it is the double layer formation that results in the ionic distribution on which the electric field acts. Let us see, in detail, how the body force comes about.

The Maxwell stress, neglecting magnetic effects, is given by [8]

$$[T] = \varepsilon \left\{ \vec{E} \otimes \vec{E} - \frac{1}{2} \vec{E} \cdot \vec{E} \, [I] \right\} \tag{1.43}$$

The body force on the fluid due to the Maxwell stress is

$$\begin{aligned} \vec{F}_E &= \nabla \cdot [T] \\ &= \varepsilon \nabla \cdot \left\{ \vec{E} \otimes \vec{E} \right\} - \varepsilon \frac{1}{2} \nabla \cdot \left(\vec{E} \cdot \vec{E} \, [I] \right) \text{ (assuming, as before, that } \varepsilon \text{ is spatially invariant)} \\ &= \varepsilon \vec{E} \left(\nabla \cdot \vec{E} \right) + \varepsilon \left(\vec{E} \cdot \nabla \right) \vec{E} - \varepsilon \tfrac{1}{2} \nabla \left(\vec{E} \cdot \vec{E} \right) \\ &= \varepsilon \vec{E} \left(\nabla \cdot \vec{E} \right) + \varepsilon \left(\vec{E} \cdot \nabla \right) \vec{E} - \varepsilon \tfrac{1}{2} \times 2 \vec{E} \cdot \nabla \vec{E} \text{ (please see footnote}^3\text{)} \\ &= \varepsilon \vec{E} \left(\nabla \cdot \vec{E} \right) \end{aligned}$$

Here, $\vec{E}$ is the total electrical field and may be expressed in terms of a potential as $\vec{E} = -\nabla \varphi$. Using this expression of $\vec{E}$, we obtain

$$\vec{F}_E = \varepsilon \nabla \varphi \nabla^2 \varphi \tag{1.44}$$

In addition to this contribution from the Maxwell stress, there is another contribution from the osmotic pressure to the body force term. The osmotic pressure is defined as the pressure required to be exerted on a solution to prevent the percolation of solvent across a semi-permeable membrane from another solution having lower concentration. In the most general case, the osmotic pressure arises whenever there is a gradient of concentration such as in the present case of non-uniform ion distribution in the EDL. In mathematical terms, osmotic pressure is defined to be

---

[3]The equality $\nabla \left(\vec{E} \cdot \vec{E} \right) = 2 \left(\vec{E} \cdot \nabla \right) \vec{E}$ is established in the context of dielectrophoresis (see (1.124)–(1.126)).

$\prod = nk_BT$. The body force on the fluid due to this osmotic pressure is $\vec{F}_O = -\nabla \prod$, and assuming isothermal conditions we obtain

$$\begin{aligned}\vec{F}_O &= -k_B T \nabla (n_+ + n_-) \\ &= -k_B T \left[-\frac{ez}{k_B T} n_+ + \frac{ez}{k_B T} n_- \right] \nabla \psi \\ &= ez \left[n_+ - n_- \right] \nabla \psi\end{aligned}$$

We use (1.41) to express $ez\,(n_+ - n_-)$ as $-\varepsilon\left(d^2\psi/dy^2\right)$. Thus,

$$\vec{F}_O = -\varepsilon \nabla^2 \psi \nabla \psi \tag{1.45}$$

We note that the potential field φ is made up of two components: φ_0 due to the externally applied electric field and ψ due to the charged substrate, i.e., $\varphi = \varphi_0 + \psi$. Moreover, the potential φ_0 satisfies the Laplace equation $\nabla^2 \varphi_0 = 0$, so that $\nabla^2 \varphi = \nabla^2 \varphi_0 + \nabla^2 \psi = \nabla^2 \psi$. Therefore, the total body force on the fluid is

$$\begin{aligned}\vec{F} &= \vec{F}_E + \vec{F}_O \\ &= \varepsilon \nabla \varphi \nabla^2 \varphi - \varepsilon \nabla^2 \psi \nabla \psi \quad \left(\nabla^2 \varphi = 0 + \nabla^2 \psi\right) \\ &= \varepsilon \nabla \varphi_0 \nabla^2 \psi \\ &= -\rho_{e,\text{free}} \nabla \varphi_0\end{aligned}$$

Finally,

$$\vec{F} = \rho_{e,\text{free}} \vec{E}_{\text{ext}} \tag{1.46}$$

Moreover, if the permittivity depends on the fluid density ρ, then an extra term gets added to the body force,[4] thus

$$\vec{F} = \rho_{e,\text{free}} \vec{E} + \nabla \left(\frac{1}{2} \vec{E} \cdot \vec{E} \rho \frac{\partial \varepsilon}{\partial \rho} \right) \tag{1.47}$$

Let us incorporate this body force in the equation of motion of the fluid flowing in a microchannel. The coordinate normal to the wall is y while the coordinate along the flow direction is x. Again for flow in a microchannel with low values of Re we neglect the inertial terms, thus

$$0 = \mu \nabla^2 \vec{u} + \rho_{e,\text{free}} \vec{E}_{\text{ext}} \tag{1.48}$$

It can be noted here that in the absence of any imposed pressure gradient and/or a gradient of zeta potential, the pressure gradient terms get dropped in the equation of

[4]The derivation of this extra term is beyond the scope of this book. However, it may be read in [12].

motion depicted by (1.48). The case with non-zero pressure gradient can be analyzed by using the Stokes equation without any electrokinetic body force term. The flow occurring under the combined effect of an imposed pressure gradient and an external electric field may be analyzed by considering the resultant effect to be dictated by a linear superposition of the following: (i) Stokes equation with a pressure gradient term but without an electrokinetic body force term and (ii) Stokes equation with an electrokinetic body force term but without a pressure gradient term. In case the osmotic pressure gradient term is not considered as a separate body force term in the momentum equation, its contribution may be clubbed with the traditional pressure gradient term so that one may write $-\nabla P = -\nabla(p+\Pi)$ in which case the electrical body force must be written as $\rho_{e,free}\vec{E}$, where $\vec{E}$ is the total electric field due, and not $\rho_{e,free}\vec{E}_{ext}$. To delineate the flow characteristics due to pure electroosmotic flow (governed by (ii) mentioned as above), we solve (1.48).

For the special case of flow in a slit-type microchannel, the flow can be suitably modeled as one-dimensional. Thus,

$$0 = \mu\frac{d^2u}{dy^2} + \rho_{e,\text{free}}E_x$$

where E_x is the externally applied electrical field parallel to the flow direction. Using (1.41) for ρ_e, we get from (1.48)

$$0 = \mu\frac{d^2u}{dy^2} - \varepsilon\frac{d^2\psi}{dy^2}E_x \tag{1.49}$$

Now, using the boundary conditions $du/dy = 0$ and $d\psi/dy = 0$ at $y=H$ and $u=0$ and $\psi = \zeta$ at $y=0$, we obtain the velocity profile as

$$u = -\frac{1}{\mu}\varepsilon\zeta E_x\left(1 - \frac{\psi}{\zeta}\right) \tag{1.50}$$

The combination $-\varepsilon\zeta E_x/\mu$ has the dimension of velocity and is referred to as the *Helmholtz–Smoluchowski velocity*, u_{HS}. However, the velocity cannot be determined just yet because the distribution of ψ is unknown. To find ψ, we use the Poisson equation (1.41) together with the Boltzmann distribution (1.31) to obtain the celebrated Poisson–Boltzmann distribution for a *z*:*z* symmetric electrolyte:

$$\nabla^2\psi = -n_0ez\left(\exp\left(-\frac{ez\psi}{k_{\text{B}}T}\right) - \exp\left(\frac{ez\psi}{k_{\text{B}}T}\right)\right) \tag{1.51}$$

Considering y as the coordinate normal to the substrate over which the EDL is established, (1.51) reduces to

$$\frac{d^2\psi}{dy^2} = 2n_0ez\sinh\left(\frac{ez\psi}{k_{\text{B}}T}\right) \tag{1.52}$$

Equation (1.52) may be solved subject to various boundary conditions. A few of these are exemplified in the following cases, considering EDL development adjacent to the solid substrate of a parallel plate without having any EDL overlap:

Case I: The zeta potential ζ at both walls (y=0 and y=2H) is specified. ζ is small such that $\left|ez\zeta/k_{\mathrm{B}}T\right| \ll 1$. In this condition,

$$\sinh\left(\frac{ez\zeta}{k_{\mathrm{B}}T}\right) \approx \frac{ez\zeta}{k_{\mathrm{B}}T}$$

This artifice is called the Debye–Hückel linearization. In practice, this linearization is invoked up to $|\zeta| = 25$ mV (for T=300 K). Accordingly, (1.52) gets simplified to

$$\frac{d^2\psi}{dy^2} = \frac{2n_0e^2z^2}{\varepsilon k_BT}\psi \tag{1.53}$$

Here, $\left(2n_0e^2z^2/\varepsilon k_{\mathrm{B}}T\right)^{-1/2} = \lambda$, called the Debye length, is a characteristic thickness of the EDL. Equation (1.53) together with the boundary conditions $\psi = \zeta$ at y=0 and y=2H is used to obtain

$$\psi = \zeta\frac{\cosh\left[\frac{y-H}{\lambda}\right]}{\cosh\left[\frac{H}{\lambda}\right]} \tag{1.54}$$

Case II: Instead of the zeta potential, as in Case I, the surface charge (per unit wall area), i.e., surface charge density σ, is specified.

In the case of the Debye length being much smaller than half the channel height, the EDL distribution may be considered to be independent at each wall. In such a case, to maintain the condition of electroneutrality, the surface charge on a wall must be equal and opposite to the total (unbalanced) charge in the EDL near that wall. This means

$$\sigma|_{y=0} = -\int_0^H \rho_e dy$$

$$= \int_0^H \varepsilon\frac{d^2\psi}{dy^2}dy \quad \text{(using the Poisson equation (1.41))}$$

$$= \varepsilon\cancelto{0}{\left.\frac{d\psi}{dy}\right|_{y=H}} - \varepsilon\left.\frac{d\psi}{dy}\right|_{y=0} \quad \text{(from symmetry condition at the centre-line)}$$

Finally,

$$\sigma = -\varepsilon\left.\frac{d\psi}{dy}\right|_{y=0} \tag{1.55}$$

Similarly,

$$\sigma|_{y=2H} = \varepsilon \frac{d\psi}{dy}\bigg|_{y=2H} \tag{1.56}$$

We again invoke the Debye–Hückel linearization artifice in order to be able to use the simplified (1.53). Using this equation, together with the conditions in (1.55) and (1.56), we obtain

$$\psi = \left(\frac{\sigma\lambda}{\varepsilon}\right)\frac{\cosh\left[\frac{y-H}{\lambda}\right]}{\sinh\left[\frac{H}{\lambda}\right]} \tag{1.57}$$

The zeta potential can be calculated to be $\zeta = \psi\,(y = 0, 2H) = \sigma\lambda/\varepsilon$. So the restriction within which the Debye–Hückel linearization is valid is

$$\left|\frac{\zeta ez}{k_{\mathrm{B}}T}\right| = \left|\frac{\sigma\lambda ez}{\varepsilon k_{\mathrm{B}}T}\right| \ll 1$$

Case III: Again, as in Case I, the zeta potential is specified but we no more consider it to be small so that the Debye–Hückel linearization cannot be used. In such a case, the governing equation to be used is the nonlinearized Poisson equation (one-dimensional):

$$\frac{d^2\psi}{dy^2} = \frac{2n_0ez}{\varepsilon}\sinh\left(\frac{ez\psi}{k_{\mathrm{B}}T}\right) \tag{1.58}$$

Equation (1.58) is subject to the boundary conditions $\psi = \zeta$ at $y = 0$ and $d\psi/dy = 0$ at $y = H$. In order to solve (1.58) analytically, we first non-dimensionalize it using the scheme $\bar{\psi} = ez\psi/4k_{\mathrm{B}}T$ and $\bar{y} = y/H$. Then,

$$\frac{1}{H^2}\frac{d^2\bar{\psi}}{d\bar{y}^2} = \frac{1}{4\lambda^2}\sinh\left(4\bar{\psi}\right)$$

or,

$$\frac{d^2\bar{\psi}}{d\bar{y}^2} = \frac{K^2}{4}\sinh\left(4\bar{\psi}\right) \tag{1.59}$$

where $K = H/\lambda$ is a ratio denoting the relative half-height of the channel to the EDL characteristic thickness. Now, multiplying both sides by $2\left(d\bar{\psi}/d\bar{y}\right)$, we get

$$2\frac{d\bar{\psi}}{d\bar{y}}\frac{d^2\bar{\psi}}{d\bar{y}^2} = \frac{K^2}{4}2\frac{d\bar{\psi}}{d\bar{y}}\sinh\left(4\bar{\psi}\right)$$

or,

$$\frac{d}{d\bar{y}}\left(\left(\frac{d\bar{\psi}}{d\bar{y}}\right)^2\right) = \frac{K^2}{8}\frac{d}{d\bar{y}}\left(\cosh\left(4\bar{\psi}\right)\right) \tag{1.60}$$

On integrating (1.60) once, we get

$$\left(\frac{d\bar{\psi}}{d\bar{y}}\right)^2 = \frac{K^2}{8}\cosh\left(4\bar{\psi}\right) + C_1$$

For most microfluidic applications, the electrical potential is zero at some distance slightly beyond the EDL region. This is not true for the case when the EDL originating at the two walls overlaps in the mid-channel region. Therefore, for the case when there is no overlap, the boundary conditions specified above can be extended to $\bar{\psi} = 0$ at $\bar{y} = 1$.

Using this extra condition gives $C_1 = -K^2/8$. Thus, we have

$$\left(\frac{d\bar{\psi}}{d\bar{y}}\right)^2 = \frac{K^2}{8}\left[\cosh\left(4\bar{\psi}\right) - 1\right] \tag{1.61}$$

Now, using the identity $\cosh(2x) = 1 + 2\sinh^2(x)$, we obtain, from (1.61)

$$\left(\frac{d\bar{\psi}}{d\bar{y}}\right)^2 = \frac{K^2}{4}\sinh^2\left(2\bar{\psi}\right)$$

or,

$$\left(\frac{d\bar{\psi}}{d\bar{y}}\right) = -\sqrt{\frac{K^2}{4}\sinh^2\left(2\bar{\psi}\right)}$$

or,

$$\frac{d\bar{\psi}}{d\bar{y}} = -\frac{K}{2}\sinh\left(2\bar{\psi}\right) \tag{1.62}$$

The negative sign indicates that if the zeta potential is negative, the potential will increase from the zeta potential at the shear plane to zero in the bulk solution. Proceeding to solve the differential equation (1.62), we get

$$\frac{d\bar{\psi}}{\sinh\left(2\bar{\psi}\right)} = -\frac{K}{2}d\bar{y}$$

or,

$$\frac{\cosh^2(\bar{\psi}) - \sinh^2(\bar{\psi})}{2\sinh(\bar{\psi})\cosh(\bar{\psi})} d\bar{\psi} = -\frac{K}{2} d\bar{y}$$

(using the identity $\cosh^2(x) - \sinh^2(x) = 1$)

or,

$$\frac{1 - \tanh^2(\bar{\psi})}{\tanh(\bar{\psi})} d\bar{\psi} = -K d\bar{y}$$

or,

$$\frac{\text{sech}^2(\bar{\psi})}{\tanh(\bar{\psi})} d\bar{\psi} = -K d\bar{y}$$

(using the identity $\text{sech}^2(x) + \tanh^2(x) = 1$)

or,

$$\frac{d(\tanh(\bar{\psi}))}{\tanh(\bar{\psi})} = -K d\bar{y}$$

or,

$$\tanh(\bar{\psi}) = C_2 \exp(-K\bar{y})$$

Using the boundary condition $\bar{\psi} = \bar{\zeta}$ at $\bar{y} = 0$, we get

$$\tanh(\bar{\psi}) = \tanh(\bar{\zeta}) \exp(-K\bar{y})$$

Finally,

$$\bar{\psi} = \tanh^{-1}\left[\tanh(\bar{\zeta}) \exp(-K\bar{y})\right] \tag{1.63}$$

In the case when the EDLs are thick, the two electrostatic potentials generated by the two plates may be taken as additive:

$$\bar{\psi}(\bar{y}) = \tanh^{-1}\left[\tanh(\bar{\zeta}) \exp(-K\bar{y})\right] + \tanh^{-1}\left[\tanh(\bar{\zeta}) \exp(-2K + K\bar{y})\right] \tag{1.64}$$

The cases discussed here, though not exhaustive, are certainly very significant ones from the point of modeling real situations. Another realm not touched upon here is that under the case of EDL overlap. When the EDL charge distribution is such that the ratio $K = H/\lambda < 4$, the center-line condition $\psi(y = H) = 0$ is no longer valid. In that case, no analytical closed form solution of (1.58) can be found.

It is important to remember that in the description of the EDL through Poisson–Boltzmann equation, the ions are assumed to be point charges and non-interacting.

This means that the real finite size of ions is neglected or, in other words, steric effects are not considered. Indeed, the Boltzmann distribution of (1.31) predicts an unbounded and physically impossible growth in counterion concentration. This aphysical picture is particularly manifested at high values of zeta potential. To overcome these limitations of the Poisson–Boltzmann equation, modifications incorporating the finite size of the ions have been invoked in the literature, leading to modified Poisson–Boltzmann equations [10]. The concerned derivations are not reproduced in this introductory reading.

Having an idea about the potential profile in the channel, we can proceed to find the velocity profile using (1.50). We consider the simple case corresponding to Case I discussed previously. The non-dimensional velocity profile in that case is shown in Fig. 1.2. Note that the velocity u has been non-dimensionalized by the Helmholtz–Smoluchowski velocity $-\varepsilon\zeta E/\mu$, while the coordinate y along the channel height has been non-dimensionalized by the half-channel height H. The non-dimensional parameter $K = H/\lambda$ represents the relative height of the channel (half-height) compared to the characteristic thickness of the EDL.

It can be seen from the figure that the velocity profile is a plug-like one (except for a thin region near the walls) with magnitude equal to u_{HS}. Thus, in the case of thin EDL, the electroosmotic flow can be thought of as having a constant profile with a slip length equal to u_{HS}. In sharp contrast to this, the velocity profile in traditional pressure gradient-driven flow is a parabolic one. The immediate advantage of this plug-like EOF profile over the parabolic case is that shear-induced axial dispersion is significantly reduced. Furthermore, the EOF velocity is not dependent on the geometrical dimensions unlike the pressure gradient-driven velocity which decreases with the second power of channel size. This, along with the elimination of moving components as well as easy integrability with the electrical/electronic

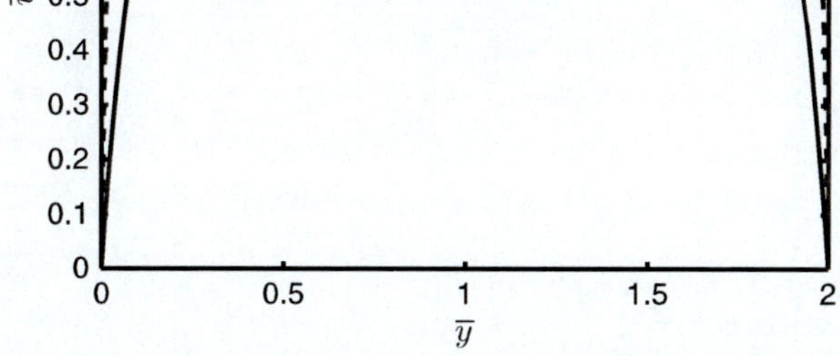

Fig. 1.2 Non-dimensional velocity in electroosmotic flow

circuitry in lab-on-a-chip devices, is the primary motivation of using EOF in miniaturized systems. However, EOF is not without its share of disadvantages: It has a strong dependence on the surface chemistry and the physico-chemical properties of the solution. The plug-like velocity profile of ideal EOF may be disturbed when stray pressure gradients are induced due to non-uniformities on the surface leading, in turn, to non-negligible axial dispersions. Not only that, EOF can give rise to significant Joule heating problems that may destabilize thermally labile biological samples. Be that as it may, EOF continues to be the most popular alternative to traditional pressure gradient-driven flow in the microfluidic realm.

3.2 Induced Charge Electroosmosis (ICEO)

It was discussed at the beginning of Section 3.1 that the genesis of EOF lies in a cloud of charge getting formed and the application of an electric field in the proper direction to induce motion of the charges (in this cloud) and, concomitantly, the suspending fluid. How the charge cloud comes about does not dictate the existence of EOF. With this fresh insight, let us look at a "different" kind of EOF.

When a conductor is placed in an electric field $\vec{E}$ as shown in Fig. 1.3, the conduction electrons reorient themselves in such a way that the net electric field within the conductor is zero, and all the induced charges appear on the surface of the conductor. The net effect is that a conductor placed in an electric field develops a surface charge (satisfying, of course, the initial electroneutrality condition within the whole conductor body). Now, if an aqueous solution of ions is present around this conductor, these surface charges will attract the oppositely charged ions, leading to the development of a cloud of ions that screens the surface charge on the conductor. It is extremely important to understand that if there were no ions present in the fluid (ideally), the electric lines of force would have met the surface of the conductor at right angles. But the presence of the oppositely charged screening cloud of ions "deflects" these lines of force such that very near to the surface these lines are almost tangential. The gross effect is that the distribution of the electric lines of force around the conductor together with its screening cloud of ions appears to be identical to that in the case of an ideal non-conductor.

If we now focus on the region in the close vicinity of the conductor, the situation is not much unlike what was present in the traditional EOF. We have a screening

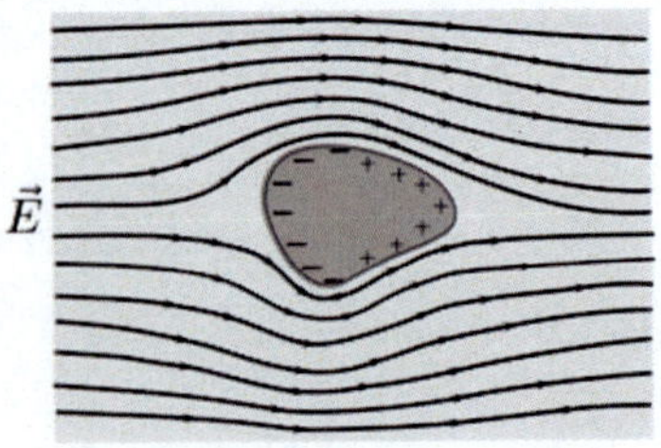

Fig. 1.3 Schematic of induced charge electroosmosis

cloud of ions (predominantly charged opposite to the surface charge) together with an electric field that is tangential to the surface. In this situation, it is intuitive to expect that the ions in the screening cloud will be actuated to move parallel to the surface of the conductor, inducing, in turn, the suspending liquid to move with them due to drag force. And that is exactly what happens, indeed.

The foregoing qualitative understanding of the "different" EOF has been developed by drawing analogies with the "traditional" EOF. The fluid actuation mechanism is, indeed, the same physically. Be that as it may, it would be naïve not to appreciate the differences between the two situations. In the case of the "traditional" EOF, the surface charging mechanism and the distribution of the screening cloud of charges (the EDL) was independent of the application of the external electric field. In fact, the same EDL distribution is established (given the same surface and solution conditions) irrespective of the fact that the external electric field is applied or not. However, in the present situation, the surface charging, and thus the development of the screening cloud, is totally dependent on the applied electric field. In simple terms, it might be said that the electric field actuates what it itself induces. And that is a clue to why this kind of flow is called induced charge electroosmosis (ICEO). Even from this qualitative description, it may be appreciated that since the surface charge and the screening cloud distribution is induced by the external electric field, the quantitative description would result in this distribution being a function of the external field. Again, since the flow actuation is brought about by the influence of the very same electric field, the resultant flow is to be expected to have a nonlinear dependence on the external field. This is indeed so, and thus, ICEO comes under the purview of nonlinear electrokinetic phenomena.

3.3 Streaming Potential

It is not necessary to apply an external electric field in order for electrokinetic effects to play a significant role in micro-scale flows. Even in purely pressure-driven flows in the presence of a charge distribution in the form of a double layer, electrokinetic effects are manifested. As the ions are advected along with the flow, they build up in the reservoirs and create a back potential. The current that is generated as a result of the advection of the ions is called streaming current. The back potential that is developed as a result of this streaming current is called *streaming potential*. This is depicted schematically in Fig. 1.4. Again, a current is generated in response to this streaming potential and it is called the conduction current. The direction of the conduction current is opposite to that of the streaming current. At equilibrium, since no external electric field is applied, the net ionic current in the system must be zero. Indeed, it is this condition of zero ionic current that is utilized to calculate the electric field associated with the streaming potential. Let us see how.[5]

We start with the general Boltzmann distribution from (1.31) with the additional consideration of a symmetric electrolyte:

---

[5] We follow the mathematical treatment presented in [6].

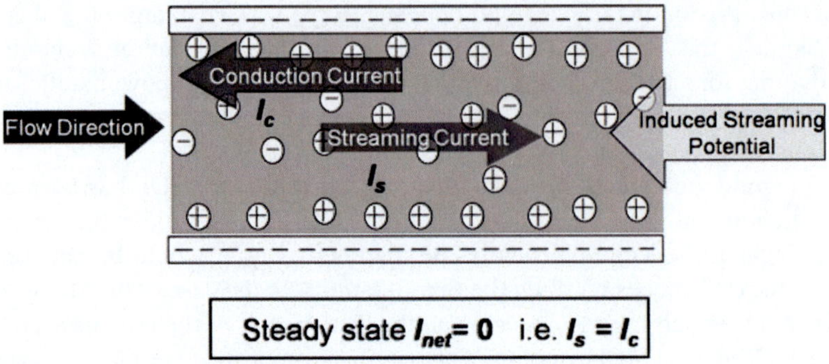

Fig. 1.4 Generation of streaming potential

$$n_\pm = n_0 \exp\left(\mp \frac{ez\psi}{k_B T}\right)$$

This, in turn, can be employed to estimate the total ionic current in a micro/nano-channel of height $2H$ (with y varying from 0 to $2H$) as

$$I_{\text{ionic}} = e \int_0^{2H} (u_+ z_+ n_+ + u_- z_- n_-)\, dy \tag{1.65}$$

For a z:z symmetric electrolyte such that $z_+ = -z_- = z$,

$$I_{\text{ionic}} = ez \int_0^{2H} (u_+ n_+ - u_- n_-)\, dy \tag{1.66}$$

Here, u_+ (u_-) refer to the axial velocities of the cations (anions), expressed as

$$u_\pm = u \pm \frac{zeE}{f_\pm} \tag{1.67}$$

Under the assumption of identical values of cationic and/or anionic friction coefficient of charge $f = f_+ = f_-$, the expression for I_{ionic} from (1.66) simplifies to

$$I_{\text{ionic}} = ez \int_0^{2H} (n_+ - n_-)\, u dy + \frac{e^2 z^2 E}{f} \int_0^{2H} (n_+ + n_-)\, dy \tag{1.68}$$

As discussed previously, for pure pressure-driven transport, I_{ionic} becomes identically zero; the corresponding value of E is called the streaming field E_s.

The velocity field is the result not only of the pressure gradient drive but also of the streaming potential. If we disregard the origin of this back potential, the situation is just like a flow taking place under the combined effect of a pressure gradient and an electric field (albeit, an induced one). In such a situation, we can take a cue from (1.49) and write the governing equation for the velocity as

$$0 = -\frac{dp}{dx} + \mu\frac{\partial^2 u}{\partial y^2} + \rho_e E_s \tag{1.69}$$

We make use of the Poisson equation, to obtain from (1.69)

$$\mu\frac{\partial^2 u}{\partial y^2} = \frac{dp}{dx} + \varepsilon\frac{\partial^2 \psi}{\partial y^2}E_s \tag{1.70}$$

The boundary conditions to be applied are, as before, u=0 and $\psi = \zeta$ at y=0 and u=0 and $\psi = \zeta$ at y=2H. Using these, we get

$$\begin{aligned} u &= \underbrace{-\frac{1}{2\mu}\frac{dp}{dx}\left(2Hy - y^2\right)}_{u_p} \underbrace{-\frac{\varepsilon\zeta E_s}{\mu}\left(1 - \frac{\psi}{\zeta}\right)}_{u_e} \\ &= u_p + u_e \end{aligned} \tag{1.71}$$

When we substitute this expression of u in (1.68) and set I_{ionic} to zero, we can solve for the expression for the induced streaming potential field E_s as

$$E_s = \frac{n_0 ez \int_0^{2H} u_p \sinh\left(\frac{ez\psi}{k_B T}\right) dy}{\frac{n_0 e^2 z^2}{f}\int_0^{2H} \cosh\left(\frac{ez\psi}{k_B T}\right) dy + \frac{n_0 ez\varepsilon\zeta}{\mu}\int_0^{2H}\left(1 - \frac{\psi}{\zeta}\right)\sinh\left(\frac{ez\psi}{k_B T}\right) dy} \tag{1.72}$$

Using the expression of u_p from (1.71) in (1.72), we get

$$E_s = \frac{\frac{-f}{2ez\mu}\frac{dp}{dx}\int_0^{2H}\left(2Hy - y^2\right)\sinh\left(\frac{ez\psi}{k_B T}\right) dy}{\int_0^{2H} \cosh\left(\frac{ez\psi}{k_B T}\right) dy + \frac{f\varepsilon\zeta}{\mu ez}\int_0^{2H}\left(1 - \frac{\psi}{\zeta}\right)\sinh\left(\frac{ez\psi}{k_B T}\right) dy} \tag{1.73}$$

We use the non-dimensional scheme $\bar{y} = y/H$, $\bar{\psi} = ez\psi/4k_B T$ to obtain the following form of the streaming potential field:

$$E_s = \frac{\frac{-fH^2}{2ez\mu}\frac{dp}{dx}\int_0^2\left(2\bar{y} - \bar{y}^2\right)\sinh\left(4\bar{\psi}\right) d\bar{y}}{\int_0^2 \cosh\left(4\bar{\psi}\right) d\bar{y} + \frac{4f\varepsilon k_B T}{\mu e^2 z^2}\bar{\zeta}\int_0^2\left(1 - \frac{\bar{\psi}}{\bar{\zeta}}\right)\sinh\left(4\bar{\psi}\right) d\bar{y}}$$

or,

$$E_s = \frac{E_0 I_1}{I_2 + \bar{R} I_3} \tag{1.74}$$

where

$$I_1 = \int_0^2 (2\bar{y} - \bar{y}^2)\sinh(4\bar{\psi})\,d\bar{y}, \quad I_2 = \int_0^2 \cosh(4\bar{\psi})\,d\bar{y}$$

$$I_3 = \int_0^2 \bar{\zeta}\left(1 - \frac{\bar{\psi}}{\bar{\zeta}}\right)\sinh(4\bar{\psi})\,d\bar{y}, \quad \bar{R} = \frac{4f\varepsilon k_B T}{\mu e^2 z^2}, \text{ and } E_0 = -\frac{fH^2}{2ez\mu}\frac{dp}{dx}$$

The ionic friction coefficient of charge f may be expressed as $f = e^2 N_A / F^2 \Lambda$ where N_A is the Avogadro number, F is the Faraday constant, and Λ is the ionic mobility. f may also be expressed as $f = k_B T / D$ where D is the same diffusivity used in the Nernst–Planck equation (1.34).

It is important to note that the streaming potential is responsible for inhibiting the flow. The decreased volumetric flow rate may be attributed, on a gross scale, to an increased viscosity of the fluid, due to the electrokinetic effects. This is often referred to as the electroviscous effect. Since the electric field due to streaming potential, E_s, is a function of the applied pressure gradient, the velocity, explicitly considering the streaming potential effects, can be expressed as

$$u_1 = -\frac{1}{2\mu}\frac{dp}{dx} f_e(y)$$

If the streaming potential effects are incorporated within an effective increased viscosity (the electroviscous effect), then

$$u_2 = -\frac{1}{2\mu_{ev}}\frac{dp}{dx}\left(2Hy - y^2\right)$$

The volumetric rate should be equal in both these for them to be equivalent. Thus,

$$\int_0^{2H} u_1 dy = \int_0^{2H} u_2 dy$$

$$\frac{\mu_{ev}}{\mu} = \frac{2H^3/3}{\int_0^{2H} f_e dy} \tag{1.75}$$

Just as a back electric potential develops in a purely pressure-driven flow, a back pressure gradient may develop when a purely electroosmotic flow takes place between two reservoirs due to the changing liquid levels in the two reservoirs. At equilibrium, the flow due to the back pressure is equal to the flow due to the application of the external electric field and there is no net flow. This is known as the finite reservoir size effect.

3.4 Electrophoresis

It is not difficult to understand electrophoresis if one has grasped the idea of electroosmosis. Both these phenomena refer to the relative movement of a charged surface and a liquid. In electroosmosis, the charged surface is held stationary with the liquid moving past it, while in electrophoresis it is the charged surface that moves relative to the liquid. In both the cases, an external electric field is the actuator of the motion. The similarity between electroosmosis and electrophoresis becomes striking if we fix our frame of reference on the moving electrophoretic particle. From such a viewpoint the relative motion manifests itself just like electroosmosis of the liquid past the particle. Indeed, this is what is done in the mathematical treatment of electrophoresis as will be clear in the analysis that follows.

The electrophoretic particle can be any charged body – a colloid, a macromolecule, or even a microorganism. This charge may be intrinsic (like in a DNA molecule) or it may be an induced one (like EDL formation). Consequently, in the most general case, the particle can be of any irregular shape with charge that can be intrinsic or induced. However, as a body representative of many real physical situations and for the ease of mathematical analysis,[6] we will restrict ourselves to the electrophoresis of a spherical particle. We will also restrict ourselves to a situation where the charge on this spherical particle is brought about due to EDL formation.

The electrical potential around a spherical particle in a polar medium can be found from the Poisson's equation (reduced form due to spherical symmetry):

$$\frac{1}{r^2}\frac{d}{dr}\left(r^2\frac{d\psi}{dr}\right) = -\frac{\rho_e}{\varepsilon} \tag{1.76}$$

or,

$$\frac{1}{r^2}\frac{d}{dr}\left(r^2\frac{d\psi}{dr}\right) = \frac{2n_0 ez}{\varepsilon}\sinh\left(\frac{ez\psi}{k_B T}\right) \tag{1.77}$$

Invoking the Debye–Hückel linearization for small potential values, we get

$$\begin{aligned}\frac{1}{r^2}\frac{d}{dr}\left(r^2\frac{d\psi}{dr}\right) &= \frac{2n_0 e^2 z^2}{\varepsilon k_B T}\psi \\ \frac{1}{r^2}\frac{d}{dr}\left(r^2\frac{d\psi}{dr}\right) &= \frac{1}{\lambda^2}\psi\end{aligned} \tag{1.78}$$

where $1/\lambda^2 = 2n_0 e^2 z^2/\varepsilon k_B T$. In order to solve (1.78), we make the substitution $\xi = r\psi$. Using this, we obtain from (1.78)

[6] We follow the structure of the development presented in [15].

$$\frac{1}{r}\frac{d^2\xi}{dr^2} = \frac{1}{\lambda^2}\frac{\xi}{r}$$

or,

$$\frac{d^2\xi}{dr^2} = \frac{\xi}{\lambda^2} \tag{1.79}$$

The general solution of (1.79) is of the form $\xi = A\exp(r/\lambda) + B\exp(-r/\lambda)$. The boundary condition $\psi \to 0$ as $r \to \infty$ implies ξ is finite when $r \to \infty$. From this, we get $A = 0$. Again, the boundary condition $\psi = \zeta$ when $r = R$ implies $\xi = \zeta R$ when r=R. From this, we get $B = R\zeta\,\exp(R/\lambda)$. Therefore,

$$\psi = \frac{\zeta R}{r}\exp\left(-\frac{r-R}{\lambda}\right) \tag{1.80}$$

The total surface charge on the spherical particle is (by the condition of electroneutrality) negative of the total charge distribution in the double layer. This condition is used to find an expression for the surface charge:

$$q_{\text{surface}} = -q_{\text{EDL}} = -\int_R^\infty 4\pi r^2 \rho_e dr \tag{1.81}$$

From (1.76),

$$\rho_e = -\varepsilon\frac{1}{r^2}\frac{d}{dr}\left(r^2\frac{d\psi}{dr}\right)$$

So, from (1.81), we get

$$q_{\text{surface}} = 4\pi\varepsilon\int_R^\infty \frac{d}{dr}\left(r^2\frac{d\psi}{dr}\right)dr = 4\pi\varepsilon\left[\cancelto{0}{r^2\frac{d\psi}{dr}\bigg|_\infty} - r^2\frac{d\psi}{dr}\bigg|_R\right] = -4\pi\varepsilon R^2\frac{d\psi}{dr}\bigg|_R \tag{1.82}$$

From (1.80), we can see that

$$\frac{d\psi}{dr}\bigg|_R = -\zeta\left(\frac{1}{R}+\frac{1}{\lambda}\right)$$

So,

$$q_{\text{surface}} = 4\pi\varepsilon R^2\zeta\left(\frac{1}{R}+\frac{1}{\lambda}\right) \tag{1.83}$$

Case I: $\lambda \gg R$

In the case where the Debye length λ is large compared to R, the particle may be treated as a point charge. The electrical force may, then, be equated with the Stokes drag on the particle to find its velocity (for steady motion). Thus,

$$q_{\text{surface}}E_x = 6\pi\mu RU \tag{1.84}$$

where U is the velocity of steady motion of the particle and E_x is the unperturbed electric field.

Using (1.84) in (1.83) and simplifying, we obtain

$$U = \frac{2}{3}\frac{\zeta\varepsilon\left(1 + {}^{R}\!/_{\lambda}\right)E_x}{\mu} \tag{1.85}$$

Since we are analyzing the situation where $R \ll \lambda$, we have

$$U \approx \frac{2}{3}\frac{\zeta\varepsilon E_x}{\mu} \tag{1.86}$$

This is known as the *Hückel equation.*

Case II: $\lambda \ll R$

When the Debye length is very small compared to the radius of the particle, in regions very close to the surface of the particle, the curvature of the particle can be neglected so that the EDL may be viewed just as it was in the case of the planar substrate of an EOF. Indeed, drawing an analogy with EOF taking place in a parallel plate geometry where the plug-like flow profile (for very thin EDL) magnitude is given by the Helmholtz–Smoluchowski velocity, we may write the fluid velocity parallel to the particle surface as

$$U_{||} = -\frac{\varepsilon\zeta E_{||}}{\mu} \tag{1.87}$$

where $E_{||}$ represents the electric field just close to the particle surface and $U_{||}$ is the tangential velocity of the fluid. Of course, it is important to note that $U_{||}$ and $E_{||}$ vary along the surface of the particle. Moreover, this velocity of the fluid is written from a reference frame fixed on the particle, so that from this frame, the fluid appears to flow past the surface. In order to find the full solution of the velocity we will follow a subtle logic. The rigorous mathematical treatment of this logical deduction can be read in [14]. We first consider the electric field distribution as it stands with the particle embedded in the field. The electric field may be represented by the potential as $\vec{E} = -\nabla\phi$. Since there are no free charges ($\rho_e = 0$), the application of Poisson's equation $\nabla \cdot \vec{E} = -\rho_e/\varepsilon$ gives

$$\nabla^2\phi = 0 \tag{1.88}$$

The boundary conditions which this standard Laplace equation is subjected to are the no penetration $\hat{n} \cdot \nabla\phi = 0$ (i.e., $\partial\phi/\partial n = 0$) and the "far-stream" ϕ_∞ (corresponding to the unperturbed electric field in the region far away from the particle). In order to solve this equation, we start with the expanded form of the Laplace equation (note the reduction to two variables from symmetry considerations):

$$\frac{\partial}{\partial r}\left(r^2\frac{\partial\phi}{\partial r}\right) + \frac{1}{\sin\theta}\frac{\partial}{\partial\theta}\left(\sin\theta\frac{\partial\phi}{\partial\theta}\right) = 0 \tag{1.89}$$

This equation may be solved using separation of variables. Let us assume a solution of the form $\phi = G(r)\,H(\theta)$. Then,

$$\frac{\partial}{\partial r}\left(r^2\frac{dG}{dr}H\right)+\frac{1}{\sin\theta}\frac{\partial}{\partial\theta}\left(G\sin\theta\frac{dH}{d\theta}\right)=0 \tag{1.90}$$

or,

$$\frac{\partial}{\partial r}\left(r^2\frac{dG}{dr}H\right)=-\frac{1}{\sin\theta}\frac{\partial}{\partial\theta}\left(G\sin\theta\frac{dH}{d\theta}\right)=K\,(\text{say}) \tag{1.91}$$

From (1.91) we get the following two equations:

$$\frac{1}{\sin\theta}\frac{d}{d\theta}\left(\sin\theta\frac{dH}{d\theta}\right)+KH=0 \tag{1.92}$$

and

$$\frac{1}{G}\frac{d}{dr}\left(r^2\frac{dG}{dr}\right)=K \tag{1.93}$$

Let us assume that K can be denoted as $n(n+1)$. Then, from (1.93), we get

$$\left(r^2G'\right)'=n\,(n+1)\,G$$

$$\Rightarrow r^2G''+2rG'-n\,(n+1)\,G=0 \tag{1.94}$$

This equation is in the Euler–Cauchy form. Thus, we can assume a solution of the form $G=r^{\alpha}$. Substituting this form in (1.94), we get

$$\alpha\,(\alpha-1)+2\alpha-n\,(n+1)=0 \tag{1.95}$$

The solutions of (1.95) are $\alpha = n,\ -1-n$. Therefore, the two solutions of (1.94) are $G_1\,(r)=r^n$ and $G_2\,(r)=1/r^{n+1}$.

Now, let us turn our attention to the solution of (1.92). Let $\cos\theta = w$, then $\sin^2\theta = 1-w^2$ and

$$\frac{d}{d\theta}=\frac{d}{dw}\frac{dw}{d\theta}=-\sin\theta\frac{d}{dw}$$

Therefore,

$$\frac{1}{\sin\theta}\frac{d}{d\theta}\left(\sin\theta\frac{dH}{d\theta}\right)+KH=0$$

or,

$$-\frac{d}{dw}\left[\sin\theta\left(-\sin\theta\frac{dH}{dw}\right)\right]+KH=0$$

or,

$$-\frac{d}{dw}\left[\left(1-w^2\right)\frac{dH}{dw}\right]+n\left(n+1\right)H=0$$

or,

$$\left(1-w^2\right)\frac{d^2H}{dw^2}-2w\frac{dH}{dw}+n\left(n+1\right)H=0 \tag{1.96}$$

This is the form of Legendre's equation. In order that the solution (together with its derivatives) of the Laplace equation (1.89) be continuous, it is necessary to restrict n to integer values (following Kreyszig [11]).

For n=0, 1, ... the Legendre polynomials $H=P_n\left(w\right)=P_n\left(\cos\theta\right)$ are solutions of Legendre's equation. Thus, there are two series of solution of (1.96):

$$\phi_{n1}\left(r,\theta\right)=A_nr^nP_n\left(\cos\theta\right)\ \text{and}\ \ \phi_{n2}\left(r,\theta\right)=B_n\frac{1}{r^{n+1}}P_n\left(\cos\theta\right)$$

Thus, the general solution can be expressed as

$$\phi\left(r,\theta\right)=\sum_{n=0}^{\infty}A_nr^nP_n\left(\cos\theta\right)+\sum_{n=0}^{\infty}B_n\frac{1}{r^{n+1}}P_n\left(\cos\theta\right) \tag{1.97}$$

The boundary conditions that must be satisfied are

$$\lim_{r\to\infty}\phi\left(r,\theta\right)=-E_xr\cos\theta\quad\text{and}\quad\left.\frac{\partial\phi\left(r,\theta\right)}{\partial r}\right|_{r=R}=0$$

The first boundary condition gives us a clue that ϕ should be of the form

$$\phi=-E_xr\cos\theta+\sum_{n=0}^{\infty}B_n\frac{1}{r^{n+1}}P_n\left(\cos\theta\right)$$

meaning $A_1=-E_x$ and A_n=0 for n=0 and $\forall n\geq 2$. And, from the second boundary condition,

$$-E_x\cos\theta-\frac{B_0P_0\left(\cos\theta\right)}{R^2}-\frac{2B_1P_1\left(\cos\theta\right)}{R^3}-\cdots=0 \tag{1.98}$$

We note that $P_0\left(x\right)=1$, $P_1\left(x\right)=x$, and so on. Using this in (1.98) and equating the like powers of $\cos\theta$, we get

$$B_0 = 0, \quad B_1 = \frac{1}{2}A_1R^3 = -\frac{1}{2}E_xR^3, \text{ and } \quad B_n = 0 \quad \forall n \geq 2$$

Therefore, from (1.97), we obtain

$$\phi = -E_x r\cos\theta - \frac{1}{2}E_xR^3\frac{\cos\theta}{r^2} = -E_x\cos\theta\left(r + \frac{R^3}{2r^2}\right) \tag{1.99}$$

From this solution, we can see that

$$E_{||} = \frac{1}{r}\frac{\partial\phi}{\partial\theta}\bigg|_{r=R} = -\frac{3}{2}E_x\sin\theta$$

together with $E_\perp = 0$ (true to the boundary condition utilized).

The velocity profile should be such that it satisfies the Navier–Stokes equations along with the slip boundary condition (1.87) and ensures that the flow does not exert any force or moment on the particle. It has been shown by Morrison [14] that irrotational flow satisfies these conditions. An irrotational flow velocity field is derivable from a potential $\vec{u} = -\nabla\Phi$. Furthermore, the velocity should satisfy the no penetration boundary condition $\hat{n}\cdot\vec{u} = 0$ or $\hat{n}\cdot\nabla\Phi = 0$. It may, now, be immediately observed that the differential equations and the boundary conditions are identical for the electric potential and the velocity potential. With the appreciation of this fact, we can take a clue from (1.87) (relating the slip velocity to the tangential electrical field) and write the relation between the potential Φ corresponding to the velocity field and the potential ϕ corresponding to the electrical field as

$$\Phi = -\frac{\varepsilon\zeta}{\mu}\phi \tag{1.100}$$

Writing this in terms of the velocity and electric field, we get

$$U = -\frac{\varepsilon\zeta E_x}{\mu} \tag{1.101}$$

So, the velocity of the particle is, in a reference fixed in space,

$$U_{\text{particle}} = \frac{\varepsilon\zeta E_x}{\mu} \tag{1.102}$$

This is the *Helmholtz–Smoluchowski equation.*

We have, until now, considered two special cases – first, when the Debye length is much larger than the radius of the particle and, second, when the Debye length is much smaller than the radius. These are the extreme cases of a general situation where the Debye length and the radius of the particle are of comparable dimensions. We will not provide the mathematical treatment of this general case, instead direct the interested reader to [15].

3.5 Dielectrophoresis

The force acting on a charged particle placed in an electric field, resulting in electrophoresis, could be intuitively understood. However, a force may also act on an uncharged but polarizable particle if it is placed in a non-uniform electric field, and this force actuates the translational motion of the particle. This phenomenon is called dielectrophoresis. It is important to note that the nature of the dielectrophoretic force is strongly dependent on the non-uniformity of the electric field and the dielectric properties of both the particle and the surrounding medium. Before proceeding to a full-fledged mathematical treatment of dielectrophoresis, let us first understand qualitatively the physical picture. Under the action of an electric field a polarizable particle will polarize so that a dipole is induced in it. If the electric field were uniform, no net force would have acted on the particle. But, since the applied electric field is spatially non-uniform, different forces act on the two ends of the "dipole" so that a net electric force does act on the particle.

Following arguments similar to the case of electrophoresis, we consider a spherical particle as shown in Fig. 1.5 and present a mathematical analysis of the dielectrophoretic force acting on it. Let the permittivity of the particle be ε_2 and that of the medium be ε_1. The electric potential in the region interior to the particle is ϕ_i while that in the surrounding medium is ϕ_e. The radius of the spherical particle is R. The governing differential equation, just like in the case of electrophoresis, is the Laplace equation $\nabla^2\phi = 0$ ($\phi\,(r \leq R) \equiv \phi_i$ and $\phi\,(r \geq R) \equiv \phi_e$). The boundary conditions that must be satisfied are[7]

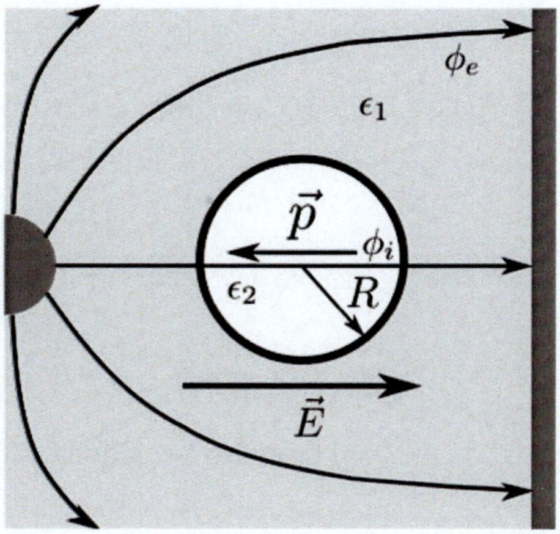

Fig. 1.5 Dielectrophoresis of a spherical particle

[7] We start with the presentation structure found in the chapter on dielectrophoresis in [2]; however, for pedagogical reasons, we give, here, a more detailed mathematical derivation.

$$\phi_i\,(0,\theta)\text{ is finite} \tag{1.103}$$

$$\phi_i\,(R,\theta) = \phi_e\,(R,\theta) \tag{1.104}$$

$$\varepsilon_2 \frac{\partial \phi_i\,(R,\theta)}{\partial r} = \varepsilon_1 \frac{\partial \phi_e\,(R,\theta)}{\partial r} \tag{1.105}$$

and

$$\lim_{r\to\infty} \phi_e\,(r,\theta) = -E_0 r\cos\theta \tag{1.106}$$

From inspection, it is clear that in order to satisfy the fourth boundary condition (1.106), the solution has to be of the form

$$\phi_e = -E_0 r\cos\theta + \sum_{n=0}^{\infty} B_{ne} \frac{1}{r^{n+1}} P_n\,(\cos\theta)$$

meaning $A_{1e} = -E_0$ and $A_{ne} = 0$ for $n = 0$ and $\forall n \geq 2$. And, from the first boundary condition (1.103) that ϕ_i should be finite at $r = 0$ gives us $\phi_i = \sum_n^\infty A_{ni} r^n P_n\,(\cos\theta)$ meaning $B_{ni} = 0\ \forall n = 0, 1, \ldots$. Now, using the second boundary condition (1.104), we get

$$\begin{aligned} -E_0 r\cos\theta + B_{0e}\frac{1}{R}P_0\,(\cos\theta) + B_{1e}\frac{1}{R^2}P_1\,(\cos\theta) + \cdots = A_{0i}P_0\,(\cos\theta) \\ + A_{1i}RP_1\,(\cos\theta) + \cdots \end{aligned} \tag{1.107}$$

We know that $P_0\,(x) = 1$, $P_1\,(x) = x$, and so on. Utilizing this in (1.107), we get

$$-E_0 r\cos\theta + B_{0e}\frac{1}{R} + B_{1e}\frac{1}{R^2}\cos\theta + \cdots = A_{0i} + A_{1e}R\cos\theta + \cdots$$

Comparing the like powers of $\cos\theta$, we get

$$A_{0i} = 0, A_{1i} = -E_0 + \frac{B_{1e}}{R^3}, B_{0e} = 0, B_{ne} = 0\ \forall n \geq 2,\text{ and } A_{ni} = 0\ \forall n \geq 2 \tag{1.108}$$

Using the third boundary condition (1.105)

$$\varepsilon_1 \left.\frac{\partial \phi_e}{\partial r}\right|_{r=R} = \varepsilon_2 \left.\frac{\partial \phi_i}{\partial r}\right|_{r=R}$$

or,

$$\varepsilon_1 \left[A_{1e}P_1\,(\cos\theta) - \frac{2B_{1e}}{R^3}P_1\,(\cos\theta)\right] = \varepsilon_2 A_{1i} P_1\,(\cos\theta)$$

or,

$$\frac{\varepsilon_1}{\varepsilon_2}\left[A_{1e} - \frac{2B_{1e}}{R^3}\right] = A_{1i} \tag{1.109}$$

Comparing (1.109) and the second equality in (1.108), we get

$$-E_0 + \frac{B_{1e}}{R^3} = \frac{\varepsilon_1}{\varepsilon_2}\left[-E_0 - \frac{2}{R^3}B_{1e}\right]$$

or,

$$B_{1e} = E_0 R^3 \frac{\varepsilon_2 - \varepsilon_1}{\varepsilon_2 + 2\varepsilon_1}$$

Therefore, using (1.109), we get

$$A_{1i} = -E_0 + E_0\frac{\varepsilon_2 - \varepsilon_1}{\varepsilon_2 + 2\varepsilon_1} \quad \text{or} \quad A_{1i} = -E_0\frac{3\varepsilon_1}{\varepsilon_2 + 2\varepsilon_1} \tag{1.110}$$

Thus, the potential ϕ is given by

$$\phi = \begin{cases} -E_0 r\cos\theta + E_0\dfrac{R^3}{r^2}\left(\dfrac{\varepsilon_2 - \varepsilon_1}{\varepsilon_2 + 2\varepsilon_1}\right)\cos\theta, & r > R \\ -E_0\dfrac{3\varepsilon_1}{\varepsilon_2 + 2\varepsilon_1}, & r < R \end{cases} \tag{1.111}$$

Let us, now, find the potential at a distance r (from the origin of a coordinate system) due to a charge distribution of density ρ_e within the domain bounded by Γ as shown in Fig. 1.6.

The permittivity of the medium is ε_1. The charge distribution domain is very small such that $\max\left[r'\right] \ll r$, where r' is the distance of an elemental volume from

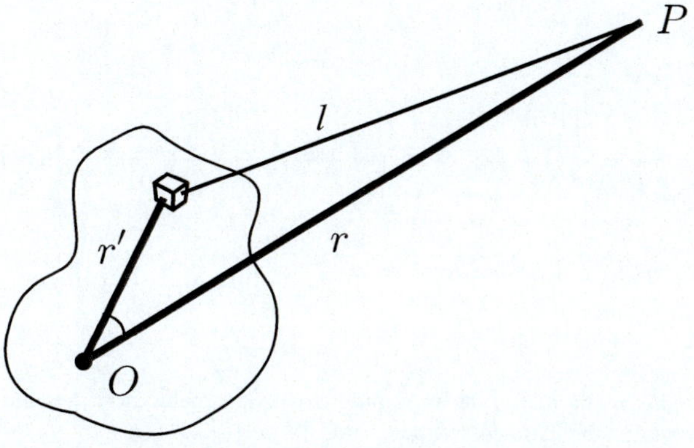

Fig. 1.6 Potential at a point due to a charge distribution in space

the origin. The potential due to this elemental volume $d\tau$ of the charge density is given by

$$dV = \frac{1}{4\pi\varepsilon_1}\frac{\rho_e d\tau}{l} \tag{1.112}$$

where l is the distance of the point of interest from the elemental volume. Therefore, the potential due to the total charge distribution is

$$V = \frac{1}{4\pi\varepsilon_1}\int \frac{\rho_e}{l} d\tau \tag{1.113}$$

From Fig. 1.6, we can write

$$\cos\theta = \frac{r^2 + r'^2 - l^2}{2rr'} \text{ (using cosine identity from trigonometry)}$$

So,

$$l^2 = r^2 + r'^2 - 2rr'\cos\theta$$

or,

$$l^2 = r^2\left[1 + \frac{r'^2}{r^2} - 2\frac{r'}{r}\cos\theta\right]$$

or,

$$l = r\sqrt{1 + \epsilon}$$

where

$$\epsilon = \frac{r'}{r}\left(\frac{r'}{r} - 2\cos\theta\right) \ll 1$$

or,

$$\frac{1}{l} = \frac{1}{r}(1 + \epsilon)^{-1/2} = \frac{1}{r}\left(1 - \frac{1}{2}\epsilon + \frac{3}{8}\epsilon^2 - \cdots\right) \text{ (using binomial expansion)}$$

Thus, from (1.113), after expanding and simplifying, we obtain[8]

[8]This is, basically, the multipole expansion method. A detailed description of this approach is found in the classical textbook of Griffiths [8].

$$V = \frac{1}{4\pi\varepsilon_1}\left[\frac{1}{r}\int \rho d\tau + \frac{1}{r^2}\int r'\cos\theta\rho d\tau + \frac{1}{r^3}\int r'^2\left(\frac{3}{2}\cos^2\theta - \frac{1}{2}\right)\rho d\tau + \cdots\right] \tag{1.114}$$

For a charge distribution, where the total charge is zero (as it is in this case), the first term called the monopole term vanishes. The contribution of the dipole term is

$$V_{\text{dip}} = \frac{1}{4\pi\varepsilon_1}\frac{1}{r^2}\int r'\cos\theta\rho d\tau \tag{1.115}$$

Here, $r'\cos\theta = \hat{r}\cdot\vec{r}'$. Therefore, from (1.115), we can write

$$V_{\text{dip}} = \frac{1}{4\pi\varepsilon_1}\frac{1}{r^2}\hat{r}\cdot\int \vec{r}'\rho d\tau \tag{1.116}$$

The integral $\vec{p} = \int \vec{r}'\rho d\tau$ is called the dipole moment of the distribution. Thus,

$$V_{\text{dip}} = \frac{1}{4\pi\varepsilon_1}\frac{\hat{r}\cdot\vec{p}}{r^2} = \frac{1}{4\pi\varepsilon_1}\frac{p\cos\theta}{r^2} \tag{1.117}$$

Interestingly, if we take a physical dipole as shown in Fig. 1.7 (as a special case of the general charge distribution with total charge zero), we find the same magnitude of the potential as just found in (1.117). Let us see how. We first note that

$$V = \frac{1}{4\pi\varepsilon_1}\left|\frac{q}{l_+} - \frac{q}{l_-}\right|$$

Now, $l_\pm^2 = r^2 + r_\pm^2 - 2rr_\pm\cos\theta$ and $l_\pm \approx \left[1 - 2\left(r_\pm/r\right)\cos\theta\right]^{1/2}$. Therefore,

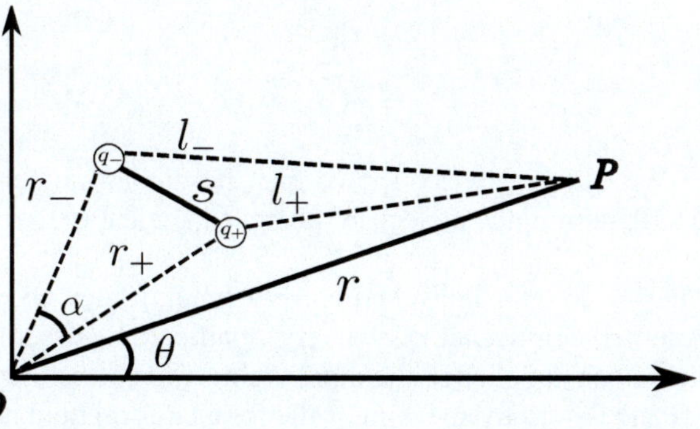

Fig. 1.7 Potential due to an actual dipole

$$V = \frac{1}{4\pi\varepsilon_1}\left|\frac{q}{r}\left(1 - 2\frac{r_+}{r}\cos\theta\right)^{-1/2} - \frac{q}{r}\left(1 - 2\frac{r_-}{r}\cos\theta\right)^{-1/2}\right|$$
$$= \frac{1}{4\pi\varepsilon_1}\frac{q}{r}\frac{|r_+ - r_-|}{r}\cos\theta \tag{1.118}$$

Now, $s^2 = r_+^2 + r_-^2 - 2r_+r_-\cos\alpha \approx r_+^2 + r_-^2 - 2r_+r_- = (r_+ - r_-)^2$. So, $s = |r_+ - r_-|$. Therefore, from (1.118), we get

$$V = \frac{qs\cos\theta}{4\pi\varepsilon_1 r^2} = \frac{q}{4\pi\varepsilon_1}\frac{\vec{s}\cdot\hat{r}}{r^2} = \frac{\vec{p}\cdot\hat{r}}{4\pi\varepsilon_1 r^2} = \frac{p\cos\theta}{4\pi\varepsilon_1 r^2} \tag{1.119}$$

Having shown that the dipole contribution of a charge distribution is equivalent to that of a physical dipole, we follow Bruus [2] to argue that if a given potential contains a component of the form $B\left(\cos\theta/r^2\right)$, then it implies that a dipole of strength $p = 4\pi\varepsilon_1 B$ is located at the center of the coordinate system. The motivation behind such a subtle argument is the form of the potential as found in (1.117) (or (1.119)).

Now, if we look at (1.111), we see that there, too, is a term $E_0R^3\left((\varepsilon_2 - \varepsilon_1)/(\varepsilon_2 + 2\varepsilon_1)\right)\left(\cos\theta/r^2\right)$ in the expression of the potential outside the spherical particle. Thus, by the artifice, just mentioned, we can say that there does exist a dipole of dipole moment

$$p = 4\pi\varepsilon_1 E_0 R^3\left(\frac{\varepsilon_2 - \varepsilon_1}{\varepsilon_2 + 2\varepsilon_1}\right) \tag{1.120}$$

The combination $K = \left(\varepsilon_2 - \varepsilon_1/(\varepsilon_2 + 2\varepsilon_1)\right)$ is called the *Clausius–Mossotti factor*.

If the electric field is non-uniform, the force on a dipole is given by

$$\begin{aligned} F &= F_+ - F_- \\ &= q(\vec{E}_{\vec{l}+d\vec{l}} - E_{\vec{l}}) \\ &= q\left[\frac{\partial\vec{E}}{\partial x}dx + \frac{\partial\vec{E}}{\partial y}dy + \frac{\partial\vec{E}}{\partial z}dz\right] \\ &= q\left(dx\hat{i} + dy\hat{j} + dz\hat{k}\right)\cdot\nabla\vec{E} = (\vec{p}\cdot\nabla)\vec{E} \end{aligned} \tag{1.121}$$

We had previously shown that the dipole moment associated with the charge distribution in the dielectric sphere is identical to that of a physical dipole. We had, of course, limited ourselves to dipole terms of the general multipole expansion. Thus, as long as the sphere is very small, it may be replaced by a physical dipole. Another significance of the very small particle assumption is that the non-uniform field does not change the dipole moment expression of (1.120) which was derived for and is, strictly speaking, valid for a uniform field. Now, we may proceed to write the expression for the dielectric force on the small dielectric spherical particle, using (1.121):

$$\vec{F}_{\mathrm{DEP}} = (\vec{p} \cdot \nabla)\vec{E} \tag{1.122}$$

with the understanding that $\vec{p}$ is the dipole moment of the charge distribution:

$$\vec{F}_{\mathrm{DEP}} = 4\pi\varepsilon_1 \frac{\varepsilon_2 - \varepsilon_1}{\varepsilon_2 + 2\varepsilon_1} R^3 \left(\vec{E}_0 \cdot \nabla\right)\vec{E}_0 \tag{1.123}$$

It is important to note that we have written $\vec{E} = \vec{E}_0$ by neglecting the higher order terms in the Taylor expansion containing the gradient terms.

Now, we simplify the $\left(\vec{E}_0 \cdot \nabla\right)\vec{E}_0$ term in (1.123) using the vector identity

$$\left(\vec{A} \cdot \nabla\right)\vec{B} = \nabla\left(\vec{A} \cdot \vec{B}\right) - (\vec{B} \cdot \nabla)\vec{A} - \vec{A} \times (\nabla \times \vec{B}) - \vec{B} \times \left(\nabla \times \vec{A}\right) \tag{1.124}$$

In this case, $\vec{A} = \vec{E}_0$ and $\vec{B} = \vec{E}_0$. Therefore,

$$\left(\vec{E}_0 \cdot \nabla\right)\vec{E}_0 = \nabla\left(\vec{E}_0 \cdot \vec{E}_0\right) - \left(\vec{E}_0 \cdot \nabla\right)\vec{E}_0 - \vec{E}_0 \times \left(\nabla \times \vec{E}_0\right) - \vec{E}_0 \times \left(\nabla \times \vec{E}_0\right) \tag{1.125}$$

We know that $\nabla \times \vec{E}_0 = 0$ because E_0 is the gradient of a scalar, and curl of the gradient of a scalar identically vanishes. Therefore, from (1.125), we obtain

$$2\left(\vec{E}_0 \cdot \nabla\right)\vec{E}_0 = \nabla\left(\vec{E}_0 \cdot \vec{E}_0\right)$$

or,

$$\left(\vec{E}_0 \cdot \nabla\right)\vec{E}_0 = \frac{1}{2}\nabla\left(\vec{E}_0 \cdot \vec{E}_0\right) \tag{1.126}$$

Using (1.126) in (1.123), we can write

$$\vec{F}_{\mathrm{DEP}} = 2\pi\varepsilon_1 \frac{\varepsilon_2 - \varepsilon_1}{\varepsilon_2 + 2\varepsilon_1} R^3 \nabla\left(\vec{E}_0 \cdot \vec{E}_0\right) \tag{1.127}$$

Up to this point, we have been considering ideal dielectrics (both particle and the suspending medium) having zero conductivities. But, for real dielectrics, we have to take into consideration the conductivities. This is done by incorporating the conductivities within a complex permittivity. Thus,

$$\hat{\varepsilon}_n = \varepsilon_n - i\frac{\sigma_n}{\omega} \quad \text{with} \quad n = 1, 2 \tag{1.128}$$

where σ_n is the conductivity and ω is the angular frequency. Then,

$$\vec{F}_{\mathrm{DEP}} = 2\pi \operatorname{Re}\left[\frac{\hat{\varepsilon}_1\left(\hat{\varepsilon}_2 - \hat{\varepsilon}_1\right)}{\hat{\varepsilon}_2 + 2\hat{\varepsilon}_1}\right] R^3 \nabla\left(\vec{E}_0 \cdot \vec{E}_0\right) \tag{1.129}$$

where Re denotes the real part.

Using Stokes law, the hydrodynamic drag force on a particle of radius R far away from any wall is given by

$$F_{\mathrm{HYD}} = 6\pi \mu R U \tag{1.130}$$

where U is the speed of the particle through the medium having coefficient of viscosity μ. To find this speed U we equate the dielectrophoretic force to the hydrodynamic force, thus

$$F_{\mathrm{DEP}} = F_{\mathrm{HYD}}$$

or,

$$2\pi \varepsilon_1 \frac{\varepsilon_2 - \varepsilon_1}{\varepsilon_2 + 2\varepsilon_1} R^3 \nabla\left(\vec{E}_0 \cdot \vec{E}_0\right) = 6\pi \mu R U \tag{1.131}$$

considering ideal dielectric particle and suspending medium. Finally, we get

$$U = \frac{1}{3\mu}\varepsilon_1 \frac{\varepsilon_2 - \varepsilon_1}{\varepsilon_2 + 2\varepsilon_1} R^2 \nabla\left(\vec{E}_0 \cdot \vec{E}_0\right) \tag{1.132}$$

This is the velocity of a non-accelerating spherical particle undergoing dielectrophoresis.

4 Surface Tension-Driven Flows

It was mentioned in Section 1 that surface effects become progressively dominant as dimensions scale down. We discussed the electrokinetic surface effects in the previous section. In this section, we discuss fluid flow actuation and control through the manipulation of surface tension forces. There are a number of agents which may be used to bring about this manipulation, namely hydrodynamic, thermal, chemical, electrical, or optical. Pertinently, the possibility of fluid actuation exploiting surface tension is contingent on the existence of an interface – be it a free surface or a liquid–fluid interface.

Surface tension (γ) is the force per unit length acting along the interface of immiscible phases. The microscopic origin of surface tension is most easily illustrated in the case of a liquid–gas system. Molecules in the liquid bulk (sufficiently distant from any surface so as not to "feel" its presence) experience equal forces (from the neighboring molecules) in all directions. In contrast to this, at the liquid–gas interface, the molecules experience a higher force from the liquid side compared to the vapor side. The net force on the interface molecules is thus directed inward toward the liquid side. In order that the interface is sustained, it must possess a certain amount of energy to overcome this net force. This energy per unit interfacial area is known as the surface energy.

4.1 Interfaces – Young and Laplace Equation

In many practical situations, there are liquid–gas interfaces present adjacent to a solid phase. To characterize such three-phase systems, reference is made to the contact angle as shown in Fig. 1.8.

A simple force balance along the solid surface gives

$$\gamma_{\mathrm{sl}} + \gamma_{\mathrm{lv}} \cos\theta = \gamma_{\mathrm{sv}}$$

or,

$$\cos\theta = \frac{\gamma_{\mathrm{sv}} - \gamma_{\mathrm{sl}}}{\gamma_{\mathrm{lv}}} \tag{1.133}$$

This is the famous Young's equation. Considering the liquid to be water, the contact angle may be used to characterize the surface. Specifically, when $0 \le \theta \le 90^\circ$, the solid is termed hydrophilic. When $\theta > 90^\circ$, the substrate is termed hydrophobic. The derivation of Young's equation may give the impression that the vertical component of the force is left unbalanced but in reality this is balanced by the normal stress in the solid substrate.

Furthermore, from the perspective of fluid mechanics, the presence of an interface creates a jump in pressure across a curved interface. The pressure is higher on the concave side. A quantitative estimate of this pressure difference follows:

Let us consider a portion *ABCD* of a curved surface as shown in Fig. 1.9. This portion is generated when two sets of mutually perpendicular planes cut the surface. The intersection of each of these panes with the surface is an arc. The radii of curvature of the arcs of length x and y as seen in the figure are, respectively, R_1 and R_2. Now, let us consider a movement of this curved surface outward by a small amount

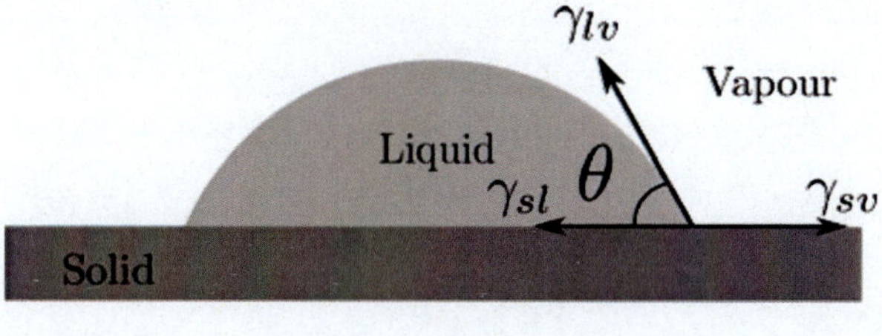

Fig. 1.8 Evaluation of contact angle

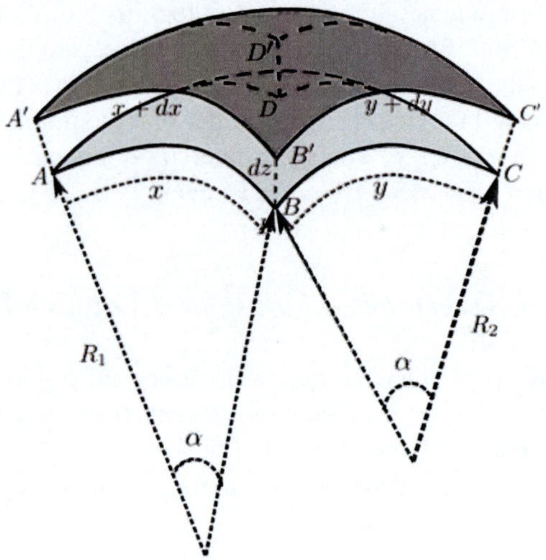

Fig. 1.9 Stretching of a surface – evaluation of the pressure difference across a curved interface

dz such that in this new position the portion which was $ABCD$ becomes $A'B'C'D'$. The arc lengths now become $x+dx$ and $y+dy$. The increase in area on moving from $ABCD$ to $A'B'C'D'$ is

$$dA = (x + dx)\,(y + dy) - xy$$

$$\approx xdy + ydx \tag{1.134}$$

Now, the change in interfacial free energy in the process is

$$dG = \gamma dA = \gamma\,(xdy + ydx) \tag{1.135}$$

Equating the change in energy with the work done due to the pressure differential Δp, it follows that

$$dG = dw$$

or,

$$dG = \Delta p dV$$

or,

$$\gamma\,(xdy + ydx) = \Delta pxydz \tag{1.136}$$

We notice from Fig. 1.9 that

$$x = R_1 d\alpha$$

and

$$x + dx = (R_1 + dz)\,\alpha$$

Therefore,

$$\frac{x + dx}{R_1 + dz} = \frac{x}{R_1}$$

or,

$$\frac{dx}{xdz} = \frac{1}{R_1} \tag{1.137}$$

Similarly,

$$\frac{y + dy}{R_2 + dz} = \frac{y}{R_2}$$

or,

$$\frac{dy}{ydz} = \frac{1}{R_2} \tag{1.138}$$

Using (1.137) and (1.138) in (1.136), we get

$$\gamma\left(xy\frac{dz}{R_2} + yx\frac{dz}{R_1}\right) = \Delta p dx dz$$

or,

$$\gamma\left(\frac{1}{R_1} + \frac{1}{R_2}\right) = \Delta p \tag{1.139}$$

This is the famous Laplace equation.

The mathematical development we have presented until now is restricted to cases where bulk internal forces due to gravity, electric field, and so on can be neglected. In order to analyze a situation which includes such forces, we take recourse to the thermodynamic description of the droplet from a fundamental viewpoint for illustration. The general form of the free energy of a droplet is

$$E = \sum_{i \neq j} A_{ij}\gamma_{ij} + \sum_{k} U_k - \lambda V \tag{1.140}$$

where V is the constant droplet volume, λ is a Lagrange multiplier to enforce the constant volume constraint, A_{ij} is the interfacial area that demarcates the phases i and j, with the corresponding surface energy being γ_{ij}, U_k is the contribution to the free energy from the kth external force (such as gravity and electric field). If the coordinates used to represent the droplet are q_m then the Young's equation (1.133) and Laplace equation (1.139) can be recovered by minimizing the free energy (1.140). In most common situations (where complex surface morphologies are absent), the condition of extremization are usually sufficient:

$$\frac{\partial E}{\partial q_m} = 0 \tag{1.141}$$

For a more complete description of this approach considering a spherical droplet, the interested reader is referred to [5].

We have, hitherto, been concerned with the equilibrium description of a droplet. There are multifarious applications of droplets in the microfluidic realm, and these necessitate the use of more advanced mathematical treatments addressing the dynamics of droplets. However, beyond droplets, surface tension effects are also important for fluid flow through channels and tubes at the micro-scale. Recognizing this importance, we present the theory of fluid transport in a microtube/capillary under the action of surface tension in the following section.

4.2 Surface Tension-Driven Flow in Microchannels/Capillary

The equilibrium height to which a liquid will rise in a vertical capillary can be found by equating the vertical component of the surface tension force on the liquid meniscus to the body force on the liquid column due to gravity:

$$2\pi a\gamma \cos\theta = \pi a^2 H\rho g$$

or,

$$H = \frac{2\gamma \cos\theta}{\rho g \alpha} \tag{1.142}$$

where H is the equilibrium height reached by the capillary.

Now, it is also important to track how this equilibrium height is reached with time (see Fig. 1.10). This means we need to record the transient conditions in the process of achieving the equilibrium height. It may first be assumed that at any instant, the liquid flow is a fully developed Poiseuille flow, in which case

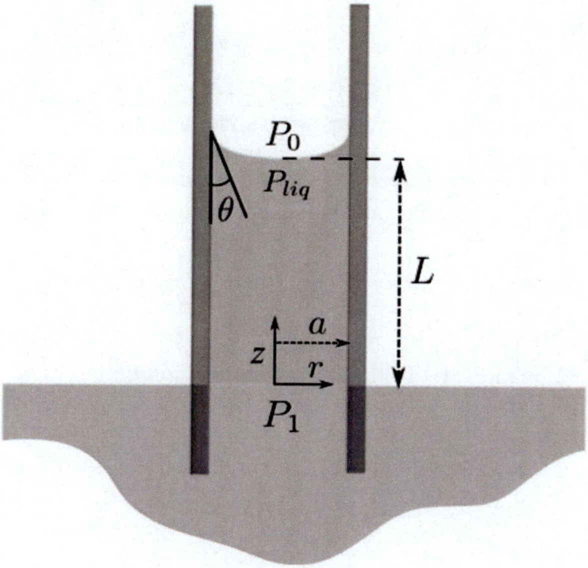

Fig. 1.10 Rise of fluid due to capillary action

$$\frac{u}{\bar{u}} = 2\left(1 - \frac{r^2}{a^2}\right) \tag{1.143}$$

where $\bar{u}$ is the average velocity. The wall shear stress is given by:

$$\sigma_w = -\mu \frac{du}{dr}\bigg|_{r=a} = \frac{4\mu}{a}\bar{u} \tag{1.144}$$

A force balance on the liquid column, which has a length $L(t)$ at any instant of time, and neglecting inertial effects reads:

$$\sigma_w 2\pi a L - 2\pi a \gamma \cos\theta + \pi a^2 \rho L g = 0. \tag{1.145}$$

Furthermore, noting that $\bar{u} = dL/dt$, we have from (1.144) and (1.145)

$$\frac{dL(t)}{dt} = \frac{a^2}{8\mu L(t)}\left[\frac{2\gamma\cos\theta}{a} - \rho g L(t)\right]$$

or,

$$\frac{dL(t)}{dt} = \frac{\gamma}{8\mu}\left[\frac{2a\cos\theta}{L(t)} - \frac{\rho g a^2}{\gamma}\right]$$

or,

$$\frac{dL(t)}{dt} = \frac{\rho g H a^2}{8\mu L} - \frac{\rho g a^2}{8\mu} \text{ (using (1.142))} \tag{1.146}$$

This is the Lucas–Washburn equation. When t is very small, $L \ll H$, i.e., $H/L \gg 1$, and the Lucas–Washburn equation reduces to

$$\frac{dL(t)}{dt} = \frac{\rho g a^2}{8\mu}\left(\frac{H}{L}\right) \tag{1.147}$$

Equation (1.147) has the following solution:

$$L(t) = \sqrt{\frac{\rho g H}{4\mu}}\sqrt{t} \tag{1.148}$$

i.e.,

$$L(t) \propto \sqrt{t}$$

Although this solution gives a good estimate, it suffers from a fundamental discrepancy. For $t \to 0$, $dL(t)/dt \to \infty$, implying an infinite velocity, when the liquid just starts rising in the capillary. When t is very large, $L \to H$ so that $L = H - \delta L$, where $\delta L \ll H$. Then,

$$\frac{dL(t)}{dt} = \frac{\rho g a^2}{8\mu}\left[\frac{H}{H - \delta L} - 1\right]$$

or,

$$-\frac{d(\delta L)}{dt} = \frac{\rho g a^2}{8\mu}\frac{\delta L}{H - \delta L}$$

or,

$$-\frac{d(\delta L)}{dt} \approx \frac{\rho g a^2}{8\mu}\frac{\delta L}{H} \tag{1.149}$$

The solution of (1.149) is

$$\delta L(t) = C \exp\left(-\frac{\rho g a^2}{8\mu H} t\right) \tag{1.150}$$

where C is an arbitrary constant of integration. Using (1.150) in the expression $L = H - \delta L$, we get

$$L(t) = H - C \exp\left(-\frac{\rho g a^2}{8\mu H} t\right) \tag{1.151}$$

This means that, at large times, the capillary height $L(t)$ approaches the equilibrium height H asymptotically as an exponential saturation.

It is important to realize that the formulation of the Lucas–Washburn equation stems from a quasi-steady approximation – an assumption which precludes the use of any inertial effect. A model, including the inertial terms, was presented by Zhmud et al. [17]. It is a statement of Newton's second law of motion as applied to the liquid moving up under the combined forces of surface tension, viscous forces, and gravity. Thus,

$$\frac{d}{dt}(MV_c) = F_{\text{Surface tension}} + F_{\text{Viscous}} + F_{\text{Gravity}} \tag{1.152}$$

where $M = \rho\pi a^2 L$, $V_c = dL/dt$, $F_{\text{Surface tension}} = 2\pi a\gamma\cos\theta$, $F_{\text{Viscous}} = -\Delta p\pi a^2$, and $F_{\text{Gravity}} = \rho g L\pi a^2$. Now, as before, we assume a fully developed flow profile, such that $Q = \left(\pi a^4/8\mu\right)\left(\Delta p/L\right)$ or $\left(\pi a^2\right)\Delta p = 8\mu\pi L\left(dL/dt\right)$. Therefore, $F_{\text{Viscous}} = -8\mu\pi L\left(dL/dt\right)$. Using these in (1.152) we obtain

$$\frac{d}{dt}\left(\rho\pi a^2 L\frac{dL}{dt}\right) = 2\pi a\gamma\cos\theta - 8\mu\pi L\frac{dL}{dt} + \rho g L\pi a^2$$

$$\rho\left[L\frac{d^2L}{dt^2} + \left(\frac{dL}{dt}\right)^2\right] = 2\frac{\gamma}{a}\cos\theta - \frac{8}{a^2}\mu L\frac{dL}{dt} - \rho g L \tag{1.153}$$

Now, although the natural initial condition to consider might seem to be $L(0) = 0$ and $dL(0)/dt = 0$, such conditions give rise to an ill-posed problem with infinite initial acceleration. However, such a fundamental drawback can be circumvented by considering an added mass $\rho\pi a^2\lambda$ inducted initially so that (1.153) becomes

$$\rho\left[(L+\lambda)\frac{d^2L}{dt^2} + \left(\frac{dL}{dt}\right)^2\right] = 2\frac{\gamma}{a}\cos\theta - \frac{8}{a^2}\mu L\frac{dL}{dt} - \rho g L \tag{1.154}$$

Here, λ is obtained from potential flow solution for a droplet of radius r ready to enter a capillary and its value depends on the capillary geometry. For the present case of a cylindrical tube $\lambda \approx 3\rho\pi r^3/8$.

The complete physical argument behind this artifice of considering an added mass can be found in Zhmud et al. [17]. Furthermore, the steps required to solve this equation numerically are also elaborated in the same reference.

We have only scratched the surface of this important phenomenon and the models pertaining to various microfluidic applications – electrowetting, electrocapillary, thermocapillary, to name just a few. The detailed considerations of these are certainly beyond the scope of this book. Yet, the fundamentals of surface tension effects as described here remain fundamentally immutable. The extra considerations of electrokinetic, thermal, optical, and other physico-chemical effects exploit these fundamentals of surface energy to manipulate, actuate, and control fluid flow.

5 Non-Newtonian Fluids

We have, hitherto, been discussing the mechanics of Newtonian fluids, i.e., fluids in which the shear stress varies linearly with the velocity gradient. But, there are certain fluids which do not follow this behavior. Even fluids which are otherwise Newtonian in nature may show non-Newtonian flow characteristics because of the presence of suspensions in them. Typical examples are biofluids like blood, which are common in microfluidics applications for medical diagnostics.

Although there do not exist any fundamental constitutive equation to universally model non-Newtonian fluids, many empirical relations have been proposed. For many engineering applications, these relations may be adequately represented by the power law model. For a one-dimensional flow, the power law model is

$$\tau = k\left(\frac{du}{dy}\right)^n \tag{1.155}$$

where k is the consistency index and n is the flow behavior index.

In order to draw a parallel with Newtonian fluid constitutive behavior, the shear stress is written as

$$\tau = k\left|\frac{du}{dy}\right|^{n-1}\left(\frac{du}{dy}\right) \tag{1.156}$$

The absolute value is used to ensure that τ has the same sign as du/dy. Continuing the parallel with the Newtonian fluid, the term $k\left|du/dy\right|^{n-1}$ is given a special name: apparent viscosity. There can be three cases:

(i) $n<1$: In this case, the apparent viscosity decreases as the rate of deformation increases. These are called pseudoplastic fluids, of which, probably, the most significant example is blood. The apparently simplistic power law model gives us an important insight into the in vivo blood flow biophysics. Blood flows through extremely narrow "tubes" within the body. Within these tubes, the gradient of velocity is extremely high at the near-wall regions. This means that the apparent viscosity is substantially low in these regions – this fact is tremendously significant from the point of view of expediting the blood pumping.

(ii) $n=1$: This case reduces to the Newtonian fluid.

(iii) $n>1$: In this case, the apparent viscosity increases as the rate of deformation increases. These are called dilatant fluids.

There is a certain class of non-Newtonian fluid which deforms under applied stress but returns, partially, to its original shape after the applied stress is released. Since it exhibits characteristics of both a viscous fluid and an elastic medium, it is termed viscoelastic. A number of biofluids come under the purview of this class; hence, viscoelasticity is particularly significant from the point of view of microfluidics.

A widely accepted mathematical model for viscoelastic fluids is the Phan-Thien–Tanner model.

$$f(\tau_{kk})[\tau] + \lambda\left[\tau^{\nabla}\right] = 2\mu[D] \tag{1.157}$$

where $D = \frac{1}{2}\left(\nabla \otimes \vec{u}^T + \nabla \otimes \vec{u}\right)$, λ is the relaxation time of the fluid, $\left[\tau^{\nabla}\right]$ is upper convected derivative of τ, defined as $\left[\tau^{\nabla}\right] = \left(D[\tau]/Dt\right) - \nabla \otimes \vec{u}^T \cdot [\tau] - [\tau] \cdot \nabla \otimes \vec{u}$, $f(\tau_{kk}) = 1 + \left(\varepsilon\lambda/\mu\right)\tau_{kk}$ is the stress coefficient function, with $\tau_{kk} = \mathrm{tr}([\tau])$. When $\varepsilon = 0$ the upper convected model is recovered.

For a fully developed flow

$$\vec{u} = \left[u(y)\ 0\right]^T$$

Therefore,

$$\nabla \otimes \vec{u} = \begin{bmatrix} \dfrac{\partial}{\partial x} \\ \dfrac{\partial}{\partial y} \end{bmatrix} \left[u(y)\ 0\right] = \begin{bmatrix} \cancelto{0}{\dfrac{\partial u(y)}{\partial x}} & 0 \\ \dfrac{\partial u(y)}{\partial y} & 0 \end{bmatrix}$$

So,

$$D = \frac{1}{2}\begin{bmatrix} 0 & \dfrac{\partial u(y)}{\partial y} \\ \dfrac{\partial u(y)}{\partial y} & 0 \end{bmatrix}$$

$$\nabla \otimes \vec{u}^T \cdot [\tau] = \begin{bmatrix} 0 & 0 \\ \dfrac{\partial u(y)}{\partial y} & 0 \end{bmatrix} \begin{bmatrix} \tau_{xx} & \tau_{xy} \\ \tau_{xy} & \tau_{yy} \end{bmatrix} = \begin{bmatrix} \tau_{xy}\dfrac{\partial u(y)}{\partial y} & \tau_{yy}\dfrac{\partial u(y)}{\partial y} \\ 0 & 0 \end{bmatrix} \tag{1.158}$$

Similarly,

$$[\tau]\cdot\nabla\otimes\vec{u}=\begin{bmatrix}\tau_{xy}\dfrac{\partial u\,(y)}{\partial y} & 0\\ \tau_{yy}\dfrac{\partial u\,(y)}{\partial y} & 0\end{bmatrix} \tag{1.159}$$

Using (1.158) and (1.159) we can write out the components from (1.157) as

$$f\,(\tau_{kk})\,\tau_{xx}-2\lambda\tau_{xy}\frac{\partial u\,(y)}{\partial y}=2\mu\times\frac{1}{2}\times 0$$

or,

$$f\,(\tau_{kk})\,\tau_{xx}=2\lambda\tau_{xy}\frac{\partial u\,(y)}{\partial y} \tag{1.160}$$

Similarly,

$$f\,(\tau_{kk})\,\tau_{yy}=0$$

or,

$$\tau_{yy}=0 \tag{1.161}$$

and

$$f\,(\tau_{kk})\,\tau_{xy}-\lambda\cancelto{0}{\tau_{yy}}\,\frac{\partial u(y)}{\partial y}=2\mu\times\frac{1}{2}\frac{\partial u(y)}{\partial y} \tag{1.162}$$

Dividing (1.160) by (1.162), we obtain:

$$\frac{f\,(\tau_{kk})\,\tau_{xx}}{f\,(\tau_{kk})\,\tau_{xy}}=\frac{2\lambda\tau_{xy}}{\mu}$$

or,

$$\tau_{xx}=\frac{2\lambda}{\mu}\tau_{xy}^2 \tag{1.163}$$

The argument of the stress coefficient function, now, becomes:

$$\tau_{kk}=\tau_{xx}+\cancelto{0}{\tau_{yy}}=\tau_{xx} \tag{1.164}$$

Now, Cauchy's equation for a steady flow (with negligible inertia terms) is

$$-\nabla p+\nabla\cdot[\tau]=0$$

Since the flow is fully developed, the equation along the x (axial)-direction is

$$-\frac{\partial p}{\partial x}+\frac{\partial \tau_{xy}}{\partial y}=0 \tag{1.165}$$

Equation (1.165), in conjunction with the boundary condition $\tau_{xy}=0$ at the center-line (y=0), gives

$$\tau_{xy}=\frac{\partial p}{\partial x}y \tag{1.166}$$

Using (1.166) in (1.163) we get

$$\tau_{xx}=\frac{2\lambda}{\mu}\left(\frac{\partial p}{\partial x}\right)^2 y^2 \tag{1.167}$$

Now, from (1.162), we have

$$f\left(\tau_{kk}\right)\tau_{xy}=\mu\frac{\partial u\left(y\right)}{\partial y}$$

or,

$$\left(1+\frac{\varepsilon\lambda}{\mu}\tau_{xx}\right)\tau_{xy}=\mu\frac{\partial u\left(y\right)}{\partial y} \tag{1.168}$$

Using (1.167) in (1.168), we get

$$\frac{\partial u\left(y\right)}{\partial y}=\left[1+\frac{2\varepsilon\lambda^2}{\mu^2}\left(\frac{\partial p}{\partial x}\right)^2 y^2\right]\frac{\partial p}{\partial x}\left(\frac{y}{\mu}\right) \tag{1.169}$$

This equation may be readily solved by using the boundary conditions relevant for a particular problem.

The fact that surface effects become important with increasing levels of miniaturization is not changed by the consideration of non-Newtonian fluids. Thus, in flows of non-Newtonian fluids, too, electrokinetic and capillary effects have a significant role to play [3, 7].

6 Acoustofluidics

Acoustofluidics is the application of acoustics in microfluidics. When acoustic waves are propagating in a fluid there arise rapidly oscillating pressure and velocity fields in it. Simultaneously, there also arises a slow non-oscillating velocity component. It is true that under normal circumstances, in keeping with our macroscopic intuition, such effects are of minuscule significance. But, at the micro-scale even such small effects can have non-negligible consequences. More importantly, such effects can be usefully exploited for fluidic actuation.

The slow non-oscillating velocity component that arises in periodic acoustically driven flows is one example of the general class of streaming flows. The physics that lies, fundamentally, at the heart of such phenomena is something that we have been neglecting in our mathematical development – albeit on the basis of sound physical justifications with contextual relevance. These are the nonlinearity and the compressibility. In the particular case of acoustofluidics, however, such physics can no longer be viably neglected.

We start with the understanding that the changes in pressure, density, and velocity brought about by the acoustic actuation are only small perturbations. Thus, we proceed, for the sake of analysis, with perturbation theory. Specifically, we consider perturbations up to second order. It will be seen that the time-invariant streaming velocities are achieved at this order. In the mathematical treatment that follows, we follow the structure of development presented in Bruus [2]. We first express the pressure, density, and velocity as asymptotic expansions in terms of the small parameter ε such that $\varepsilon \ll 1$.

Equations of motion in acoustofluidics may be developed on the basis of second-order asymptotic expansion of the field variables appearing in the Navier–Stokes equation. The corresponding expanded variables have the following mathematical structures:

$$p = p_0 + \varepsilon p_1 + \varepsilon^2 p_2 \tag{1.170}$$

$$\rho = \rho_0 + \varepsilon \rho_1 + \varepsilon^2 \rho_2 \tag{1.171}$$

and

$$\vec{v} = \vec{0} + \varepsilon \vec{v}_1 + \varepsilon^2 \vec{v}_2 \tag{1.172}$$

Here, p_0 and ρ_0 denote the values of the fluid pressure and density, respectively, in the undisturbed state with zero velocity.

Let us, first, consider the general form of the mass conservation equation (1.5):

$$\frac{\partial \rho}{\partial t} = -\nabla \cdot (\rho \vec{v})$$

Using (1.171) and (1.172), and neglecting terms with coefficients having powers of ε higher than 2, we get

$$\frac{\partial \rho_0}{\partial t} + \varepsilon \frac{\partial \rho_1}{\partial t} + \varepsilon^2 \frac{\partial \rho_2}{\partial t} = -\nabla \cdot \left[\varepsilon \rho_0 \vec{v}_1 + \varepsilon^2 \rho_0 \vec{v}_2 + \varepsilon^2 \rho_1 \vec{v}_1 \right] \tag{1.173}$$

From (1.173) it is clear that

$$\text{Zeroth-order equation: } \frac{\partial \rho_0}{\partial t} = 0 \tag{1.174}$$

$$\text{First-order equation: } \frac{\partial \rho_1}{\partial t} = -\nabla \cdot (\rho_0 \vec{v}_1) \tag{1.175}$$

$$\text{Second-order equation: } \frac{\partial \rho_2}{\partial t} = -\nabla \cdot (\rho_0 \vec{v}_2 + \rho_1 \vec{v}_1) \tag{1.176}$$

Let us, next, consider the general linear momentum conservation equation without invoking the condition of incompressibility as in (1.15):

$$\rho \frac{\partial \vec{v}}{\partial t} + \rho \vec{v} \cdot \nabla \vec{v} = -\nabla p + \mu \nabla^2 \vec{v} + \left(\mu_v + \frac{1}{3}\mu \right) \nabla (\nabla \cdot \vec{v}) \tag{1.177}$$

Using (1.170), (1.171), and (1.172) in (1.177) and neglecting terms with coefficients having powers of ε higher than 2, we get for the various terms

$$\begin{aligned}
\rho \frac{\partial \vec{v}}{\partial t} &= \varepsilon \rho_0 \frac{\partial \vec{v}_1}{\partial t} + \varepsilon^2 \rho_0 \frac{\partial \vec{v}_2}{\partial t} + \varepsilon^2 \rho_1 \frac{\partial \vec{v}_1}{\partial t} \\
\rho \vec{v} \cdot \nabla \vec{v} &= \varepsilon^2 \rho_0 \vec{v}_1 \cdot \nabla \vec{v}_1 \\
\nabla p &= \nabla p_0 + \varepsilon \nabla p_1 + \varepsilon^2 \nabla p_2 \\
\mu \nabla^2 \vec{v} &= \vec{0} + \varepsilon \nabla^2 \vec{v}_1 + \varepsilon^2 \nabla^2 \vec{v}_2 \quad \text{and} \\
\nabla (\nabla \cdot \vec{v}) &= \varepsilon \nabla (\nabla \cdot \vec{v}_1) + \varepsilon^2 \nabla (\nabla \cdot \vec{v}_2)
\end{aligned}$$

Equating the terms having as coefficients the same powers of ε, we get

$$-\nabla p_0 = 0 \tag{1.178}$$

$$\rho_0 \frac{\partial \vec{v}_1}{\partial t} = -\nabla p_1 + \mu \nabla^2 \vec{v}_1 + \left(\mu_v + \frac{1}{3}\mu \right) \nabla (\nabla \cdot \vec{v}_1) \tag{1.179}$$

and

$$\rho_0 \frac{\partial \vec{v}_2}{\partial t} + \rho_1 \frac{\partial \vec{v}_1}{\partial t} + \rho_0 \vec{v}_1 \cdot \nabla \vec{v}_1 = -\nabla p_2 + \mu \nabla^2 \vec{v}_2 + \left(\mu_v + \frac{1}{3}\mu \right) \nabla (\nabla \cdot \vec{v}_2) \tag{1.180}$$

Next, considering the pressure, and using Taylor's expansion about the equilibrium pressure p_0, we obtain

$$
\begin{aligned}
p &= p_0 + \left.\frac{\partial p}{\partial \rho}\right|_0 \frac{\rho-\rho_0}{1!} + \left.\frac{\partial^2 p}{\partial \rho^2}\right|_0 \frac{(\rho-\rho_0)^2}{2!} + \text{higher order terms} \\
&= p_0 + \left.\frac{\partial p}{\partial \rho}\right|_0 \left(\varepsilon\rho_1 + \varepsilon^2\rho_2\right) + \left.\frac{\partial}{\partial \rho}\left(\frac{\partial p}{\partial \rho}\right)\right|_0 \frac{\left(\varepsilon\rho_1 + \varepsilon^2\rho_2\right)^2}{2} + \text{higher order terms} \\
&= p_0 + \varepsilon c_0^2\rho_1 + \varepsilon^2 c_0^2\rho_2 + \varepsilon^2 \frac{1}{2}\left.\frac{\partial c^2}{\partial \rho}\right|_0 \rho_1^2 + \text{higher order terms}
\end{aligned}
\tag{1.181}
$$

On comparing (1.181) with (1.170), we get

$$
p_1 = c_0^2\rho_1 \tag{1.182}
$$

Finally, from (1.181), (1.176), and (1.180), the second-order equations are, respectively,

$$
p_2 = c_0^2\rho_2 + \frac{1}{2}\left.\frac{\partial c^2}{\partial \rho}\right|_0 \rho_1^2 \tag{1.183}
$$

$$
\frac{\partial \rho_2}{\partial t} = -\rho_0 \nabla\cdot\vec{v}_2 - \nabla\cdot(\rho_1\vec{v}_1) \tag{1.184}
$$

and

$$
\begin{aligned}
\rho_0\frac{\partial \vec{v}_2}{\partial t} &= -c_0^2\nabla\rho_2 - \frac{1}{2}\left.\frac{\partial c^2}{\partial \rho}\right|_0 \nabla\rho_1^2 - \rho_1\frac{\partial \vec{v}_1}{\partial t} \\
&\quad -\rho_0\vec{v}_1\cdot\nabla\vec{v}_1 - \mu\nabla^2\vec{v}_2 + \left(\mu_v + \frac{1}{3}\mu\right)\nabla\left(\nabla\cdot\vec{v}_2\right)
\end{aligned}
\tag{1.185}
$$

We assume that the time dependence of all first-order fields is harmonic, $\exp(-i\omega t)$. Then, we take the time average of each of (1.184) and (1.185). Thus,

$$
\left\langle\frac{\partial \rho_2}{\partial t}\right\rangle = \langle -\rho_0\nabla\cdot\vec{v}_2 - \nabla\cdot(\rho_1\vec{v}_1)\rangle
$$

$$
\cancelto{0}{\frac{\partial}{\partial t}\langle\rho_2\rangle} = -\rho_0\nabla\cdot(\langle\vec{v}_2\rangle) - \nabla\cdot(\langle\rho_1\vec{v}_1\rangle)
$$

or,

$$\nabla \cdot \langle \vec{v}_2 \rangle = -\frac{1}{\rho_0} \nabla \cdot \langle \rho_1 \vec{v}_1 \rangle \tag{1.186}$$

and

$$\begin{aligned}
\left\langle \rho_0 \frac{\partial}{\partial t} (\nabla \cdot \vec{v}_2) \right\rangle &= \left\langle -c_0^2 \nabla^2 \rho_2 - \nabla \cdot \rho_1 \frac{\partial \vec{v}_1}{\partial t} - \rho_0 \nabla \cdot [(\vec{v}_1 \cdot \nabla) \vec{v}_1] \cdots \right. \\
&\quad \left. - \frac{1}{2} \frac{\partial c^2}{\partial \rho}\bigg|_0 \nabla^2 \rho_1^2 + \mu \nabla^2 (\nabla \cdot \vec{v}_2) + \left(\mu_v + \frac{1}{3}\mu \right) \nabla^2 (\nabla \cdot \vec{v}_2) \right\rangle \\
\Rightarrow \frac{1}{c_0^2} \rho_0 \frac{\partial}{\partial t} (\nabla \cdot \langle \vec{v}_2 \rangle) &= -\nabla^2 \langle \rho_2 \rangle - \nabla \cdot \left\langle \frac{\rho_1}{c_0^2} (-i\omega) \vec{v}_1 \right\rangle - \frac{\rho_0}{c_0^2} \nabla \cdot \langle (\vec{v}_1 \cdot \nabla) \vec{v}_1 \rangle \\
&\quad - \frac{1}{2} c_0^2 \frac{\partial c^2}{\partial \rho}\bigg|_0 \nabla^2 \left\langle \rho_1^2 \right\rangle + \frac{1}{c_0^2} \left(\mu_v + \frac{4}{3}\mu \right) \nabla^2 (\nabla \cdot \langle \vec{v}_2 \rangle) \\
\Rightarrow \nabla^2 \langle \rho_2 \rangle &= \frac{i\omega}{c_0^2} \nabla \cdot \langle \rho_1 \vec{v}_1 \rangle - \frac{\rho_0}{c_0^2} \nabla \cdot \langle (\vec{v}_1 \cdot \nabla) \vec{v}_1 \rangle \\
&\quad - \frac{1}{2c_0^2} \frac{\partial c^2}{\partial \rho}\bigg|_0 \nabla^2 \left\langle \rho_1^2 \right\rangle - \frac{(\mu_v + 4/3\mu)}{\rho_0 c_0^2} \nabla^2 (\nabla \cdot \langle \rho_1 \vec{v}_1 \rangle)
\end{aligned} \tag{1.187}$$

Equations (1.186) and (1.187) denote, respectively, the streaming components of the velocity and density that arise due to the system oscillations. From the point of view of microfluidics, these streaming components are of great significance because they can be exploited to bring about transport at such micro-scales.

7 Conclusions

We have, in each of the sections, described models of phenomena which are individually responsible for enriching the mechanics of fluid flow. However, beyond these preliminary considerations, in real microfluidic applications, these various cases might be significant in a combined manner. The corresponding mathematical models of these intricate systems would necessitate the incorporation of the individual models in a cohesive manner. But, in the general case even this is perhaps an oversimplified statement. It must be appreciated that the co-existence of more than one effect – for example, electrokinetics together with surface tension – might imply not just a straightforward superimposition. Indeed, the simultaneous presence of two or more effects might lead to results that do not follow from intuitive expectations. Genuine insights into such intertwined phenomena can be gained only if one has truly grasped the fundamentals of flow physics as we have tried to delineate in a very rudimentary way here in this chapter. Furthermore, it must be appreciated that the models of flow physics are strongly rooted in the developments of various physico-chemical hydrodynamics developed over the last couple of centuries or perhaps even more. Yet, as so often happens that the technology drives the development of science, the special considerations of microfluidic applications have led to new and fundamental insights especially in the last couple of decades. This, in turn, has motivated fruitful innovations, thus making the realm of micro-scale flow physics a truly potent and fertile field. As the continual feedback between theory and application goes on, the story of micro-scale flow physics has spiraled into an epic – and it is far from over!

References

1. Acheson, D. J., *Elementary Fluid Dynamics*, Oxford University Press, New York, NY (2003).
2. Bruus, H., *Theoretical Microfluidics*, Oxford University Press, Oxford (2008).
3. Chakraborty, S., *Electroosmotically Driven Capillary Transport of Typical Non-Newtonian Biofluids in Rectangular Microchannels*, Analytica Chimica Acta, 605, 175–184 (2007).
4. Chakraborty, S., *Electric Double Layer*, in Li, D. *Encyclopedia of Microfluidics and Nanofluidics* (pp. 444–453), Springer, New York, NY (2008).
5. Chakraborty, S., *Surface Tension Driven Flow*, in Li, D., *Encyclopedia of Microfluidics and Nanofluidics* (pp. 1955–1968), Springer, New York, NY (2008).
6. Chakraborty, S., Das, S., *Streaming-Field-Induced Convective Transport and Its Influence on the Electroviscous Effects in Narrow Fluidic Confinement Beyond the Debye-Hückel Limit*, Physical Review E, 77, 037303 (2008).
7. Das, S., Chakraborty, S., *Analytical Solutions for Velocity, Temperature and Concentration Distribution in Electroosmotic Microchannel Flows of a Non-Newtonian Bio-Fluid*, Analytica Chimica Acta, 559, 15–24 (2006).
8. Griffiths, D., *Introduction to Electrodynamics*, Prentice Hall, New Delhi (1998).
9. Hunter, R. J., *Zeta Potential in Colloid Science*, Academic, New York, NY (1981).
10. Kilic, M. S., Bazant, M. Z., Ajdari, A., *Steric Effects in the Dynamics of Electrolytes at Large Applied Voltages. II. Modified Poisson-Nernst-Planck Equations*, Physical Review E, 75, 021503 (2007).
11. Kreyszig, E., *Advanced Engineering Mathematics*, 8th ed., Wiley, Singapore (2004).
12. Landau, L. D., Lifshitz, E. M., *Electrodynamics of Continuous Media*, Pergamon, Oxford (1984).
13. Li, D., *Electrokinetics in Microfluidics*, Elsevier, London (2004).
14. Morrison, F. A., *Electrophoresis of a Particle of Arbitrary Shape*, Journal of Colloid and Interface Science, 34, 2 (1970).
15. Probstein, R. F., *Physicochemical Hydrodynamics*, Wiley-Interscience, Hoboken, NJ (2003).
16. Wall, S., *The History of Electrokinetic Phenomena*, Current Opinion in Colloid and Interface Science, 15, 119–124 (2010).
17. Zhmud, B. V., Tiberg, F., Hallstensson, K., *Dynamics of Capillary Rise*, Journal of Colloid and Interface Science, 228, 263–269 (2000).

Chapter 2
Scaling Laws

Amitabha Ghosh

Abstract It is a very well-known fact that when size diminishes to extremely small levels, Newtonian mechanics fails. In such situations one has to use quantum mechanics for studying physical systems of such extremely small dimensions. However, it is less readily recognized that even before reaching such extremely small dimensions leading to the breakdown of Newtonian mechanics there are many special aspects one needs to take care of for successful analysis and design of small systems. The counterintuitive features arise not due to breakdown of Newtonian mechanics but due to the changes in the order of predominance of physical phenomena caused by drastic reduction in size from the scales we are familiar with in our daily experience. In simple language this is called scaling effect and the laws which govern such effects are called scaling laws. While dealing with microsystems our normal engineering intuition fails and it needs to be replaced by special "microintuition" for developing microsystems which is becoming a very important engineering activity with the turn of the century. This chapter presents the rudiments of the common scaling laws and their importance.

Keywords Scaling · Microsystems · Micromechanics

1 Introduction

Once I was visiting the residence of one of my closest friends and colleague who had just returned from a summer trip abroad. He brought a toy train for his elder son who was about 11 years old. He had a younger son also, who received a battery-driven nice toy racing car as a gift. When I reached their home in the evening of a weekend, I found that his elder son was pestering his father for a clarification. The boy was observant and was asking his father why the toy train and the toy car were not destroyed when they derailed or collided whereas a real train or car gets destroyed completely on such occasions. I noticed with some amusement the

A. Ghosh (✉)
Indian National Science Academy, New Delhi, India; Indian Institute of Technology (IIT), Kanpur, India; Bengal Engineering and Science University, Howrah 711103, India
e-mail: amitabha@iitk.ac.in

S. Chakraborty (ed.), *Mechanics Over Micro and Nano Scales*,
DOI 10.1007/978-1-4419-9601-5_2,

problem my friend was facing to satisfy his son's query. He was a professor of economics. I could understand that he was intuitively getting the feel of why this happens, but could not satisfy his son's scientific mind. When I took a very small and a long piece of the same mild steel wire lying in the courtyard and demonstrated the much stiffer/stronger behavior of the smaller piece both the father and the son got some idea about what we call the scaling effect. From our daily experience we all have some intuition about the change of properties and characteristics of objects with drastically different sizes but of similar shapes. However, a formal study of the change of characteristics of objects and systems with change in scales is essential in modern engineering, particularly because of the emergence of microsystems technology.

Though the phenomenon of the effect of scaling on the properties and characteristics of objects and systems is known since a long time the emergence of "scaling laws" as a serious branch of systematic study is relatively recent. This is primarily due to the increased dependence of modern technology on miniaturization. At the same time using the scaling effects to predict the behavior and properties of a large system by experimenting on a small-sized scale model is a very useful tool. Physicists have also used the basic ideas behind "scaling laws" under a different name "dimensional analysis." Many times the characteristics or properties of a system can be expressed through combinations of various parameters so that each group is dimensionless. Thus, any change in size, i.e., scale, does not affect the magnitudes of these quantities and the performance of a system can be predicted from the results obtained on the performance of a similar system, but of different size. In more recent times the trend of miniaturization in technology has brought back "scaling laws" to the center stage of related design and fabrication activities.

1.1 Trend of Miniaturization

It may not be out of place to look into the reasons behind the ongoing technological revolution based on miniaturization. Figure 2.1 shows the current tendency to make systems and devices more intelligent and autonomous. To achieve higher degree of intelligence it is essential to drastically increase sensory data (by many orders of magnitude). This demands that the sensors be miniaturized so that a large number of these can be accommodated in small areas and at the same time neither the cost nor the energy consumption exceeds acceptable limits. Similarly for activating such machines/devices it is essential to employ a large number of actuators of miniaturized size (to effectively use the scaling effect of some physical laws as will be seen later) working in parallel.

Actuators in futuristic machines and devices will not only be numerous in number but also be distributed over the whole system instead of remaining confined to a few number of locations of the systems as is currently the practice. This will provide the required dexterity to the moving elements of the machine/system. In fact, the engineers and technologists are often taking lessons from the living world

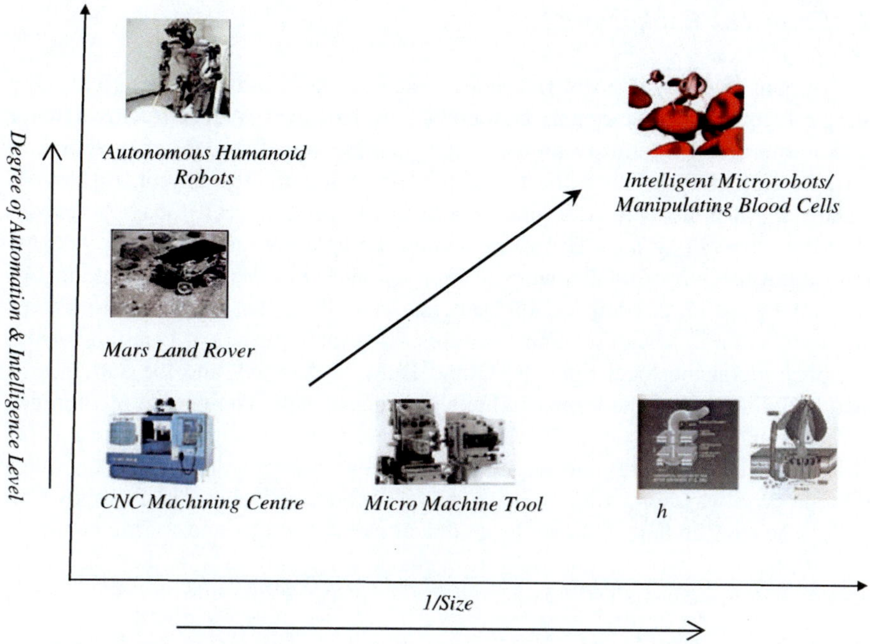

Fig. 2.1 Trend of miniaturization

for conceiving ultramodern machines. All living objects are dependent on miniaturization and massive parallelism for sensing and actuation. This also influences the technique of fabrication, and already there is a tendency to depend on "bottom-up" approaches instead of the traditional "top-down" approach. In the "bottom-up" approach the required shape, size, and characteristic features are achieved through material manipulation at the micro, nano, and even molecular levels. Such processes have been given a new name: "fabrionics." Quite often, therefore, it becomes necessary to utilize the phenomenon of "self-assembly" to make such fabrications technologically feasible and economically viable.

All these new and emerging revolutionary concepts are going to pave the way for the next turning point in human history – the Third Industrial Revolution. In all aspects of futuristic engineering, the principle of life science will play an extremely important role. In fact a completely new branch of engineering – "synthetic biology" – has slowly started to emerge where machines and devices will be artificially created following new emerging techniques of "fabrionics" but which will often function using the phenomena and principles of life science.

It is clear that understanding and developing such new era machines and devices can be possible only when one applies the knowledge and intuition acquired through experience in the macroscopic world keeping in mind the scaling laws.

1.2 Historical Background

As it happens in many areas of mechanics, the scientific study of scaling laws and scaling effects was first started by Galileo Galilei. However, it all started by a major mistake committed by him in his early life. Galileo was a student of medicine at the University of Pisa; but as he found that his major interest was in mathematics and natural philosophy (the old term used for what we call today "science") he left the university after a few years without completing his studies. He became a self-taught teacher of mathematics at Florence and was desperately looking forward to the position of professor of mathematics at Pisa. He was invited to deliver two public lectures when the chair of professor of mathematics at Pisa fell vacant. To impress upon the local ruler the Grand Duke of Tuscany and the audience in general, Galileo chose the topic of Dante's model of hell. The people of Florence loved to hear about the topic. One of the models was by Antonio Manetti, a former member of the Florentine Academy. A competing theory was by Alessandro Vellutello who was not a Florentine. Manetti's model of Dante's hell is shown in Fig. 2.2a where the hell is shown as a conical cavity with its apex at the center of the earth; the base was a circle with its center at Jerusalem. Vellutello's inferno is comparatively much smaller as indicated in Fig. 2.2b. Vellutello's one main reason

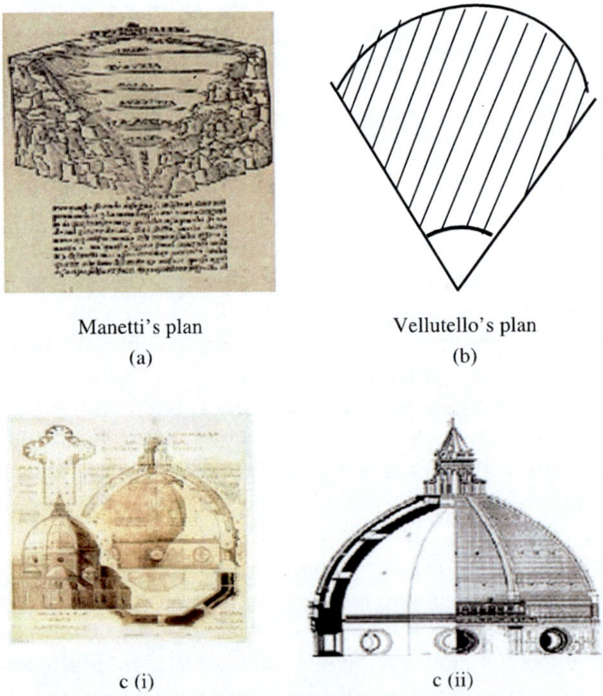

Fig. 2.2 Plan of hell in Galileo's lecture and Brunelleschi dome

for a smaller cavity was because he felt the cover of Manetti's hell was too thin. Galileo ridiculed Vellutello's model as too small to accommodate the sinners. He demolished Vellutello's objection to Manetti's model having a cover too thin to support itself by demonstrating the example of the Brunelleschi dome of Florentine Cathedral. He argued that relatively speaking the dome thickness (Fig. 2.2c) was equivalent to that of the cover of the inferno but it can also hold itself against falling under its own weight. This argument was based upon scale invariance of strength under self-weight. The audience went into rapture on seeing the rival model being ruthlessly demolished by Galileo. And, obviously, Galileo got the job of professor of mathematics at Pisa.

Later Galileo shifted to the university at Padua, very close to Venice, and his most active and productive years were spent there. The Republic of Venice had to depend upon a strong navy for obvious reasons and designing and building ships was an activity of prime importance. Being a very practical scientist it is conceivable that Galileo was in close collaboration with the Venetian arsenal and studied many aspects of ship design. This led him to study the basic aspects of strength of beams and structures. It was at that time he realized the major mistake he had made while demolishing the model of hell on the comparative argument based on the dome of the cathedral and the terrestrial cover of the hell. He kept his finding to his heart but started a thorough scientific study of the scaling laws. (This was perhaps to show his preparedness with the correct science in case someone else discovered the mistake in his lecture which was acclaimed enthusiastically and he got the job at Pisa.)

Galileo realized that large ships break under their own weight when out of water but a scale model of the same ship made of the same wood behaved much stronger. He also noticed that a thin square board can float in water (even though the board material is heavier than water) when the size of the piece is small enough. He rightly argued that when the piece is downsized its area (proportional to its weight) decreases faster than the rate at which the perimeter (which receives the support from waters' surface tension) decreases. Galileo also observed that animals cannot be simply scaled up. As the weight increases at third power of the size scale the bone's supporting cross-sectional area has to be disproportionately larger. Figure 2.3a explains the matter.

If system 1 is three times bigger than system 2 the weight ratio is $W_1/W_2 = 27$. Hence the cross-sectional area of the supporting column is $A_1/A_2 = 27$

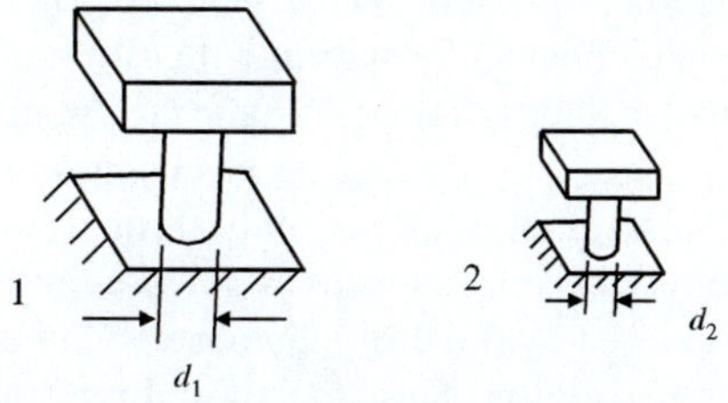

Fig. 2.3a Scaling of supporting columns

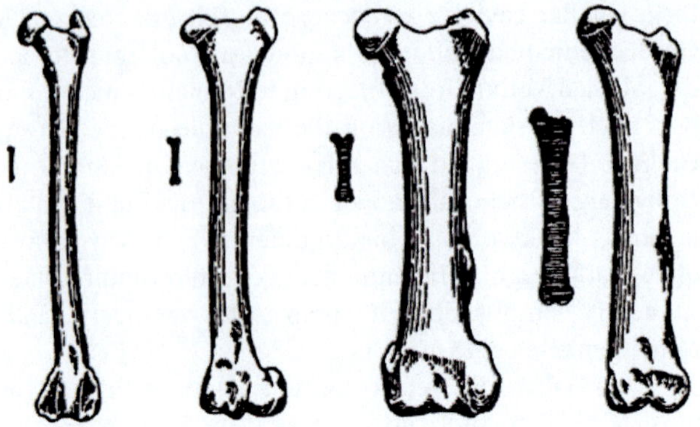

Fig. 2.3b Bone shapes with varying size

So, $A_1 = 27A_2$, since $A_1 \propto {d_1}^2$ and $A_2 \propto {d_2}^2$.

$$d_1 = 5.2d_2 > 3d_2$$

This way Galileo explained why the bones of larger animals are much thicker than those of small animals as shown in an illustration from his famous book *Dialogues on The New Sciences* (Fig. 2.3b).

2 Scaling Laws and Their Importance

The scaling laws are proportionality relations of any parameter associated with an object (or system) with its length scale. For example, the volume of an object varies as cubic length (i.e., as l^3); on the other hand, its surface area scales as l^2. Therefore, a smaller object possesses larger surface area to its volume when compared with a bigger object with similar geometrical shape. There are primarily two types of scaling laws. One is related to the scaling of physical size of objects. The other type is related to the scaling of a phenomenological behavior of an object/machine.

The first thing that a modern engineer or scientist requires for designing a miniaturized device is an understanding of the scaling laws. When all aspects of the device scale in a similar way the geometric integrity is maintained with size. Such type of scaling is called "isomorphic" (or "isometric") scaling. On the other hand, if different elements of a system with different functionalities do not scale in a similar way, the scaling is called "allometric" scaling. Scaling laws deal with the structural and functional consequences of changes in size or scale among otherwise similar structures/organisms; thus, only through the scaling laws a designer becomes aware of physical consequences of downscaling devices and systems. Human intuition is conditioned by the everyday observed phenomena around us within the common range of perception. This "macrointuition" may lead to erroneous designs of microsystems

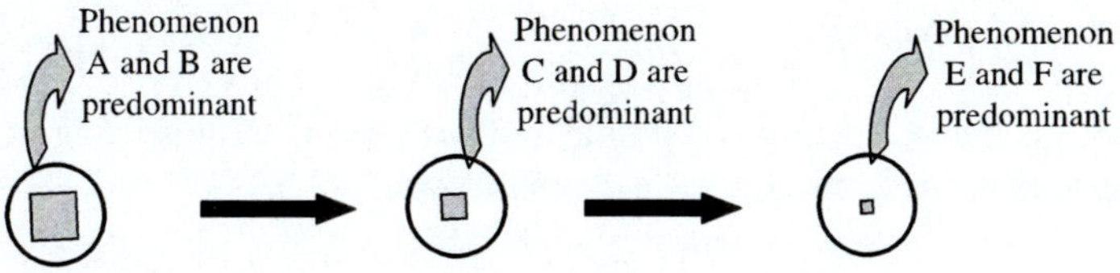

Fig. 2.4 Change in predominance of phenomena with varying scale

if the appropriate "scaling laws" are not taken into consideration, as these laws provide an idea about the system's performance at a totally different scale from that of an existing system at normal scales. Thus, the scaling laws help designers to develop a kind of "microintuition" that is essential for successful design of microsystems. In general, the performances of different subsystems of a system scale differently. This can lead to a different appearance of a system with much smaller size.

Another important aspect of scaling laws is very significant while conceiving, planning, and designing miniaturized devices and systems. When the size of an object is scaled down, different physical phenomena become predominant at different scales. Figure 2.4 shows this aspect of scaling laws.

This change of relative importance with size of a system is very important for successful design of a miniaturized system. For example, at macroscopic scale the weight of an object is predominant and it falls down under the influence of gravity. But when the same object becomes very small the weight (which scales as l^3) becomes relatively insignificant compared to air friction which depends on the surface area scaling as l^2. Therefore, even small air currents can keep it floating. When a glass full of water is overturned the water gets spilled; on the other hand, water confined to a capillary tube does not come out even when the tube is upturned. This is because the force due to surface tension becomes predominant at very small scales.

In complex systems scaling laws become relevant for understanding the interplay among various physical phenomena and geometric characteristics. Sometimes, relatively simple scaling laws, applicable to very complex systems, can provide clues to some fundamental aspects of the system. Thus, scaling laws are not only important for designing microsized systems but also very useful in understanding the basic physical principles involved in many complex phenomena.

3 Scaling Laws and Their Application

In this section important scaling laws and their use (and influence) will be taken up for discussion.

3.1 Geometric Scaling

The laws for geometric scaling are simple and well known. The scaling of various geometric parameters follows the laws given below:

Perimeter (P) $\propto l$
Area (A) $\propto l^2$
Volume (V) $\propto l^3$

where l is the length scale.

The scaling law $A \propto l^2$ can be used in geometry. This law states that the area of a geometric figure shown in Fig. 2.5a scales as l^2. If keeping the geometric shape unchanged the size is changed the area of the figure will change in proportion to the square of the length scale. So, $A_1{:}A_2{:}A_3 = 1{:}1/4{:}1/64$.

Figure 2.5b shows how the scaling law can be used to prove the Pythagoras theorem. ABC is a right-angled triangle and CD is the normal to the hypotenuse AB. It is very easy to show that the three triangles ΔABC, ΔACD, and ΔBCD are similar. So, their areas will be proportional to the square of any characteristic length (let it be the hypotenuse of each triangle).

Thus, the area of $\Delta ABC = \lambda AB^2$, the area of $\Delta ACD = \lambda AC^2$ and the area of $\Delta BCD = \lambda CB^2$, where λ is a constant of proportionality. It is obvious that

$$\text{Area of } \Delta ACD + \text{Area of } \Delta BCD = \text{Area of } \Delta ABC$$
$$\text{or, } \lambda AC^2 + \lambda BC^2 = \lambda AB^2$$
$$\text{or, } AC^2 + BC^2 = AB^2$$

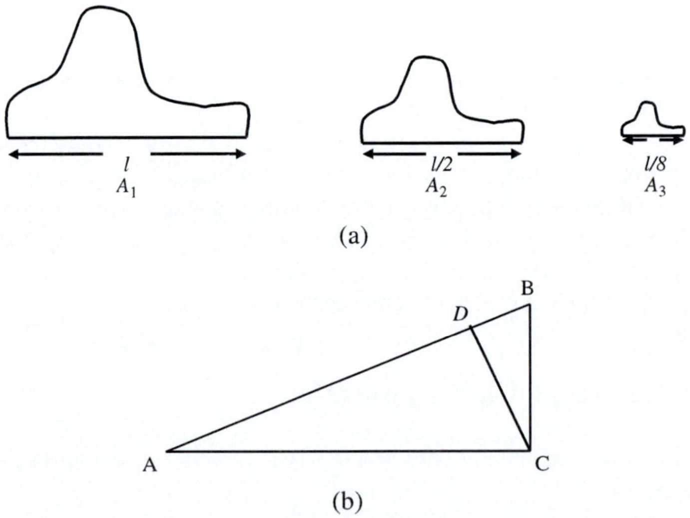

Fig. 2.5 Scaling of area and proof of Pythagoras theorem

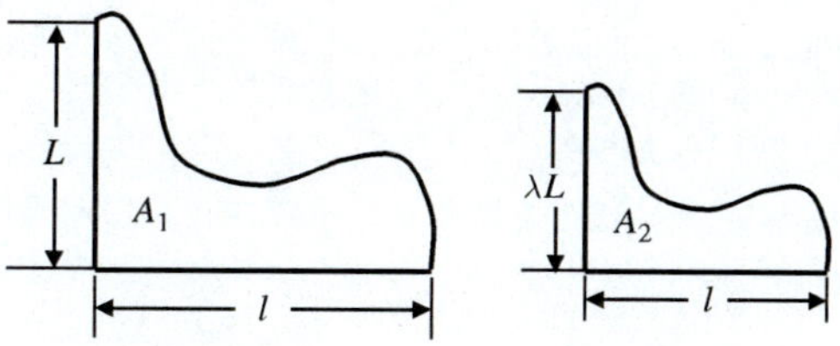

Fig. 2.6a Unsymmetrical scaling of area

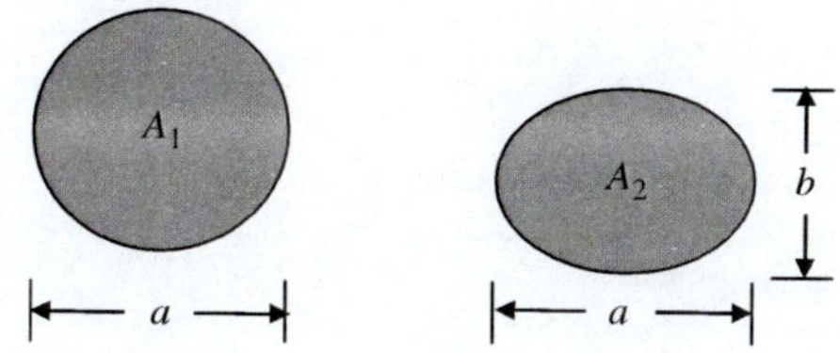

Fig. 2.6b Area of an ellipse by scaling law

If only one dimension is scaled the area will also scale as linear power of that length scale. Figure 2.6a shows two figures where only the vertical dimension is changed. So, if A_1 and A_2 be the areas of the two figures, $A_1{:}A_2 = 1{:}\lambda$.

Figure 2.6b shows a circle with diameter a and area $A_1(= \pi a^2/4)$. So, when the vertical scale is reduced and a is reduced to b the area of this is flattened circle (an ellipse with major axis a and minor axis b) A_2 will be in proportion to the vertical scale. So, $A_1{:}A_2 = 1{:}\lambda$.

where $b = \lambda a$ or $\lambda = b/a$. Thus

$$(\pi/4)a^2 : A_2 = 1{:}b/a$$

or,

$$A_2/[(\pi/4)a^2] = (b/a)/1$$

or,

$$A_2 = (\pi/4)ab$$

It is also quite obvious that Pythagoras theorem is valid for not only squares drawn on the three sides of a right-angled triangle but figures with any shape as indicated in Fig. 2.6c, where areas A, B, and C satisfy the condition $A = B + C$.

The quantities P/V and A/V scale as l^{-2} and l^{-1}, respectively. These two ratios control many important aspects. An object (like a piece of board) floating on a liquid surface experiences upward force due to the surface tension proportional to the length of the perimeter (Fig. 2.7a). So, the upward force F can be written as βP. On the other hand, the downward force due to gravity w will be ςA where ς is the weight of the sheet per unit area.

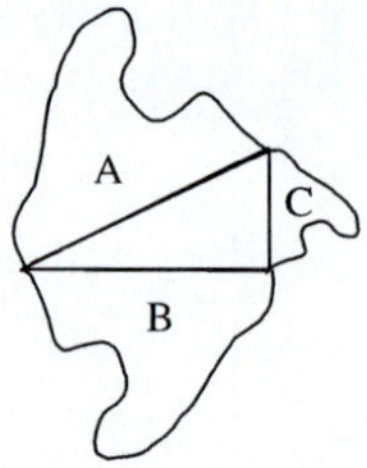

Fig. 2.6c Pythagoras theorem for arbitrary geometrical shape

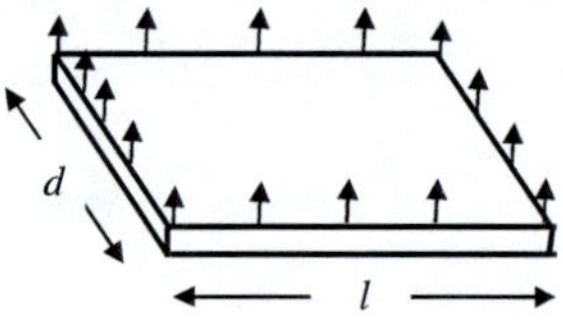

Fig. 2.7a Surface tension on the perimeter of a floating square plate

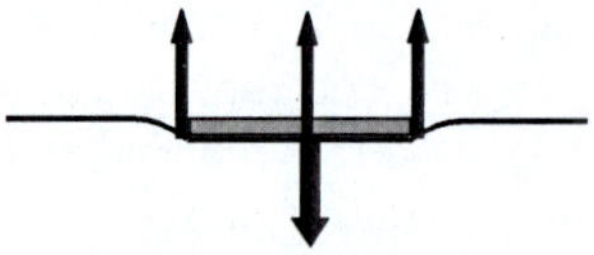

Fig. 2.7b Forces on a floating plate

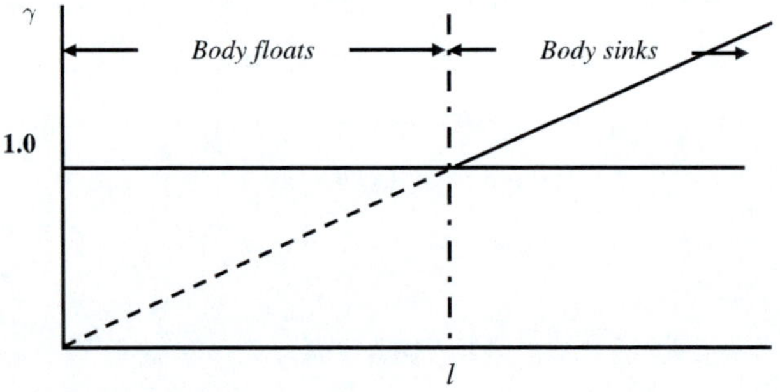

Fig. 2.7c Condition for a board to float

The ratio "γ" of the downward weight W and upward force F, W/F, scales as l^2/l, i.e., l.

Figure 2.7c shows that when the size reduces and γ becomes less than 1 the body floats. This was also discovered by Galileo.

3.2 Scaling in Mechanics

Scaling effects on problems of mechanics are very important and needs to be considered while designing systems and devices at microscopic scales. This will be elaborated with the help of a few standard problems in mechanics. To begin with, problems in solid mechanics will be taken up.

3.2.1 Cantilever Beam

Figure 2.8 shows a typical cantilever beam of length L, width b, and thickness h. If a force F acts at the tip of the beam the resulting deflection of the tip is δ. Thus the stiffness of the beam can be represented by the quantity $k = F/\delta$. It is known that

$$\delta = (FL^3)/(3EI) \tag{2.1}$$

where $I = (1/12)bh^3$. Therefore,

$$k = F/\delta = 3EI/L^3 = Ebh^3/4L^3$$

If the material remains the same, the stiffness of a cantilever beam scales as the length scale (l) of the beam. So,

$$k \propto l$$

If one has to find out the stiffness property under the beam's own weight, the deflecting force will be proportional to its weight that scales as l^3. Since I scales as l^4,

$$\delta \propto l^3 \times l^3 \times l^{-4} \propto l^2 \tag{2.2}$$

Thus, smaller beams behave stiffer than the larger ones.

Next, let the problem of the beam's ability to prevent breakage under self-weight be considered. The bending moment is maximum at the fixed end and its magnitude is $(1/2)\rho bhL^2$, where ρ is the density of the beam material. The maximum stress $\sigma_{\max}$ at this end is given by

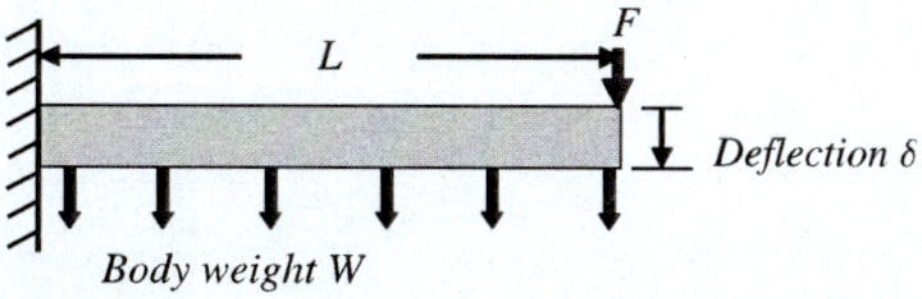

Fig. 2.8 Forces on a cantilever beam

$$\sigma_{\max} = (3\rho L^2)/h \tag{2.3}$$

Hence,

$$\sigma_{\max} \propto l$$

Figure 2.9a shows the nature of dependence of the maximum stress developed on the length scale (i.e., the size of the beam keeping the proportions among various dimensions unchanged). It is seen that beyond a particular size the beam breaks under self-weight as $\sigma_m \geq \sigma_u$, the limiting stress the beam material can withstand.

3.2.2 Simply Supported Beam

A similar analysis for a simply supported beam (shown in Fig. 2.9b) results in $M_{\max} = (1/8)\chi bhL^2$ at the middle of the beam. And the corresponding maximum stress is given by

$$\sigma_{\max} = (3/4)\rho L^2/h \propto l \tag{2.4}$$

In this case also when $\sigma_{\max}$ reaches the breaking stress of the material with the increase in the size of the beam, it breaks under self-weight. This is the reason why a real ship breaks when it is brought out of water though a scale model behaves strongly.

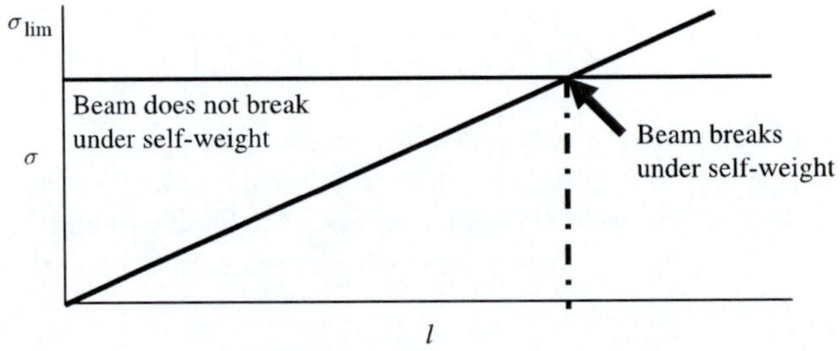

Fig. 2.9a Dependence of maximum stress due to self-weight of a cantilever beam on size

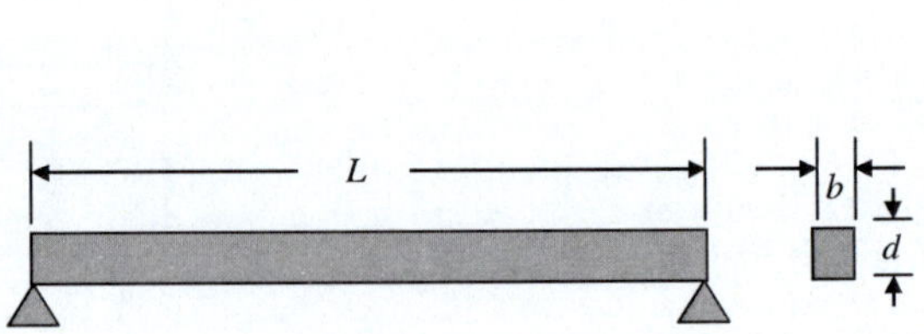

Fig. 2.9b A simply supported beam

3.2.3 Columns

In both the above cases a single phenomenon, bending of beams, played the predominant role. Sometimes, with scale, the deciding factor for failure may shift from one phenomenon to another.

Figure 2.10 shows a solid spherical object of diameter D mounted on a circular column of height h and diameter d. If the diameter of the column be just right to withstand the compressive load then

$$\sigma_u = \{\rho(1/6)\pi D^3\}/\{(\pi/4)d^2\} = (2/3)\rho(D^3/d^2)$$

Hence,

$$d^2 = (2/3)(\rho/\sigma_u)D^3$$

or,

$$d = \sqrt{(2/3)(\rho/\sigma_u)D^3} \propto l^{3/2} \tag{2.5}$$

Thus d has to scale as $l^{3/2}$ to prevent compression failure. When designing such a system it is noticed that with scale the relative dimensions change. For example, in this case D and h may scale as l but d has to scale as $l^{3/2}$, i.e., for larger systems the column is disproportionately thicker as shown in Fig. 2.11.

This is noticed in the case of many trees. The stems of saplings are relatively much slender compared to the trunk of the tree when it grows and becomes big. But there is another important point that needs to be noted. With the change in size even the mode of failure may change from compressive failure of the column to the buckling or instability of the column.

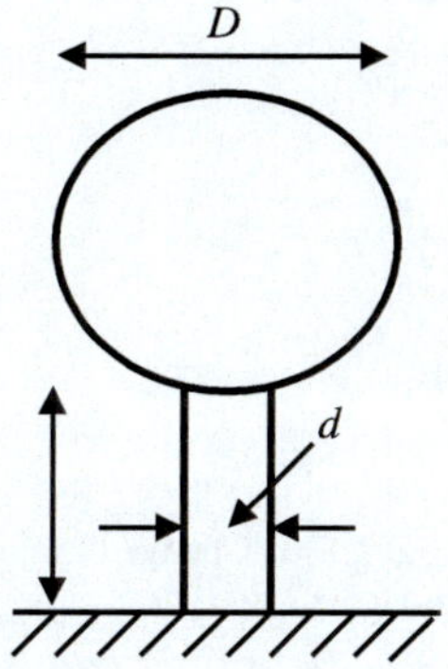

Fig. 2.10 A spherical object supported by a column of circular cross section

Fig. 2.11 Allometric scaling of a sphere on a column

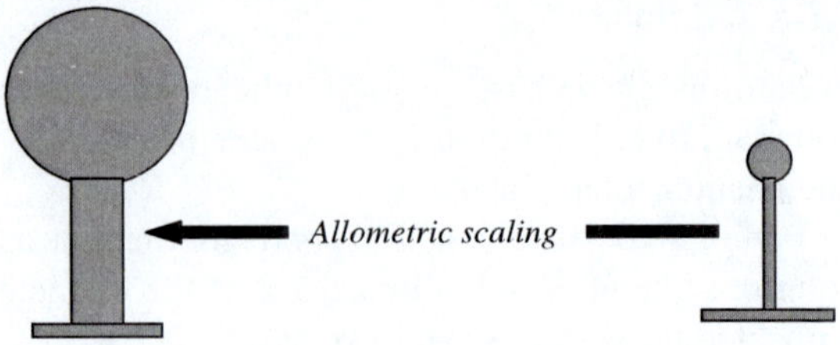

The condition for buckling of a column, whose upper end is free and the lower end is fixed, is given by $F_\mathrm{b} = (\pi^2 EI)/4h^2$, where F_b is the vertical force on the tip of the column to cause instability and buckling failure. Thus $F_\mathrm{b} \propto l^2$.

But actual load acting F_a at the tip is the weight of the sphere that is proportional to its volume. So, $F_\mathrm{a} \propto l^3$. Hence, Fig. 2.12a indicates how the failure mode may change from compression failure to buckling as the length scale reaches a critical value l^* (h may be also taken as a measure of the length scale).

If the height of the column h be χD then column diameter d_b to prevent buckling is obtained from the fallen equation:

$$\{\pi^2 E(1/64)\pi d_\mathrm{b}^4\}/(4\alpha\chi^2 D^2) = (1/6)\pi D^3 \rho$$

or,

$$d_\mathrm{b}^4 = \{(128\pi^2/3)(\chi^2 \rho/E)\} D^5 \qquad (2.6)$$

Again the diameter of the column d_c to prevent compressive failure is obtained from

$$[\{(1/6)\pi D^3 \rho\}/\{(\pi/4)d_\mathrm{c}^2\}] = \sigma_u$$

or,

$$d_\mathrm{c}^2 = \{(2/3)(\rho/\sigma_u)\} D^3 \qquad (2.7)$$

so, $d_\mathrm{b} \propto D^{5/4}$ and $d_\mathrm{c} \propto D^{3/2}$ (Fig. 2.12b).

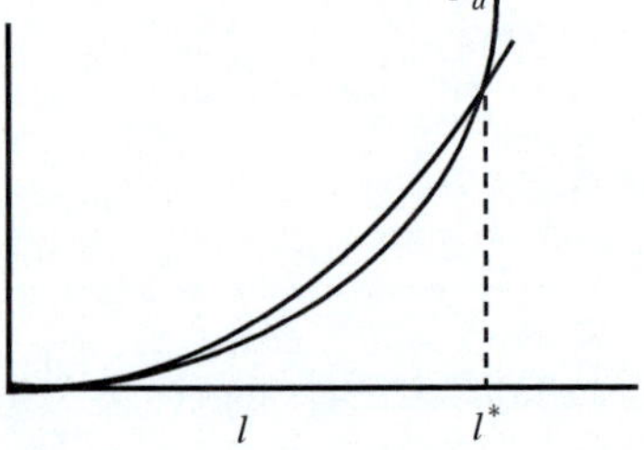

Fig. 2.12a Change of failure mode

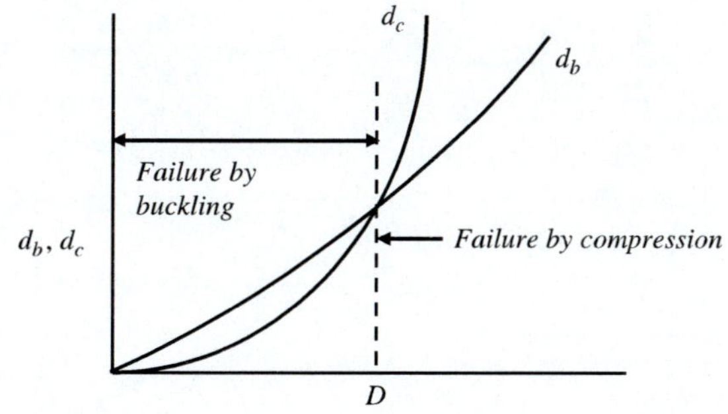

Fig. 2.12b Critical column diameter for change of failure mode

Therefore, at very small scales stability of the supporting columns becomes a problem instead of compressive failure.

3.3 Micromechanisms

In the microscopic scales, mechanisms and machines are generally of monolithic construction. This is primarily to facilitate fabrication. At that small scale it is next to impossible to assemble micron-sized parts and develop revolute hinges and prismatic joints without increasing the cost of the device to prohibitive levels. For providing the capacity of relative motion among various members localized compliances are used. Figure 2.13 shows how a hinge can be replaced by a localized compliance. Though in the second case the range of relative motion is somewhat limited in most situations it does not impose any serious problems or limitation.

Figure 2.14 shows the different types of arrangements normally used to simulate revolute, prismatic, and spherical joints. The maximum possible deflection $\delta_{\max}$ achievable with a particular material as a function of the dimensions is also given in the figure. To proceed further, some relationships are assigned to the dimensions as follows:

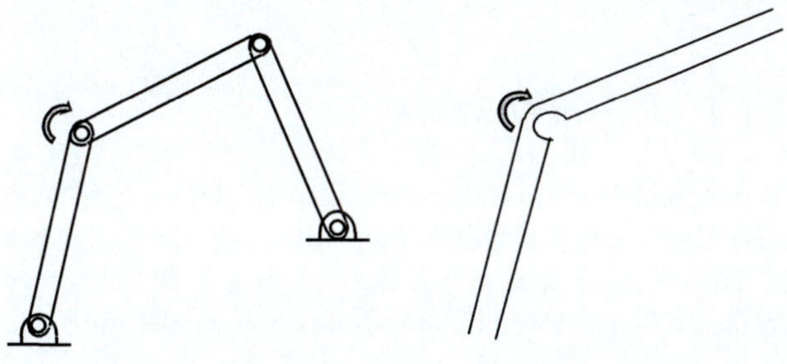

Fig. 2.13 Replacement of a hinge by localized compliance

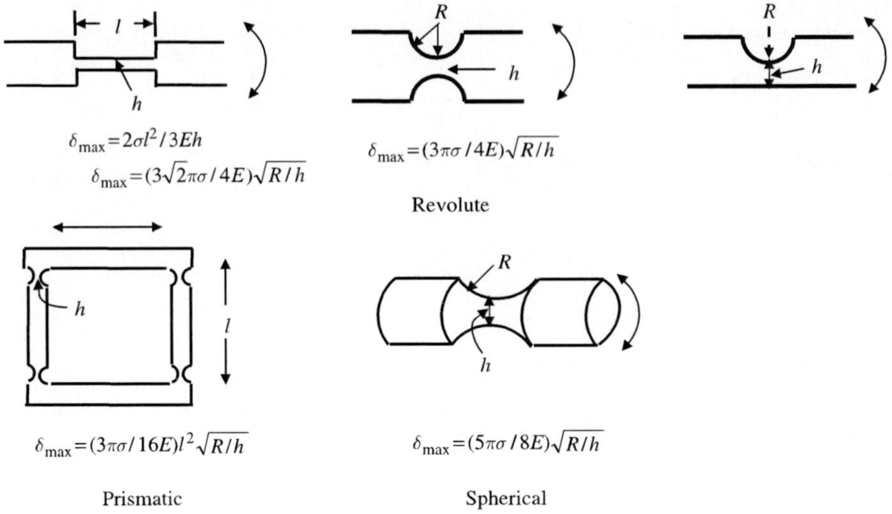

Fig. 2.14 Different types of localized compliances

$$h = \lambda_1 l \text{ in the first case}$$
$$h = \lambda_2 R \text{ in all the other cases}$$
$$h = \lambda_3 l \text{ in the case of the prismatic joint}$$

The parameters λ_1, λ_2, and λ_3 specify the geometric shape independent of the size. Using this dependence on the size scale δ_{max} is found as follows:

$$\delta_{max} = (2\sigma/3\lambda_1 E)l \propto l^1 \text{ first case}$$
$$\delta_{max} = (3\pi\sigma/4E\sqrt{\lambda_2}) \propto l^0 \text{ second case}$$
$$\delta_{max} = (3\sqrt{2}\pi\sigma/4E\sqrt{\lambda_2}) \propto l^0 \text{ third case}$$
$$\delta_{max} = (3\pi\sigma/16E)(\sqrt{3/\lambda_2\lambda_3})l^2 \propto l^2 \text{ fourth case}$$
$$\delta_{max} = (5\pi\sigma/8E)\sqrt{\lambda_2} \propto l^0 \text{ fifth case}$$

Thus, it is clear that except for the first and fourth cases the deflection is independent of the size. Both in the first and fourth cases the deflection increases with size and the systems behave much stiffer in the microscopic scales.

3.4 Scaling in Dynamics

In dynamics the scaling effects can be determined using the laws of motion. The important quantities which are involved in dynamical problems are discussed below.

The displacement scales as l, velocity as lt^{-1} and acceleration as lt^{-2}. Since mass scales as l^3 and force is mass times the acceleration, dynamic force scales as l^4t^{-2}.

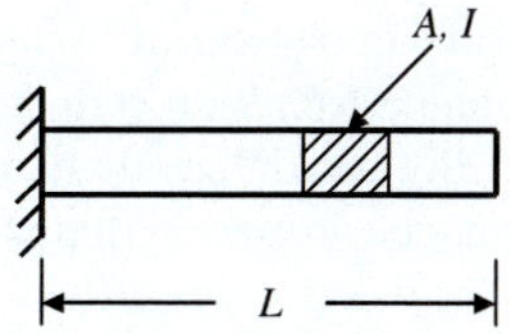

Fig. 2.15 A vibrating cantilever beam

Energy and work done by a force have the same dimension which is force times displacement. Hence, energy scales as l^5t^{-2}. Mass moment of inertia has dimension of mass times length square; thus it scales as l^5. Angular displacement is dimensionless; angular velocity and acceleration have dimensions of t^{-1} and t^{-2}, respectively. Moment, being mass moment of inertia times angular acceleration, scales as l^5t^{-2}. As mass moment of inertia scales as l^5 it is extremely small in microsystems and that is why micromotors can be both started and stopped almost instantaneously. Both mass and mass moment of inertia depend very heavily on the scale and, therefore, their importance in systems of very small size can be ignored. As a result a kinetostatic analysis is good enough in case of micromachines and micromechanisms. Such systems are under static equilibrium conditions at all instants.

Elastic bodies' dynamic characteristics are also influenced by scaling. The example of free vibration of a uniform, prismatic cantilever beam is presented here.

Figure 2.15 shows such a beam of length L, cross-sectional area A, and second moment of area I. If the modulus of elasticity and the density of the material be E and ρ, respectively, the natural frequency of transverse vibration is

$$\omega_n \propto (1/L^2)\sqrt{EI/A\rho} \tag{2.8}$$

So, if the material is kept unchanged and the size of the beam is changed (keeping dimensional proportions intact) it can be seen that $\omega_n \propto l^{-1}$.

As a result the system's natural frequency increases with miniaturization. Typical frequency of microsystems and devices can be in MHz (or even GHz) range. The response time of mechanical systems in the microrange can be as low as electronic devices. This is an important point to be remembered while designing microsystems.

3.5 Scaling in Fluid Mechanics

Matters related to fluid flow are severely affected by scaling effects. Very common daily occurrences like a speck of dust floating around without falling to the ground due to gravity and the reluctance of a fluid in a capillary tube to come out due to gravity, when the tube is kept vertical, are all examples of these scaling effects. The force due to surface tension becomes predominant at microscales as it scales as l in comparison with weight that scales as l^3.

In micron-sized channels the Reynolds number for a fluid flow is very low and may be around 1; it is known that

$$Re \propto vd\rho/\eta$$

where ρ and η are the density and viscosity of the fluid, d is the diameter of the channel, and v is the velocity of the fluid. Since v scales as l it is clear that

$$Re \propto l^2 \tag{2.9}$$

At such low Reynolds number the flow is extremely laminar and it becomes very difficult to mix fluids in microchannels. The viscous drag on a body offered by a fluid limits the terminal velocity v_{lim}

$$v_{\text{lim}} \propto 4gd^2\rho/18\eta \propto l^2 \tag{2.10}$$

As a result v_{lim} is very small for microparticles and this causes the particles to float with the moving air instead of falling.

From Hagen–Poiseuille's equation it is known that the volumetric fluid flow rate satisfies the following equation:

$$Q = (\pi r^4 \Delta p)/(8\eta L) \tag{2.11}$$

where the quantities are explained in Fig. 2.16. So, for a given pressure drop rate (i.e., $\Delta p/L$) $Q \propto l^4$. Hence for a given volume flow rate, the pressure drop rate $(\Delta p/L) \propto l^{-4}$.

So, it is extremely difficult to push fluid through microchannels using pressure drop. This makes conventional pressure-driven pumping difficult for microchannels. Surface driving forces are more suitable as such forces scale favorably.

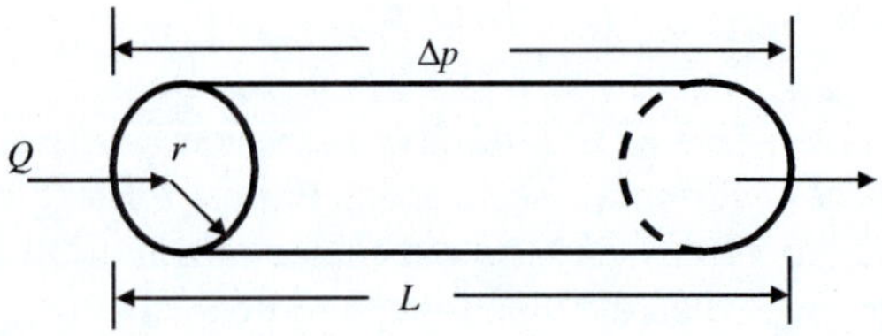

Fig. 2.16 Fluid flow in a circular pipe

3.6 Scaling of Common Forces

Solving dynamical problems involves various types of forces. The most common types of forces and their scaling are tabulated below and Fig. 2.17 shows the nature of some common types of forces.

Force	Scaling
Surface tension	l^1
Fluid force/electrostatic force	l^2
Weight/inertia force/electromagnetic	l^3
Electromagnetic force (constant current density)	l^4

For handling different types of forces in a compact manner, William Trimmer proposed a matrix representation for force scaling. This column matrix, called the "force scaling vector F" is defined as follows:

$$F = [l^F] = \begin{bmatrix} l^1 \\ l^2 \\ l^3 \\ l^4 \end{bmatrix} \tag{2.12}$$

The above matrix represents four different cases of force laws (given in the table above) which scale differently. Using this nomenclature, scaling of different parameters can be determined as explained below.

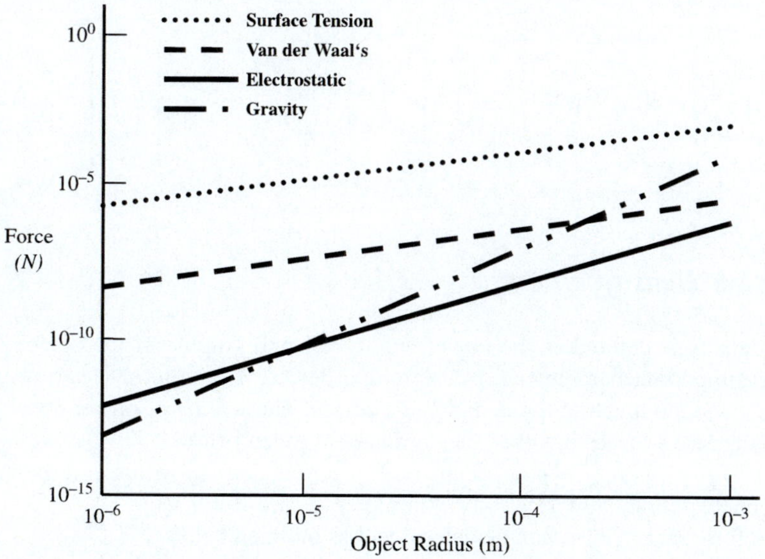

Fig. 2.17 Scaling of different types of forces

3.7 Scaling of Acceleration

From the second law of motion, $F = ma$, with m as mass and a as acceleration. So, $a = F/m$ and its scaling for different types of force can be expressed in a compact form as follows:

$$[a] = \left[l^F\right] \times \left[l^3\right]^{-1} = \begin{bmatrix} l^1 \\ l^2 \\ l^3 \\ l^4 \end{bmatrix} \times \left[l^{-3}\right] = \begin{bmatrix} l^{-2} \\ l^{-1} \\ l^0 \\ l^1 \end{bmatrix} \tag{2.13}$$

because m scales as volume, i.e., as l^{-3}.

3.8 Scaling of Time

If s be the displacement in time t (starting from rest)

$$s = (1/2)at^2$$

So,

$$t = (2s/a)^{1/2} = (2sm/F)^{1/2}$$

Thus,

$$[t] = [s]^{1/2}[m]^{1/2}[F]^{-1/2} = \left[l^1\right]^{1/2}\left[l^3\right]^{1/2}\begin{bmatrix} l^1 \\ l^2 \\ l^3 \\ l^4 \end{bmatrix}^{-1/2} = \begin{bmatrix} l^{1.5} \\ l^1 \\ l^{0.5} \\ l^0 \end{bmatrix} \tag{2.14}$$

3.9 Power Density

Power density is defined as the power supply per unit volume of the device. This is a very important parameter for designing a device. Too little of it can result in inactivity and too much of it can damage a device. The scaling of power density for different types of forces involved can be found as given below.

Let $\gamma = P/V$ where γ = power density, P = power, and V = volume of the device under consideration. Now P = (work/time) = Fs/t and $\gamma = Fs/Vt$.

Therefore, scaling of γ can be determined as outlined below:

$$[\gamma] = [F] \times [s] \times [V]^{-1} \times [t]^{-1}$$

Using (2.12) and (2.14) in the R.H.S. of the above equation and noting that $[s] = [l^1]$ and $[V] = [l^3]$ the following equation is obtained:

$$[\gamma] = \begin{bmatrix} l^1 \\ l^2 \\ l^3 \\ l^4 \end{bmatrix} [l^1] \times [l^3]^{-1} \begin{bmatrix} l^{1.5} \\ l^1 \\ l^{0.5} \\ l^0 \end{bmatrix} = \begin{bmatrix} l^{-2.5} \\ l^{-1} \\ l^{0.5} \\ l^0 \end{bmatrix} \tag{2.15}$$

The table below presents the important parameters for four different types of force law:

Force scaling	Acceleration scaling	Time scaling	Power density scaling
l^1	l^{-2}	$l^{1.5}$	$l^{-2.5}$
l^2	l^{-1}	l^1	l^{-1}
l^3	l^0	$l^{0.5}$	$l^{0.5}$
l^4	l^1	l^0	l^2

So, it is clear that devices operated by electrostatic actuators tend to have higher power density at small scales whereas for electromagnetic motors power density is low for small sizes. Thus miniaturized machines should be operated by electrostatic forces instead of electromagnetic drives.

3.10 Scaling of Electrical Parameters

Besides mechanics scaling laws have important consequences for electrical systems also. The three passive electrical elements, resistors, capacitors, and inductors, are taken up for determining the scaling effects.

3.10.1 Resistance

The resistance of a given conductor (Fig. 2.18a) of length L, cross-sectional area A, and specific resistivity ρ is given by

$$R = L\rho/A = [l^1] \times [l^2]^{-1} = l^{-1}$$

So, for a given material electrical resistance scales as l^{-1}.

3.10.2 Capacitance

For a parallel plate capacitor (Fig. 2.18b) of plate area A, plate gap d, and ε as the permittivity of gap insulation material the capacitance

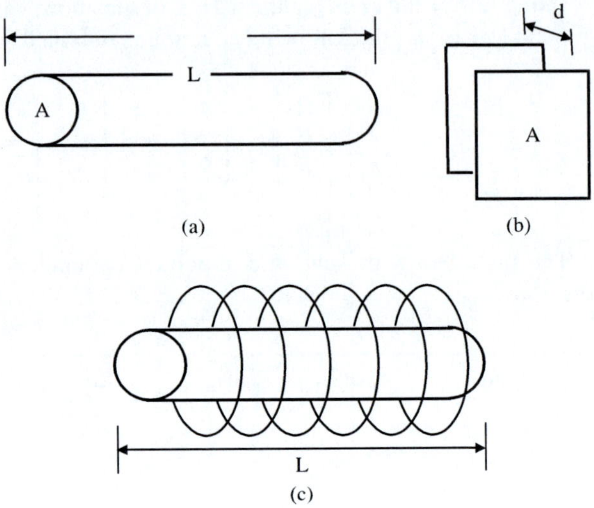

Fig. 2.18 Basic electrical elements

$$C = \varepsilon A/d = \left[l^2\right] \times \left[l^1\right]^{-1} = l^1$$

So, capacitance scales as l^1.

3.10.3 Inductance

With N as the number of coils per unit length with a coil area A and L as the length of the inductor (Fig. 2.18c), the inductance $\mathcal{L}$ is given by

$$\mathcal{L} = \mu N^2 A/L = \left[l^2\right] \times \left[l^1\right]^{-1} = l^1$$

where μ is the permeability of the material between the coils. Hence, the inductance scales as l^1.

In an electrical circuit combinations of these basic units govern the characteristics of a system. For example, the time constant of a circuit is dependent on the product RC and governs the behavior when the voltage varies. To understand how it will scale the following expression helps:

$$RC = [R] \times [C] = \left[l^{-1}\right] \times \left[l^1\right] = \left[l^0\right]$$

So time constant is independent of scaling.

3.11 Some Important Scaling Laws

A table containing the scaling exponent of different physical quantities is given for ready reference and use. Scaling laws for more complex parameters can be found out from these. If the parameter/quantity be P then scaling will be $[P] = [l^n]$

Physical quantity, P	Scaling exponent "n"
Bending stiffness	1
Mass	3
Mass moment of inertia	5
Second moment of area	4
Strength	2
Shear stiffness	1
Natural frequency	−1
Reynolds number	2
Electrical resistance	−1
Electrical capacitance	1
Inductance	1
Surface tension, van der Walls force	1
Fluid force	2
Inertia force	3, 4
Kinetic energy	3 (Const. speed), 5
Potential energy (gravitational)	4
Elastic potential energy	2
Strength to weight ratio	−1
Resistance power loss	1
Thermal time constant	2
Heat capacity	3
Electric field energy	−2
Available power	3
Power loss/power available	−2
Electromagnetic force	3
Electrostatic force	2

3.12 Scaling in Electromagnetic and Electrostatic Phenomena

When two parallel electrically conducting plates of area "A" at a distance "d" and a voltage "V" are applied as shown in Fig. 2.19, the corresponding potential energy stored in the capacitor that is formed is given by $U = -1/2CV^2$, where C is the capacitance. The capacitance C for the system is proportional to the area "A" and inversely proportional to the gap between the plates "d." Thus

$$U \propto AV^2/d \tag{2.16}$$

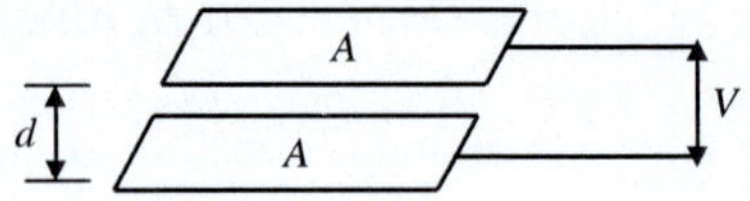

Fig. 2.19 A charged capacitor

The area A scales as l^2 and the gap d scales as l^1. To find out the scaling effect of the voltage V it may be done by considering the breakdown voltage given by Paschen's effect as shown in Fig. 2.20.

Approximately, the breakdown voltage can be taken as proportional to the gap d when $d > 10\ \mu$m. Hence, the voltage V can be considered to scale as l^1. Hence, the electrostatic potential energy

$$U \propto l^2/l^1(l^1)^2 \propto l^3$$

Scaling of electrostatic force can be found out as shown in Fig. 2.21.

At the symmetric position F_L and F_B will be zero. But when the plates are misaligned

$$F_L = -\partial U/\partial L \propto l^3/l^1 \propto l^2$$
$$F_B = -\partial U/\partial B \propto l^3/l^1 \propto l^2$$
$$F_d = -\partial U/\partial d \propto l^3/l^1 \propto l^2$$

Thus, electrostatic force scales as l^2.

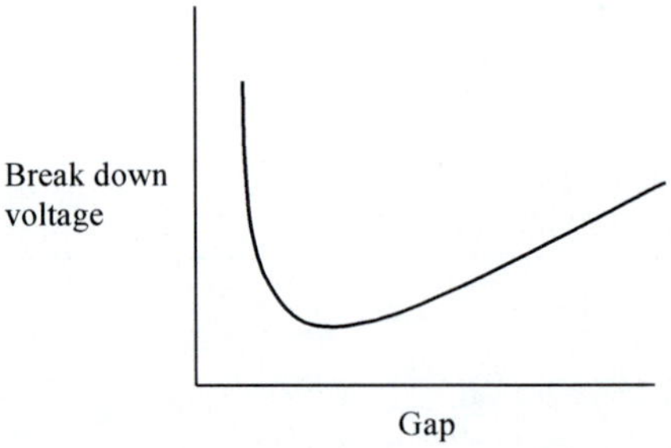

Fig. 2.20 Paschen's effect

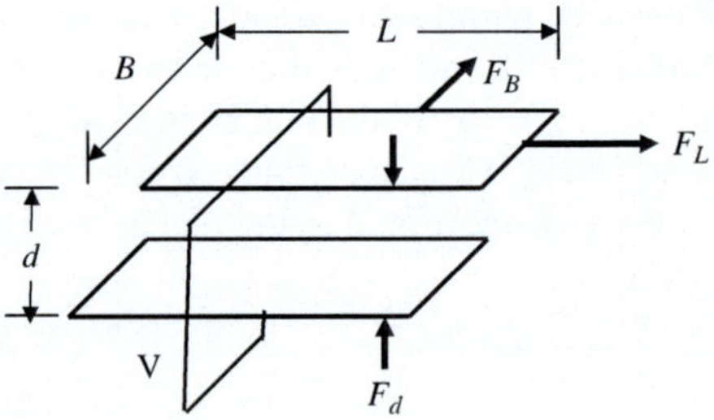

Fig. 2.21 Scaling of electrostatic force from a charged capacitor

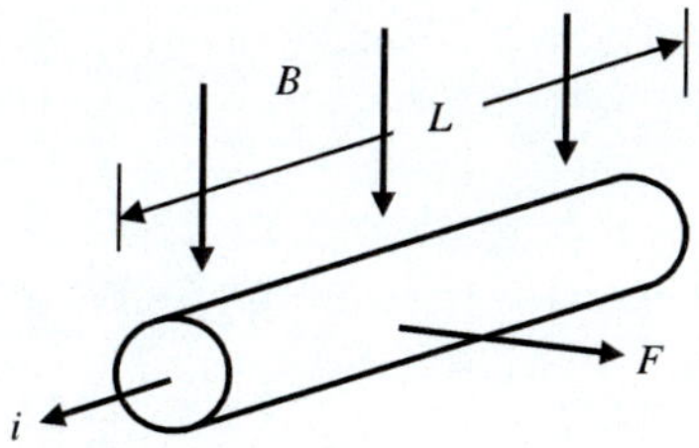

Fig. 2.22 Current-carrying wire in a magnetic field

Figure 2.22 shows a wire of length L carrying a current i in a magnetic field B. The force F on the wire is given by

$$F = iLB \tag{2.17}$$

For a given current density the current i is proportional to the cross-sectional area of the wire. So,

$$i \propto l^2$$

Again $B = \mu_0(n/L_C)i_C$, where μ_0 is the permeability, n is the number of turns in the wire, L_C is the total length of the coil, and i_C is the current through the coil producing the magnetic field. As i_C scales as the cross-sectional area of the coil area it scales as l^2 and L_C scales as l^1. Thus B scales as l.

Therefore, as $L \propto l$, $F \propto l^4$. Thus, electromagnetic force scales as l^4.

3.13 Scaling Laws Related to Surface/Volume Ratio

As the surface area (S) of an object scales as l^2 and the volume (V) as l^3 the ratio S/V scales as l^{-1}. This simple scaling law has many profound implications both in the living and nonliving worlds. It decides many important matters in living organisms, thermal phenomena, dynamics of particles in fluids, etc. Figure 2.23 shows the way S/V ratio varies with size.

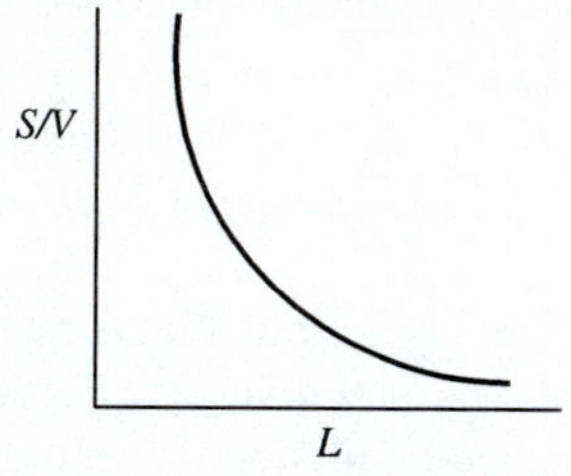

Fig. 2.23 Variation of S/V ratio with size

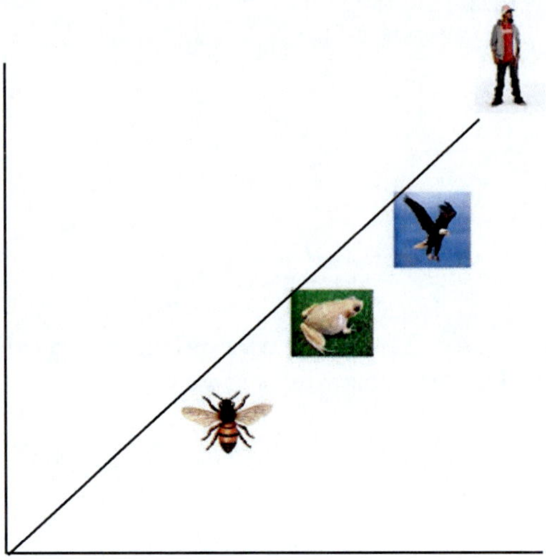

Fig. 2.24 Effect of *S/V* ratio on living objects

If one considers the phenomena of heating and cooling the influence of *S/V* ratio becomes very apparent. When a microsized chip comes out of girding operation it is white hot and brilliantly luminous. But by the time it flies away from the machining area taking only a fraction of a second it cools down to almost room temperature. Similarly during hailstorm the small-sized hails melt away very quickly but big-sized chunks of ice from ice factories take a long time to melt away. Both are effects of the scaling law applied to the *S/V* ratio. Heating or cooling rate of an object depends on the surface area but the amount of heat in the body at a given temperature is proportional to the volume. The *S/V* ratio scales as l^{-1} and, therefore, is much larger for smaller objects compared to a big-sized body. Thus, small objects (like the grinding chips) cool down (or lose the heat of the body) very fast.

In the living world also the *S/V* ratio plays a critical role. Eating food generates heat through metabolism. So, to maintain a specific temperature small-sized mammals have to eat more frequently as due to larger *S/V* ratio they tend to lose heat at a relatively faster rate. Figure 2.24 shows this phenomenon. For still smaller living organisms like insects it is not possible to maintain a fixed body temperature.

It was already mentioned how the small dust particles float around in air current (even if of small speeds). This is because of the fact that the viscous drag with air is proportional to surface area but the gravitational pull depends on the volume.

3.14 Allometric Scaling Laws in Biology

There is nothing that is more complex than life. However, despite the extreme complexity and diversity, living organisms obey certain very simple scaling laws. The general equation that represents the scaling behavior of living organisms spanning a mass range of over 21 orders of magnitude (smallest microbe of 10^{-13} g to the largest mammals and plants of mass 10^8 g) can be written as follows:

$$X = X_0 M^{\lambda} \tag{2.18}$$

where X is some observable (and quantifiable) biological parameter, X_0 is a normalizing constant, M is the mass of the organism, and λ is an exponent. An equation of this type is called allometric scaling law and λ is the allometric exponent. The table below shows various parameters along with the corresponding exponent.

Sl. No.	Parameter, X	Exponent, λ
1	Metabolic rate	3/4
2	Lifespan	1/4
3	Growth rate	−1/4
4	Heart beat rate	−1/4
5	Length of aorta, height of trees	1/4
6	Radii of aorta, radii of tree trunks	3/8

A very interesting point is that λ takes values which are simple multiplies of ¼. Figure 2.25 shows the plot of metabolic rate against body mass for the whole spectrum of living organisms. It is also interesting to note from the above table that the fundamental principles of life provide for certain invariant quantities like the total number of heart beats in a lifetime which is approximately equal to 1.5×10^9, irrespective of the size of the organism. Another interesting point regarding the metabolic rate scaling must be discussed here. It was mentioned earlier that the rate of heat loss through the surface scales as l^2. But the metabolic rate, which is proportional to the heat generation rate, scales as l^3 and is proportional to mass M. Thus, to maintain a constant temperature the rate of heat generation should be equal to the rate of heat loss. M scales as volume, i.e., l^3, and the surface area S scales as l^2, given as $(M^{1/3})^2$ or $M^{2/3}$. Thus, the metabolic rate should scale as $M^{2/3}$

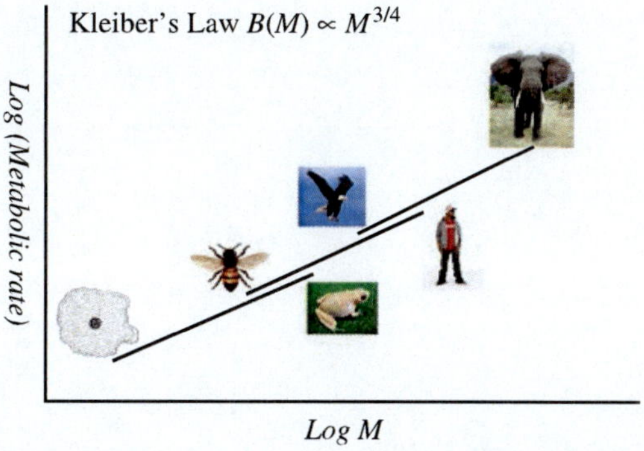

Fig. 2.25 Kleiber's law

for maintaining a fixed body temperature. But research over a long period by many researchers has established that

$$\text{Metabolic rate} = B \propto M^{3/4} \tag{2.19}$$

This is known as Kleiber's law. This controversial issue was finally explained by West more recently and the fractal nature of the fluid transportation system (blood circulatory system in animals) is the true reason behind the exponent being 3/4 instead of 2/3 as expected according to the scaling of surface to volume ratio.

As mentioned before this scaling law gave a deeper understanding of the life forms at all scales and the fractal nature of the fluid distribution system got unraveled as the most efficient form of transport.

Certain other features related to the living world also demonstrate the significance of scaling laws. The strength and ability to jump always appear to be disproportionately higher for smaller living objects. The strength always appears to be more when the size scales down as shown earlier. Certain other abilities can also be scaled.

3.14.1 Jumping

Figure 2.26 shows an animal of body mass M jumping vertically up through a height "h." So, the potential energy gained is equal to "Mgh." This energy is supplied by the muscles of the limbs and is proportional to the mass of the muscles "m." So, work done (= energy gained due to the height gained)

$$W \propto m$$

But the mass of the muscles "m" is proportional to the overall body mass "M." Hence

$$W \propto m \propto M \tag{2.20}$$

Again $W =$ gain is potential energy $= Mgh$. Hence,

$$Mgh \propto M \tag{2.21}$$

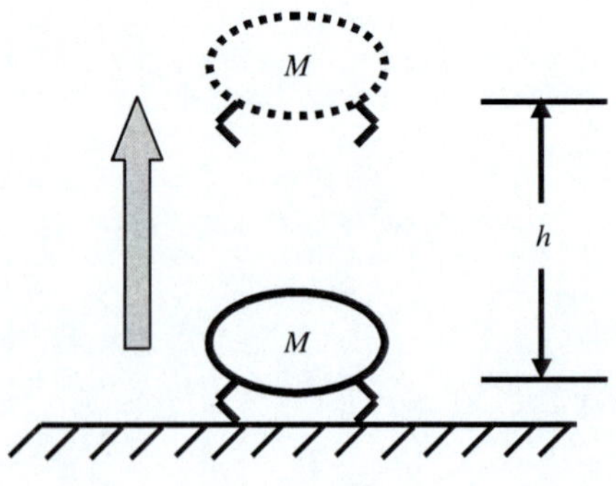

Fig. 2.26 Jumping of an animal

Equation (2.21) leads to the conclusion that "h" is independent of M. Or, in other words, the ability to gain height by leaping is independent of the size of the animal. So, smaller animals appear to have relativity higher capability to jump.

3.14.2 Running

The number of steps "n" an animal has to take for covering a given distance is inversely proportional to the size of limbs. Or,

$$n \propto L^{-1} \tag{2.22}$$

where L represents the limb size. Thus, "n" scales as l^{-1}.

Again the body mass M depends on the volume and so M scales as l^3. Or,

$$\begin{aligned} M &\propto L^3 \\ \therefore \quad L &\propto M^{1/3} \end{aligned} \tag{2.23}$$

And using (2.23) in (2.22)

$$n \propto M^{-1/3} \tag{2.24}$$

The work done per step is proportional to the body mass M. So, for covering a given distance (using "n" steps) the work done

$$W \propto nM \propto M^{1/3}\,M$$

or,

$$W \propto M^{2/3} \tag{2.25}$$

For comparing the situations using a common norm the work done to move a given distance per unit body mass (specific energy of running)

$$w = W/M \propto M^{2/3}/M \propto M^{-1/3} \tag{2.26}$$

The above equation shows that larger the animal, less is the specific energy of running. So, as the energy available for an animal is proportional to its body mass M a bigger animal can cover a larger distance.

3.14.3 Swimming

The speed of swimming or cruising for an animal or a vessel depends on the power of the entity and the fluid resistance it has to overcome. Now work done per unit time (i.e., power) is

$$W \propto \text{distance covered in unit time} \propto v$$

Again $W \propto$ fluid resistance and fluid resistance $\propto$ surface area (S). Furthermore, fluid resistance is also proportional to speed (V).

Hence compiling the above relations

$$P \propto V^2 S \tag{2.27}$$

Now the power depends on the body mass, i.e., body volume ($\propto L^3$) and scales as l^3. But, the surface area S scales as l^2 ($\propto L^2$) Thus, from (2.27) one gets

$$V^2 \propto P/S \propto L \tag{2.28}$$

And $V \propto \sqrt{L}$. Hence, swimming/cruising speed scales as $l^{1/2}$. This is known as Froude's law.

3.14.4 Flying

The capability to fly (for birds and planes) also follows nice scaling laws. If Mg be the body weight, S the wing area, V the speed, and α the lift, then

$$Mg = \text{Lift} = \alpha \propto SV^2$$

or,

$$M \propto SV^2 \tag{2.29}$$

Now if L represents a characteristic length of the flying object, $M \propto L^3$ and $S \propto L^2$. Thus, $L \propto M^{1/3}$ and hence $S \propto M^{2/3}$. Using this in (2.29)

$$M \propto M^{2/3} V^2$$

or,

$$M^{1/3} \propto V^2$$

or,

$$V \propto M^{1/6} \tag{2.30}$$

Figure 2.27 shows the cruising speed for flying objects over a range of their sizes.

3.15 Some Other Interesting Scaling Laws

Simple power laws $y(x) = y_0 x^{\lambda}$ are abundant in nature and such laws are self-similar. Thus, if "x" is scaled by multiplying it by a factor "a," $y(x)$ will be $(y_0 a^{\lambda})x^{\lambda}$ which is a similar power law.

Many phenomena of nature occur in different sizes or intensities and in most cases they follow simple scaling laws.

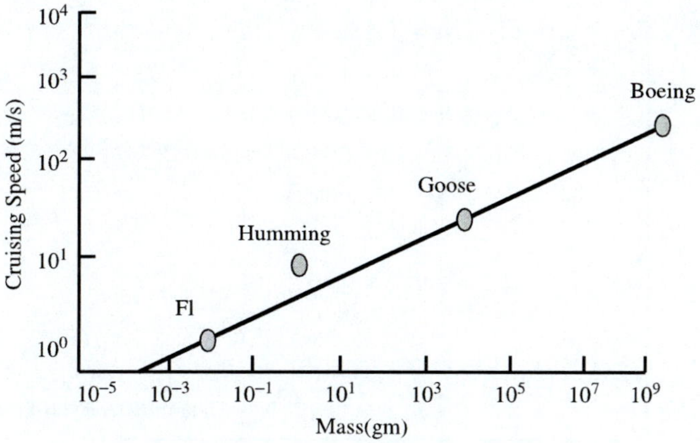

Fig. 2.27 Scaling of cruising speed

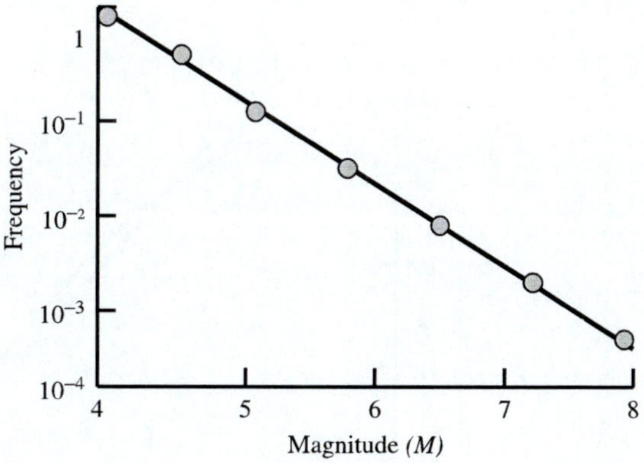

Fig. 2.28 Scaling of earthquake frequency

The frequency of occurrence of earthquakes and their respective intensities follow a very nice power law and it is known as Gutenberg–Richter scaling law as given by $\log N = a - bM$, where N is the frequency of occurrence and M is the magnitude in Richter scale. Figure 2.28 shows the situation.

The size of meteors falling on the earth and their frequency of appearance also follow the nice scaling law shown in Figure 2.29.

Another important phenomenon, number of species versus their size, also follows a scaling law to some extent but not exactly (Fig. 2.30).

3.16 Importance of Scaling Laws in Microactuation

It has already been shown in Fig. 2.17 how different types of forces scale. In the macroscopic scale mostly electromagnetic actuators are employed. Electromagnetic

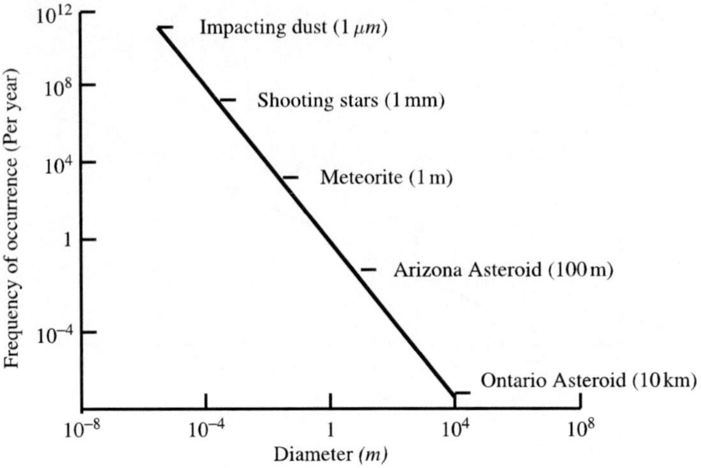

Fig. 2.29 Scaling of objects falling on the earth

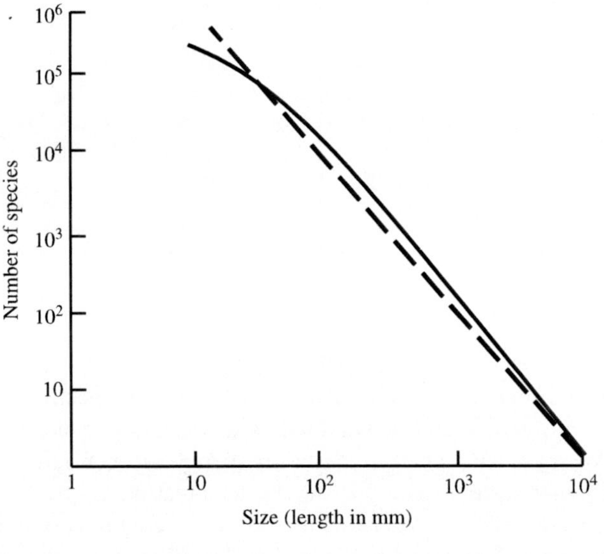

Fig. 2.30 Number of species with size

force scales as either l^3 or l^4. Hence, when the size of the device is very small actuation principle based upon electromagnetic force becomes ineffective. On the other hand, electrostatic force scales as l^2 and, therefore, actuators based upon electrostatic force are more suitable for microscopic devices.

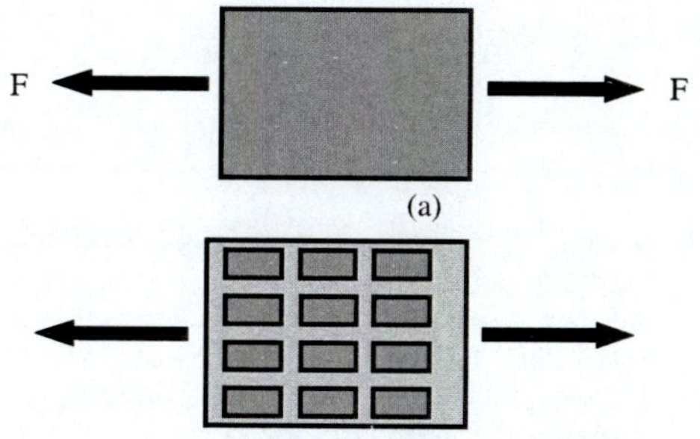

Fig. 2.31 Massively parallel miniaturized actuator

The scaling law can be very nicely used even for actuators of macroscopic size. One example can explain the principle. Figure 2.31a shows an actuator with a characteristic length L based upon electrostatic force and let the force it can develop be F. Now the actuator is miniaturized and scaled down by a factor n. Since length scale is reduced by a factor n the volume of the miniaturized device will be n^{-3} times the volume of the original actuator. Next n^3 numbers of these miniaturized actuators are placed inside a volume equal to that of the original device and arrangement is made so that all these n^3 numbers of actuators act in parallel.

Since the size of each device is reduced by a factor n each one will develop a force F/n^2 as electrostatic force scales as l^2. The n^3 numbers of these actuators, which act in parallel, will develop a total force

$$n^3 \times F/n^2 = nF$$

When $n \sim 1000$ almost 1000 times larger force can be generated by the composite actuator. This principle of miniaturized and massively parallel systems is employed by nature in many situations and the muscles are nothing but massively parallel miniaturized protein molecules acting on the basis of electrostatic force.

4 Concluding Remarks

Scaling laws play an extremely important role in designing microsized systems and devices. Macrointuition can often mislead a designer as intuition develops based on their experience in the macroworld. Scaling laws can help a designer to develop a kind of microintuition. It should also be remembered that completely different concepts may be essential for designing miniaturized systems and guidance to that can come from an understanding of the scaling laws. Since miniaturization is going be a very important phenomenon leading the world toward the eagerly awaited Third Industrial Revolution, study of scaling laws is going to acquire primary importance in the coming years.

Bibliography

1. Ananthasuresh, G. K. et al., "Micro and Smart Systems", Wiley, India, 2010
2. Dusenbery, David B., "Living at Micro Scale", Harvard University Press, Cambridge, Mass, USA, 2009
3. Madou, Marc J., "Fundamentals of Microfabrication" (2nd Edition), CRC Press, Boca Ratan, FL, USA, 2002
4. Peterson, Mark A., "Galileo's Discovery of Scaling Laws", Internet arXiv:physics/0110031v3
5. Schroeder, Manfred, "Fractals, Chaos, Power Laws", Dover, New York, NY, 1991
6. Trimmer, W., "The Scaling of Micromechanical Devices" from "Micromechanics & MEMS", IEEE Press, No. PC4390, 1991
7. West, G. B. and Brown, J. H., "The Origin of allometric scaling laws in biology from genomes to ecosystems: towards a quantitative unifying theory of biological structure and organization", The Journal of Experimental Biology, 208, 1575–1592, 2005

Chapter 3
Modeling of Two-Phase Transport Phenomena in Porous Media: Pore-Scale Approach

Puneet K. Sinha and Partha P. Mukherjee

Abstract Fundamentals of two-phase transport in a porous medium with emphasis on pore-scale physics are discussed. Lattice Boltzmann modeling and pore network modeling techniques are reviewed and are deployed to study two-phase flow in the engineered porous materials used in a polymer electrolyte fuel cell (PEFC). Computation of fluid distribution, capillary pressure, and relative permeability is discussed. Porous material microstructure–PEFC performance relation is showcased, highlighting the potential of deploying pore-scale modeling to the advancement of fundamental two-phase transport understanding and PEFC porous material development. Additionally, the need for further advancement of pore-scale models is highlighted.

Keywords Porous medium · Two-phase flow · Microstructure · Pore-scale modeling · Lattice Boltzmann · Pore network · Fuel cells · Liquid water · Gas diffusion layer · Capillary fingering · Drying · Capillary pressure · Relative permeability

1 Introduction

Two-phase transport phenomena in porous media are abundant in a wide variety of engineering applications, particularly in energy and environmental systems. The inherent heterogeneity and multiple length scales of porous media, as well as the complexity involved in the coupled multi-physical interactions, result in a significant challenge to the fundamental understanding of two-phase transport phenomena. In recent years, owing to increased awareness in clean energy and environmental issues and thrust in energy security, it is imperative to understand the underlying multi-scale and multi-physicochemical transport phenomena ubiquitous in porous media in such energy and environmental systems. Pertinent examples of recent interest include CO_2 sequestration and storage in geologic porous formations [1] and fuel cells consisting of disparate engineered porous media for clean energy conversion [2].

P.P. Mukherjee (✉)
Computer Science and Mathematics Division, Oak Ridge National Laboratory,
Oak Ridge, TN 37831, USA
e-mail: mukherjeepp@ornl.gov

S. Chakraborty (ed.), *Mechanics Over Micro and Nano Scales*,
DOI 10.1007/978-1-4419-9601-5_3,

In this chapter, we attempt to cover the fundamentals of two-phase transport modeling in porous media and emphasize on the pore-scale modeling aspects in the context of engineered porous materials in fuel cells.

1.1 Immiscible Two-Phase Transport in Porous Media

Transport of two fluid phases in porous media is inherently complex due to the underlying complicated topology of the pore space as well as the fluid/fluid and fluid/solid interfacial dynamics.

Most porous media flow modeling is based on the continuum theory, which consists of the governing equations and constitutive closure relations. The fundamental equations governing two-phase flow and transport consist of the mass balance equations for the individual phases in the continuum-scale, macroscopic modeling framework, in which the variables and properties are averaged over a representative elementary volume (REV). The concept of REV is essential, defined as being indifferent with the set of macroscopic variables used to describe a porous medium and invariant with respect to arbitrary time and space [3, 4]. By averaging, the variations due to the microscopic heterogeneity are smoothed out, and the governing equations essentially describe an equivalent homogeneous system. In applying the continuum formulations, the governing equations require constitutive closure relations to account for the influence of porous microstructure and wetting characteristics on the underlying transport.

The mass balance equation for the individual phases can be expressed as [4–6]

$$\frac{\partial}{\partial t}(\varepsilon \rho_\alpha s_\alpha) + \nabla \cdot \left(\rho_\alpha \mathbf{q}_\alpha\right) = I_\alpha + S_\alpha \tag{3.1}$$

In the above equation, ε is porosity, the subscript α denotes the fluid phase, ρ_α is the density of phase α, s_α is the phase saturation defined as the fraction of the pore volume occupied by phase α, $\mathbf{q}_\alpha$ is the flux vector of phase α, I_α and S_α denote, respectively, the interface mass transfer and source/sink terms for phase α.

Instead of formal momentum balance equations, extension of Darcy's law is used to define the phase flux. Darcy's law essentially represents a momentum balance for the control volumes containing single fluid phase characteristics of low Reynolds number flow and assumes that the pressure gradient of the individual phase is a driving force of the respective phase [7]. The extension of Darcy's law is written as

$$\mathbf{q}_\alpha = -\frac{\kappa \kappa_{r\alpha}}{\mu_\alpha}\left(\nabla P_\alpha - \rho_\alpha g\right) \tag{3.2}$$

In (3.2), κ is the intrinsic permeability of the porous medium, $\kappa_{r\alpha}$ is the relative permeability describing the reduction of the intrinsic permeability of the medium owing to the saturation of phase α, μ_α is the dynamic viscosity of phase α, P_α denotes the phase α pressure, and g is the gravitational acceleration.

Substituting (3.2) into (3.1) yields

$$\frac{\partial}{\partial t}(\varepsilon \rho_\alpha s_\alpha) - \nabla \cdot \left(\rho_\alpha \frac{\kappa \kappa_{r\alpha}}{\mu_\alpha} (\nabla P_\alpha - \rho_\alpha g) \right) = I_\alpha + S_\alpha \tag{3.3}$$

For a two-phase system, based on the wetting characteristics of the phase relative to the porous matrix phase, the phases are generally identified as wetting ($\alpha = \mathrm{w}$) and non-wetting ($\alpha = \mathrm{n}$) phases. It is important to note that (3.3) represents two equations involving nine unknowns, namely phase saturation (s_w, s_n), density ($\rho_\mathrm{w}, \rho_\mathrm{n}$), pressure ($P_\mathrm{w}, P_\mathrm{n}$), relative permeability ($\kappa_\mathrm{rw}, \kappa_\mathrm{rn}$), and intrinsic permeability (κ). Solution of the set of equations for a two-phase system, therefore, requires seven additional closure relations, described below.

1.1.1 Saturation Relation

The pore space jointly occupied by the two phases is related as

$$s_\mathrm{w} + s_\mathrm{n} = 1 \tag{3.4}$$

1.1.2 Equation of State (EOS)

EOS provides the equilibrium relationship between density, pressure, and temperature:

$$\left.\begin{aligned} \rho_\mathrm{w} &= \rho_\mathrm{w}(P_\mathrm{w}, T) \\ \rho_\mathrm{n} &= \rho_\mathrm{n}(P_\mathrm{n}, T) \end{aligned}\right\} \tag{3.5}$$

An isothermal assumption represents that the phase density depends on pressure alone. Also, often a constant density pertaining to a fluid phase is assumed, where the spatial gradients in density are small.

1.1.3 Capillary Pressure

Capillary pressure represents the difference between the non-wetting and wetting phase pressures and can be related to the phase saturation:

$$P_\mathrm{c} = P_\mathrm{n} - P_\mathrm{w} \tag{3.6}$$

1.1.4 Relative Permeability

Relative permeability can be expressed as a function of phase saturation:

$$\left.\begin{aligned} \kappa_\mathrm{rw} &= \kappa_\mathrm{rw}(s_\mathrm{w}) \\ \kappa_\mathrm{rn} &= \kappa_\mathrm{rn}(s_\mathrm{n}) \end{aligned}\right\} \tag{3.7}$$

1.1.5 Intrinsic Permeability

Intrinsic permeability depends on the pore morphology and connectivity of the porous medium:

$$\kappa = \kappa(\varepsilon, \text{connectivity, grain size, shape,} \ldots) \tag{3.8}$$

The last three constitutive relationships are briefly elaborated below.

1.2 Two-Phase Transport Parameters and Closure Relations

1.2.1 Intrinsic Permeability

Intrinsic permeability strongly and uniquely depends on the pore structure. Because of the complicated geometry of natural pore space, often empirical permeability relations are employed. Such relations have proved to be useful for porous media consisting of spherical particle packing [8]. In this regard, the Rumpf and Gupte relation [9] has received significant acceptance, derived from random sphere-packed porous bed measurements:

$$\kappa_{\text{RG}} = \frac{\bar{D}^2}{5.6}\varepsilon^{5.5} \tag{3.9}$$

$$\bar{D} = \frac{\int D^3 N(D) dD}{\int D^2 N(D) dD} \tag{3.10}$$

$\bar{D}$ is the surface average diameter and $N(D)$ the grain size distribution function. An exhaustive description of various intrinsic permeability relations is available in the book by Dullien [8].

While the Rumpf–Gupte relation is purely phenomenological, Carman–Kozeny relation [10, 11] is the most popular semi-empirical permeability model. This model considers the porous medium as an equivalent conduit, with arbitrary cross-sectional shape, but with, however, a constant average cross-sectional area [10, 11]. The generic form of the Carman–Kozeny relation for the intrinsic permeability of granular media consisting of arbitrarily shaped particles is written as

$$\kappa_{\text{CK}} = \frac{\bar{D}^2}{180}\frac{\varepsilon^3}{(1-\varepsilon)^2} \tag{3.11}$$

Despite its wide acceptance, the validity of the Carman–Kozeny relation at low and high porosities and for porous media with non-spherical particles and wide size distribution is questionable [8].

1.2.2 Interfacial Tension and Wettability

When two immiscible fluids are in contact, the interface contracts due to the difference in cohesive force between like molecules of each fluid and adhesive force between dissimilar molecules across the fluid interface. The resulting force per unit length of the interface is called interfacial tension. Commonly, the interfacial tension between a liquid and its own vapor is referred to as surface tension. It should be noted that interfacial tension depends on temperature and solute concentration. Kinetic energy of molecules increases with temperature weakening the intermolecular force in a fluid. This results in a decrease in surface tension with increase in temperature, reaching a value of zero at the critical temperature. According to Eötvös rule [4, 8], surface tension of a pure liquid can be given as

$$\sigma V_{\mathrm{m}}^{2/3} = k_{\mathrm{E}} \left(T_{\mathrm{C}} - T \right) \tag{3.12}$$

where σ is the surface tension of the pure liquid, k_{E} the Eötvös constant, and V_{m} the molar volume of the liquid. T and T_{C} represent the temperature and the critical temperature, respectively. Figure 3.1 depicts pure liquid water surface tension as a function of temperature. Additionally, in the presence of a solute, the surface tension of a liquid can increase or decrease depending on the solute properties, e.g., mixing salt increases liquid water surface tension whereas in the presence of detergent powder liquid water surface tension is reduced.

While fluid–fluid interaction can be characterized by interfacial tension, interaction of a fluid with solid surface is defined by wettability. Wettability of a fluid is quantified by contact angle, θ, defined as the angle between the solid surface and

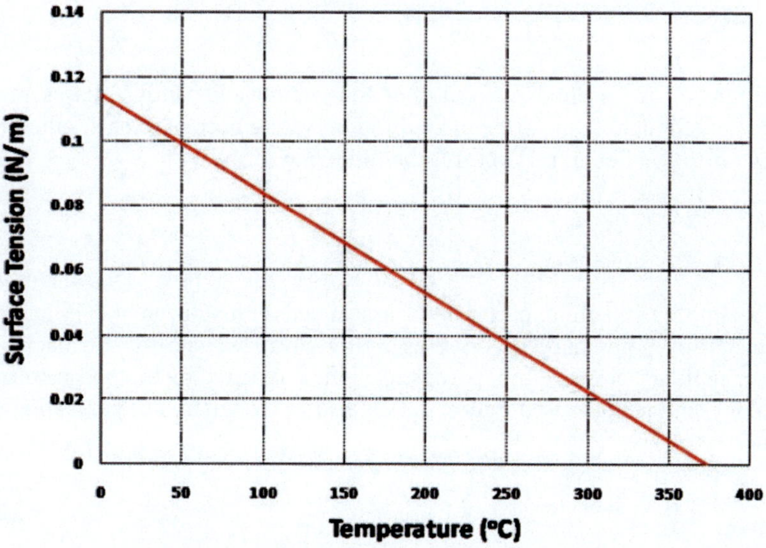

Fig. 3.1 Surface tension of liquid water as a function of temperature

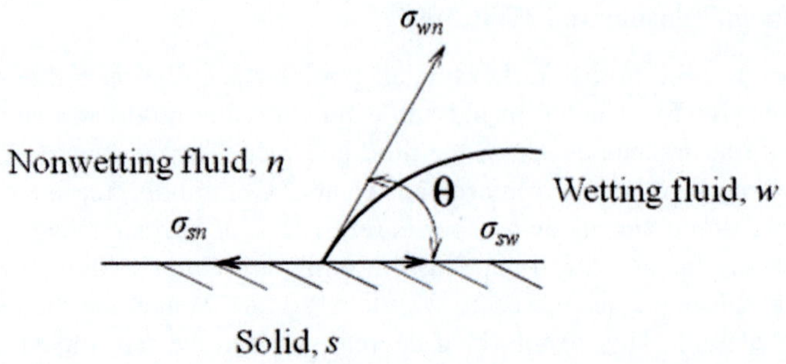

Fig. 3.2 Schematic of contact angle between a fluid–fluid interface and a solid

tangent to the fluid curvature drawn from the contact point, as depicted in Fig. 3.2. In the two immiscible fluid systems, the fluid that displays stronger affinity to the solid surface is defined as the wetting phase ($\theta < 90°$) whereas the other fluid is termed as the non-wetting phase ($\theta > 90°$).

1.2.3 Capillary Pressure–Saturation Relation

When two immiscible fluids are in contact, a curved fluid–fluid interface tends to develop. For the interface to be in equilibrium, the presence of interfacial tension renders pressure discontinuity at the interface. The pressure difference between the non-wetting and the wetting phases is termed as the capillary pressure, P_c, and can be expressed by the Young–Laplace equation [4, 8]:

$$P_c = P_n - P_w = \sigma \left(\frac{1}{R_1} + \frac{1}{R_2} \right) \tag{3.13}$$

where P_n and P_w represent the pressures of non-wetting and wetting phases, respectively. R_1 and R_2 denote the principal radii of curvature normal to each other. For a capillary tube of radius, r, (3.13) takes the following form:

$$P_c = \frac{2\sigma \cos\theta}{r} \tag{3.14}$$

In a real porous medium, each pore is associated with a different capillary pressure dictated by pore shape and size as well as local wettability. At macroscale, therefore, capillary pressure of a porous medium is dictated by the porous medium morphology, and the phase saturation, s_α, defined as the fraction of pores filled with the respective phase:

$$s_\alpha = \frac{V_\alpha}{V_{pore}} \tag{3.15}$$

where V_α and V_{pore} represent the pore volume filled with phase α and the total pore volume of a porous medium, respectively. In macroscopic modeling, where the porous medium is considered as an equivalent homogeneous system, thereby neglecting the detailed porous medium morphology, empirical correlations are typically employed to define capillary pressure–saturation relationship. Several empirical and semi-empirical relations obtained from measurements and phenomenological models often used in porous media two-phase transport are discussed at length in the classical books by Dullien [8] and Bear [4].

1.2.4 Relative Permeability–Saturation Relation

During multiphase flow in a porous medium, the available pore space is shared by multiple fluids. Reduced available cross-sectional area and varied fluid distribution in the pore space cause increased flow resistance for each fluid when compared with flow resistance in the absence of other fluids. The relative permeability of a fluid phase, $k_{r\alpha}(s)$, is defined as the ratio of effective permeability for that phase at a given saturation, $k_{\text{eff},\,\alpha}(s)$, to the absolute intrinsic permeability of the porous medium, κ:

$$k_{\text{r}\alpha}\left(s\right) = \frac{\kappa_{\text{eff},\alpha}\left(s\right)}{\kappa} \tag{3.16}$$

where s is the saturation of phase α.

The relative permeability depends on the phase distribution and saturation in the porous medium which in turn are strong functions of the microstructure and wetting characteristics. In a typical relative permeability measurement, two fluids are injected simultaneously into a core sample of a specific porous medium at a predetermined ratio. After certain time steady-state equilibrium is established. The criterion of steady state is determined by the condition that the inflow equals the outflow and/or that constant pressure drop has been reached across the sample. From the measured pressure drop across the sample, the flow rate, and the dimensions of the sample, the relative permeability can be computed according to (3.2). The test is then repeated with an increasing injection ratio to obtain the values for different saturation levels. Figure 3.3 demonstrates a typical relative permeability–saturation relationship for two-phase flow in a porous medium, where the irreducible phase saturation denotes the threshold value at which the phase is no longer connected. As non-wetting fluid invades a porous medium, progressively smaller pores are left for wetting fluid. This results in a sharp decline in relative permeability of wetting fluid with increase in the non-wetting fluid saturation. For an exhaustive discussion on different measured and phenomenological relative permeability relations, the readers are encouraged to refer to the books on porous media transport phenomena by Dullien [8] and Bear [4].

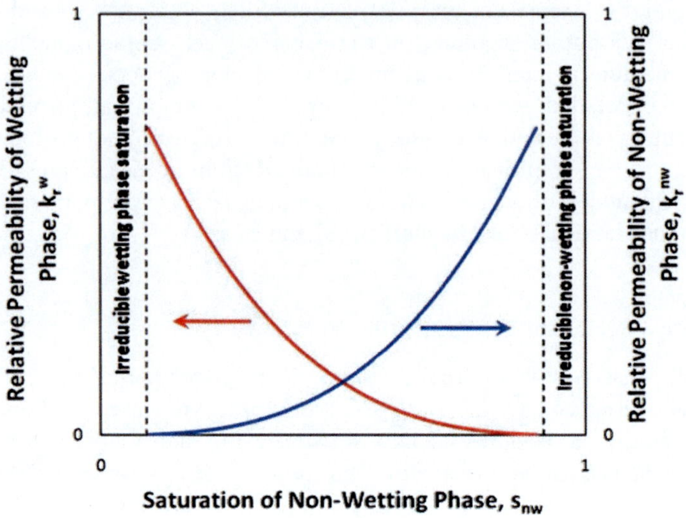

Fig. 3.3 Schematic representation of relative permeability of two phases as a function of non-wetting phase saturation

2 Pore-Scale Modeling Approaches

Most of the two-phase flow and transport simulations are based on the continuum theory, which consists of the governing equations and constitutive closure relations. The primary challenges involved in the two-phase transport modeling using the continuum approach are [12] as follows:

- Availability of appropriate constitutive two-phase closure relations is critical to the fidelity of the computational predictions. Accurate measurements are not only costly and time consuming but also faced with experimental difficulties emanating from disparate flow regimes and porous media microstructures. It is important to note that the relative permeability, which is the most important two-phase parameter, depends strongly on the phase distributions, microscopic structure of solid matrix, and wetting characteristics. The lack of fundamental understanding at the porescale precludes the possibility of obtaining meaningful constitutive relations that can capture the essential two-phase behavior at the macroscale.
- In the macroscopic modeling, solid-phase wettability and interfacial area between phases are not explicitly included, which are of paramount importance for accurate description of the phase distribution and two-phase dynamics predictions.

Pore-scale modeling provides unique opportunities to study two-phase transport because detailed information is available at the microscopic scale. Appropriate numerical experiments can be designed to evaluate constitutive closure relations (e.g., capillary pressure and relative permeability), which can be incorporated into macroscopic two-phase simulations. Furthermore, upscaling of pore-scale physical processes provides fundamental information for high-fidelity macroscopic simulations. Pore-scale models for solving multiphase flow and transport through porous media can be broadly classified into rule-based and first-principle-based models [12].

Rule-based models try to capture physical processes by incorporating adequate physics with an idealized network representation of the porous medium. Such models, based on diluted physical description, nonetheless provide useful predictions of the transport properties of a given porous medium. Simulations based on these models are generally faster since there is no need to solve large systems of equations. The most prominent among these rule-based models is the pore network (PN) modeling approach. In the PN models, a porous medium is represented by a lattice of wide pores connected by narrower constrictions called throats. The original PN model is credited to the pioneering works by Fatt [13–15], who computed capillary pressure and relative permeability in a network of interconnected pores. The PN modeling has received significant attention in investigating a variety of processes in porous media, including dispersion [16], two-phase relations [17, 18], three-phase displacement [19, 20], reactive transport [21–23], heat transfer [24], drying [25, 26], CO_2 sequestration [27], to name a few. While the PN models were primarily developed to address flows in low-porosity and low-permeability geologic porous media, Thompson [28] expanded the applicability to high-porosity and high-permeability fibrous media.

Unlike the rule-based approaches, the first-principle-based methods resolve the underlying transport processes by solving the governing equations. The governing equations can be solved by employing the fine-scale conventional computational fluid dynamics (CFD) methods, the so-called top-down approach, or the coarse-grain approach, i.e., the "bottom-up" approach [29]. The molecular dynamics, lattice gas, and lattice Boltzmann (LB) methods fall under the "bottom-up" approach category. With a given set of suitable boundary conditions, the governing differential equations can be properly discretized on a computational grid using standard CFD techniques, namely finite difference, finite volume, or finite element methods. Even though several CFD-based two-phase models, such as front-capturing [30–33] and front-tracking [34, 35] approaches, have been developed, the applicability of fine-scale CFD yet remains a challenge to simulate two-phase flow in complex porous media. The molecular dynamics approach [36, 37] takes into account the movements and collisions of all individual molecules constituting the fluid with detailed description of the intermolecular interactions and thereby providing realistic equations of state characterizing the real fluid. However, the complexity of interactions as well as the number of molecules representative of the actual fluid make the molecular dynamics models computationally prohibitive for application to macroscopic flows in porous media.

The lattice Boltzmann method, instead of tracking all the individual molecules, considers the behavior of a collection of particles, comprised of a large number of molecules, moving on a regular lattice, thereby reducing the degrees of freedom of the system and making the pore-scale simulation computationally tractable. Owing to its numerical stability and constitutive versatility, the LB method has developed into a powerful technique for simulating complex flows in recent years [38]. Unlike the conventional Navier–Stokes solvers based on the discretization of the macroscopic continuum equations, LB methods consider flows to be composed of a collection of pseudo-particles residing on the nodes of an underlying lattice structure which interact according to a velocity distribution function. The lattice Boltzmann method is an ideal scale-bridging numerical scheme which incorporates simplified kinetic models to capture microscopic or mesoscopic flow physics and yet the macroscopic averaged quantities satisfy the desired macroscopic equations [38]. The LB model incorporates phase segregation and surface tension in multiphase flows through interparticle force/interactions, which are difficult to implement in traditional methods. Therefore, compared to the conventional CFD methods, the LB modeling approach better represents the pore morphology of the actual porous medium and incorporates more rigorous physical description of the flow processes. Although in some cases it is computationally more demanding, this can be compensated by high-performance computing, since the LB algorithm is inherently parallel. Several two-phase LB models have been presented in the literature. The first immiscible multiphase LB model proposed by Gunstensen et al. [39] used red- and blue-colored particles to represent two kinds of fluids. The phase separation is then produced by the repulsive interaction based on the color gradient and color momentum. The model proposed by Shan and Chen [40, 41] imposes a non-local interaction between fluid particles at neighboring lattice sites. The interaction potentials control the form of the equation of state of the fluid. Phase separation occurs automatically when the interaction potentials are properly chosen. The free energy-based approach proposed by Swift et al. [42, 43] relies on the description of non-equilibrium dynamics, such as Cahn–Hilliard approach. The free energy model satisfies local momentum conservation. In the multiphase model proposed by He et al. [44, 45], two sets of probability distribution functions (PDFs) are employed. The first PDF set is used to simulate pressure and velocity fields and another PDF set is used to capture the interface only, which makes this approach essentially close to the interface capturing methods in spirit. This approach is thermodynamically more consistent; however, it shows numerical instability. Among the aforementioned two-phase LB models, the S–C model is widely used [46–52] due to its simplicity in implementing boundary conditions in complex porous structures; versatility in terms of handling fluid phases with different densities, viscosities, and wettabilities; as well as the capability of incorporating different equations of state.

Pore-scale modeling offers excellent versatility in the fundamental investigation of transport phenomena at the microscopic scale. In this chapter, rule-based (pore network model) and first-principle-based (lattice Boltzmann model) pore-scale modeling techniques are discussed in the context of two-phase transport phenomena prevalent in the polymer electrolyte fuel cell (PEFC) porous media.

2.1 Two-Phase Transport in PEFC Porous Media

A typical PEFC, schematically shown in Fig. 3.4, exhibits a layered architecture and consists of the anode and cathode compartments, separated by a proton-conducting polymeric membrane. The anode and cathode sides each includes gas channel, gas diffusion layer (GDL), and catalyst layer (CL). Usually, two thin catalyst layers are coated on both sides of the membrane, forming a membrane–electrode assembly. Humidified hydrogen comprises the anode feed, whereas humidified air is fed into the cathode. Hydrogen and oxygen combine electrochemically within the active

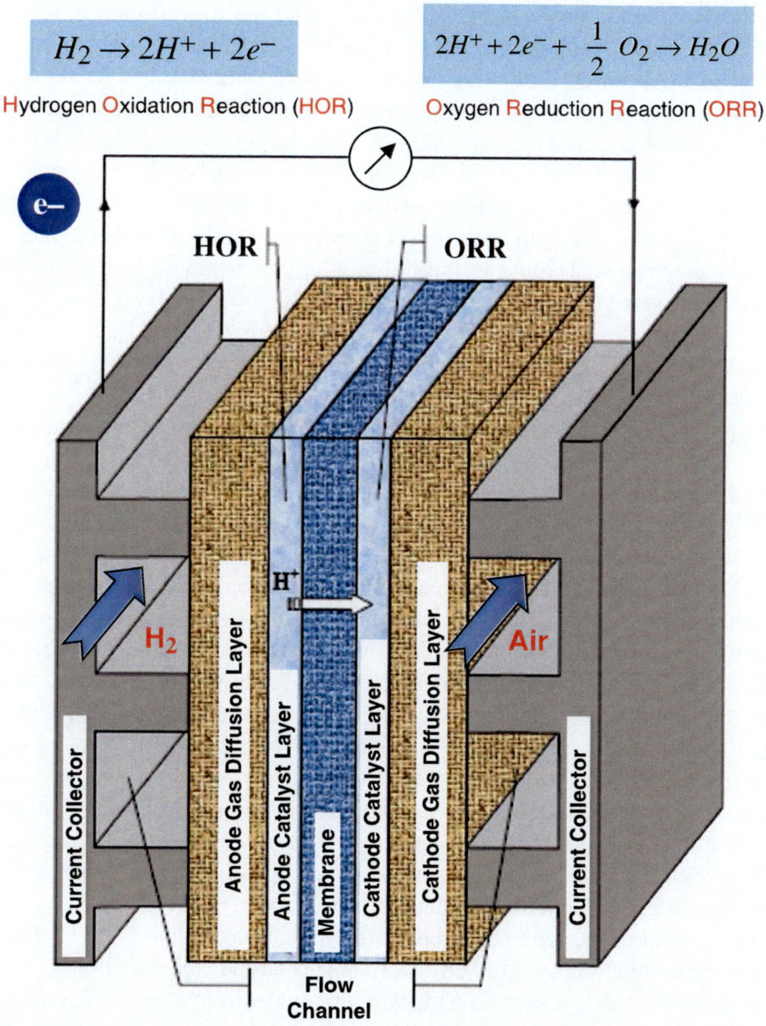

Fig. 3.4 Schematic diagram of a polymer electrolyte fuel cell

catalyst layers to produce electricity, water, and waste heat. The gas diffusion layer allows the transport of reactants to and products from the reaction sites and also conducts electrons and heat. Additionally, a microporous layer (MPL) is placed between the CL and GDL to facilitate water and heat transport. Gottesfeld and Zawodzinski [53] provided a comprehensive overview of the PEFC function and operation. The hydrogen oxidation reaction occurring at the anode catalyst layer has orders of magnitude higher reaction rate than the oxygen oxidation reaction (ORR) in the cathode catalyst layer. The ORR is therefore a potential source of large voltage loss and hence the cathode catalyst layer is the electrode of primary importance in a PEFC. Due to the acid nature of the polymer membrane and low-temperature operation, Pt or Pt alloys are the best known catalysts for PEFCs.

Despite tremendous recent progress in enhancing the overall cell performance, a pivotal performance/durability limitation in PEFCs centers on liquid water transport and resulting flooding in the constituent components [2, 54, 55]. At high current density operation, "mass transport limitations" come into play due to the excessive liquid water buildup mainly in the cathode side. Liquid water blocks the porous pathways in the CL and GDL thus causing hindered oxygen transport to the reaction sites as well as covers the electrochemically active sites in the CL thereby increasing surface overpotential. This phenomenon is known as "flooding" and is perceived as the primary mechanism leading to the limiting current behavior in the cell performance. The catalyst layer and gas diffusion layer, therefore, play a crucial role in the PEFC water management aimed at maintaining a delicate balance between reactant transport from the gas channels and water removal from the electrochemically active sites.

Water management research has received wide attention in recent years including the development of several macroscopic computational models for liquid water transport in PEFCs. Comprehensive overview of the various PEFC models is furnished by Wang [2] and Weber and Newman [55]. The macroscopic models, reported in the literature, are based on the theory of volume averaging and treat the catalyst layer and gas diffusion layer as macrohomogeneous porous layers. Due to the macroscopic nature, the current models fail to resolve the influence of the structural morphology on the underlying two-phase dynamics. Additionally, there is serious scarcity of two-phase correlations, namely capillary pressure and relative permeability as functions of liquid water saturation, as constitutive closure relations for the two-phase PEFC models tailored specifically for the CL and GDL microstructures. Current two-phase fuel cell models often employ a capillary pressure–saturation relation for modeling liquid water transport in hydrophobic gas diffusion media adapted by Pasaogullari and Wang [56] and Nam and Kaviany [57] from Udell's work [58] in the form of Leverett J-function [59]. Very recently few attempts to experimentally evaluate the capillary pressure and relative permeability relations as functions of liquid water saturation for the PEFC GDL have been reported in the literature [60–64]. Despite substantial research, both theoretical and experimental, there is serious paucity of fundamental understanding regarding the overall structure–transport–performance interactions and underlying two-phase dynamics in the catalyst layer and the gas diffusion layer.

Fig. 3.5 SEM images of typical carbon paper and carbon cloth GDL structures (reproduced from [93] with permission from Elsevier)

Owing to the complex microstructure characterized by significantly small pore size and small flow rate, the two-phase transport in the PEFC porous sructures exhibits the dominance of surface forces as compared to the gravity, viscous force, and inertia force. Representative low values of capillary number $\left(\mathrm{Ca} = \mu_n U_n / \sigma \sim 10^{-4}\right)$ and Reynolds number $\left(\mathrm{Re} = \rho_n U_n D / \mu_n \sim 10^{-6}\right)$ illustrate the strong influence of surface tension force.

In this chapter, we specifically focus on the different aspects of two-phase transport phenomena occurring in the fibrous gas diffusion layer of the PEFC. The multi-faceted functionality of a GDL includes reactant distribution, liquid water transport, electron transport, heat conduction, and mechanical support to the membrane electrode assembly. Carbon fiber-based porous materials, namely non-woven carbon paper and woven carbon cloth with thickness ~200–300 μm, have received wide acceptance as materials of choice for the PEFC GDL owing to high porosity (~70% or higher) and good electrical/thermal conductivity. Mathias et al. [65] provided a comprehensive overview of the GDL structure and functions. Figure 3.5 shows representative scanning electron microscope (SEM) images of carbon paper (non-woven) and carbon cloth (woven) GDL structures.

3 Two-Phase Lattice Boltzmann Model

The interaction potential-based two-phase lattice Boltzmann model (LBM) by Shan and Chen [40, 41], henceforth referred to as the S–C model, has been deployed in the PEFC research.

3.1 Methodology

In brief, the S–C model [40, 41] introduces k distribution functions for a fluid mixture comprising of k components. Each distribution function represents a fluid component and satisfies the evolution equation. The non-local interaction between particles at neighboring lattice sites is included in the kinetics through a set of potentials. The evolution equation for the kth component can be written as

$$f_i^k(\mathbf{x} + \mathbf{e}_i\delta_t, t + \delta_t) - f_i^k(\mathbf{x}, t) = -\frac{f_i^k(\mathbf{x}, t) - f_i^{k(\mathrm{eq})}(\mathbf{x}, t)}{\tau_k} \tag{3.17}$$

where $f_i^k(\mathbf{x}, t)$ is the number density distribution function for the kth component in the ith velocity direction at position $\mathbf{x}$ and time t and δ_t is the time increment. In the term on the right-hand side, τ_k is the relaxation time of the kth component in lattice unit and $f_i^{k(\mathrm{eq})}(\mathbf{x}, t)$ is the corresponding equilibrium distribution function. The right-hand side of (3.17) represents the collision term, which is simplified to the equilibrium distribution function $f_i^{k(\mathrm{eq})}(\mathbf{x}, t)$ by the so-called BGK (Bhatnagar–Gross–Krook) or the single-time relaxation approximation [66]. The spatio-temporal discrete form of the LB evolution equation based on the BGK approximation is given by (3.17). For a 3D 19-speed lattice ($D3Q19$, where D is the dimension and Q is the number of velocity directions), the schematic of which is shown in Fig. 3.6 with the velocity directions, the equilibrium distribution function, $f_i^{k(\mathrm{eq})}(\mathbf{x}, t)$, assumes the following form [66, 67]:

$$\left.\begin{aligned}
f_0^{k(\mathrm{eq})} &= \beta_k n_k - \frac{1}{2} n_k \mathbf{u}_k^{\mathrm{eq}} \cdot \mathbf{u}_k^{\mathrm{eq}} \\
f_i^{k(\mathrm{eq})} &= \frac{1-\beta_k}{12} n_k + \frac{1}{6} n_k \left(\mathbf{e}_i \cdot \mathbf{u}_k^{\mathrm{eq}}\right) + \frac{1}{4} n_k \left(\mathbf{e}_i \cdot \mathbf{u}_k^{\mathrm{eq}}\right)^2 - \frac{1}{12} n_k \mathbf{u}_k^{\mathrm{eq}} \cdot \mathbf{u}_k^{\mathrm{eq}} \\
&\quad \text{for } i = 1, \ldots, 6 \\
f_i^{k(\mathrm{eq})} &= \frac{1-\beta_k}{24} n_k + \frac{1}{12} n_k \left(\mathbf{e}_i \cdot \mathbf{u}_k^{\mathrm{eq}}\right) + \frac{1}{8} n_k \left(\mathbf{e}_i \cdot \mathbf{u}_k^{\mathrm{eq}}\right)^2 - \frac{1}{24} n_k \mathbf{u}_k^{\mathrm{eq}} \cdot \mathbf{u}_k^{\mathrm{eq}} \\
&\quad \text{for } i = 7, \ldots, 18
\end{aligned}\right\} \tag{3.18}$$

In the above equations, the discrete velocities, $\mathbf{e}_i$, are given by

$$\mathbf{e}_i = \begin{cases} (0,0,0), & i = 0 \\ (\pm 1,0,0), (0,\pm 1,0), (0,0,\pm 1), & i = 1-6 \\ (\pm 1,\pm 1,0), (\pm 1,0,\pm 1), (0,\pm 1,\pm 1), & i = 7-18 \end{cases} \tag{3.19}$$

The free parameter, β_k, relates to the speed of sound of a region of pure kth component as

$$\left(c_s^k\right)^2 = \frac{1-\beta_k}{2} \tag{3.20}$$

Fig. 3.6 Schematic of *D3Q19* lattice structure

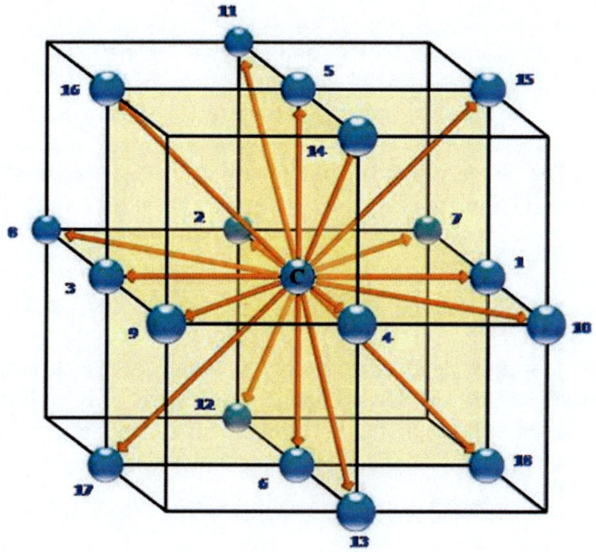

The number density of the kth component, n_k, is defined as

$$n_k = \sum_i f_i^k \tag{3.21}$$

The mass density of the kth component is defined as

$$\rho_k = m_k n_k = m_k \sum_i f_i^k \tag{3.22}$$

The fluid velocity of the kth fluid is defined through

$$\rho_k \mathbf{u}_k = m_k \sum_i \mathbf{e}_i f_i^k \tag{3.23}$$

where m_k is the molecular mass of the kth component.

The equilibrium velocity, $\mathbf{u}_k^{\mathrm{eq}}$, is determined by the expression

$$\rho_k \mathbf{u}_k^{\mathrm{eq}} = \rho_k \mathbf{u}' + \tau_k \mathbf{F}_k \tag{3.24}$$

where $\mathbf{u}'$ is a common or base velocity, which can be perceived in the context of single-component, single-phase flow, on top of which an extra component-specific velocity due to interparticle interaction is added for each component. In this particle-based approach, interparticle interaction, which manifests in terms of surface tension, wall adhesion, and external gravity force, is realized through the total force, $\mathbf{F}_k$, acting on the kth component, including fluid/fluid interaction $\mathbf{F}_{1k}$, fluid/solid interaction $\mathbf{F}_{2k}$, and external force $\mathbf{F}_{3k}$ [67], and is expressed as

$$\mathbf{F}_k = \mathbf{F}_{1k} + \mathbf{F}_{2k} + \mathbf{F}_{3k} \tag{3.25}$$

The conservation of momentum at each collision in the absence of interaction force (i.e., in the case of $\mathbf{F}_k = 0$), requires $\mathbf{u}'$ to satisfy the following relation:

$$\mathbf{u}' = \frac{\sum_{k=1}^{s} \rho_k \mathbf{u}_k / \tau_k}{\sum_{k=1}^{s} \rho_k / \tau_k} \tag{3.26}$$

A simple long-range interaction force between particles of the kth component at site $\mathbf{x}$ and the $\bar{k}$th component at site $\mathbf{x}'$ is introduced and the total fluid/fluid interaction force on the kth component at site $\mathbf{x}$ is given by

$$\mathbf{F}_{1k}(\mathbf{x}) = -\psi_k(\mathbf{x}) \sum_{\mathbf{x}'} \sum_{\bar{k}=1}^{s} G_{k\bar{k}}(\mathbf{x}, \mathbf{x}') \psi_{\bar{k}}(\mathbf{x}')(\mathbf{x}' - \mathbf{x}) \tag{3.27}$$

where $G_{k\bar{k}}(\mathbf{x}, \mathbf{x}')$ is Green's function and satisfies $G_{k\bar{k}}(\mathbf{x}, \mathbf{x}') = G_{k\bar{k}}(\mathbf{x}', \mathbf{x})$. It reflects the intensity of interparticle interaction. $\psi_{k\bar{k}}(\mathbf{x})$ is called the "effective number density" and is defined as a function of $\mathbf{x}$ through its dependency on the local number density, $\psi_k = \psi_k(n_k)$. In the $D3Q19$ lattice model, the interaction potential couples nearest and next-nearest neighbors and the Green's function is given by

$$G_{k\bar{k}}(\mathbf{x}, \mathbf{x}') = \begin{cases} g_{k\bar{k}}, & |\mathbf{x} - \mathbf{x}'| = 1 \\ g_{k\bar{k}}/2, & |\mathbf{x} - \mathbf{x}'| = \sqrt{2} \\ 0, & \text{otherwise} \end{cases} \tag{3.28}$$

where $g_{k\bar{k}}$ represents the strength of interparticle interactions between component k and $\bar{k}$. The effective number density, $\psi_k(n_k)$, is taken as n_k in the present model and other choices will give a different equation of state. The interactive force between the fluid and the wall is realized by considering the wall as a separate phase with constant number density and is given by

$$\mathbf{F}_{2k}(\mathbf{x}) = -n_k(\mathbf{x}) \sum_{\mathbf{x}'} g_{k\mathrm{w}} n_{\mathrm{w}}(\mathbf{x}')(\mathbf{x}' - \mathbf{x}) \tag{3.29}$$

where n_{w} is the number density of the wall, which is a constant at the wall and zero elsewhere, and $g_{k\mathrm{w}}$ is the interaction strength between the kth component and the wall. The interactive strength, $g_{k\mathrm{w}}$, defines the wall wettability and is positive for a non-wetting fluid and negative for a wetting fluid. It is to be noted that $\mathbf{F}_{2k}$ is perpendicular to the wall and will not affect the no-slip boundary condition. The action of a constant body force can be incorporated as

$$\mathbf{F}_{3k} = \rho_k \mathbf{g} = m_k n_k \mathbf{g} \tag{3.30}$$

where $\mathbf{g}$ is the constant body force per unit mass.

The continuity and momentum equations can be obtained for the fluid mixture as a single fluid using Chapman–Enskog expansion procedure in the nearly incompressible limit [68]:

$$\left.\begin{aligned} &\frac{\partial \rho}{\partial t} + \nabla \cdot (\rho \mathbf{u}) = 0 \\ &\rho \left[\frac{\partial \mathbf{u}}{\partial t} + (\mathbf{u} \cdot \nabla)\mathbf{u} \right] = -\nabla p + \nabla \cdot [\rho \nu (\nabla \mathbf{u} + \mathbf{u}\nabla)] + \rho \mathbf{g} \end{aligned}\right\} \tag{3.31}$$

where the total density and velocity of the fluid mixture are given, respectively, by [69]

$$\left.\begin{aligned} \rho &= \textstyle\sum_k \rho_k \\ \rho \mathbf{u} &= \textstyle\sum_k \rho_k \mathbf{u}_k + \frac{1}{2} \sum_k \mathbf{F}_k \end{aligned}\right\} \tag{3.32}$$

The pressure, which is usually a non-ideal gas equation of state, is given by [68]

$$p = \sum_k \frac{(1 - \beta_k) m_k n_k}{2} + 3 \sum_{k,\bar{k}} g_{k\bar{k}} \psi_k \psi_{\bar{k}} \tag{3.33}$$

In the present model, $m_k = 1$ and $\beta_k = 1/3$, which is commonly used in the literature. Then the equation of state can be written as

$$p = \sum_k \frac{n_k}{3} + 3 \sum_{k,\bar{k}} g_{k\bar{k}} \psi_k \psi_{\bar{k}} \tag{3.34}$$

The viscosity is given by

$$\nu = \frac{\left(\sum_k \alpha_k \tau_k - 1/2\right)}{3} \tag{3.35}$$

where α_k is the mass density concentration of the kth component and is defined as [70]

$$\alpha_k = \frac{\rho_k}{\sum_k \rho_k} \tag{3.36}$$

It should be noted that the introduction of fluid/solid interaction has no effect on the macroscopic equations since $\mathbf{F}_{2k}$ exists only at the fluid/solid interface. The relaxation time, τ_k, is estimated based on the viscosity and mass density representations given by (3.35) and (3.36) of the kth component and is detailed in [40, 68, 69]. Furthermore, in this interparticle potential model, the separation of a two-phase fluid into its components is automatic [40]. The primary physical parameters, such as the fluid/fluid and fluid/solid interaction parameters, need a priori evaluation through model calibration using numerical experiments. The fluid/fluid interaction gives rise to the surface tension force and the fluid/solid interaction manifests in the wall adhesion force. The fluid/fluid and fluid/solid interaction parameters are evaluated by designing two numerical experiments, *bubble test* in the absence of solid phase and *static droplet test* in the presence of solid wall, respectively. The details of these numerical experiments are elaborated elsewhere [47, 49, 71–73].

3.2 Two-Phase Simulation Studies

We describe a two-phase displacement numerical experiment for investigating liquid water transport and flooding dynamics in the reconstructed GDL microstructure in an ex situ setup. Additionally, no thermal effect on the two-phase transport is considered.

The numerical setup is designed to simulate a quasi-static primary drainage experiment, typically devised in the petroleum/reservoir engineering applications and detailed elsewhere in the literature [3, 10, 74], for simulating immiscible, two-phase transport in the porous microstructure. A non-wetting phase (NWP) reservoir is added to the porous structure at the front end and a wetting phase (WP) reservoir is added at the back end [73, 75–77]. These two end reservoirs added to the GDL domain in the through-plane (i.e., thickness) direction are composed of void space. It should be noted that for the primary drainage simulation in the significantly hydrophobic GDL, liquid water is the NWP and air is the WP.

The primary drainage process is simulated starting with zero capillary pressure, by fixing the NWP and WP reservoir pressures to be equal. Then the capillary pressure is increased incrementally by decreasing the WP reservoir pressure while maintaining the NWP reservoir pressure at the fixed initial value. The pressure gradient drives liquid water into the initially air-saturated GDL by displacing it. The primary objective of the quasi-static displacement simulation is to study liquid water behavior through the GDL structures and the concurrent response to capillarity as a direct manifestation of the underlying pore morphology.

3.2.1 GDL Microstructure Reconstruction

Realistic description of the microstructure is an essential prerequisite for the two-phase LB modeling. We briefly describe a stochastic reconstruction model for the GDL microstructure generation, which is employed as the input structure into the subsequent LBM simulations. The GDL microstructure reconstruction is based on the non-woven structure generation technique originally proposed by Schladitz et al. [78]. The specific assumptions made in the non-woven GDL reconstruction method, which are well justified by inspecting the corresponding carbon paper GDL SEM micrographs, include [78, 79] the following: (1) the fibers are long compared to the sample size and their crimp is negligible; (2) the interaction between the fibers

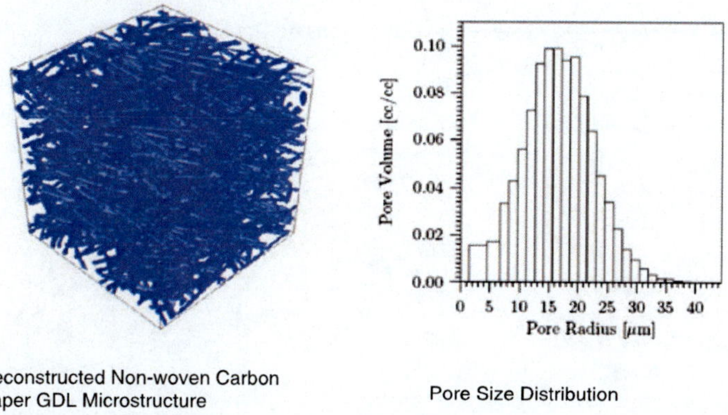

Permeability	In-plane	Through-plane	In-plane/Through-plane
Reconstructed GDL	30.9 darcy	21.1 darcy	1.46
Measured data	33 darcy	18 darcy	1.83

Fig. 3.7 Reconstructed non-woven carbon paper GDL microstructure along with the evaluated structural properties (reproduced from [79] with permission from the Electrochemical Society)

can be neglected, i.e., the fibers are allowed to overlap; and (3) the fiber system, owing to the fabrication process, is macroscopically homogeneous and isotropic in the material plane, defined as the *xy* plane. With these assumptions, the stochastic reconstruction technique is adequately described as a Poisson line process with one-parametric directional distribution where the fibers are realized as circular cylinders with a given diameter and the directional distribution provides in-plane/through-plane anisotropy in the reconstructed GDL microstructure [79]. Figure 3.7 shows the reconstructed microstructure of a typical non-woven, carbon paper GDL [79] with porosity around 72% and thickness of 180 μm along with the structural parameters in terms of the estimated pore size distribution and the anisotropy in the in-plane vs. through-plane permeability values.

3.2.2 Effect of Microstructure

Figure 3.8 shows the liquid water distribution as well as the invasion pattern from the primary drainage simulation with increasing capillary pressure in the initially air-saturated reconstructed carbon paper GDL (~72% porosity) characterized by hydrophobic wetting characteristics with a static contact angle of 140° [75]. The reconstructed GDL structure used in the two-phase simulation consists of $100 \times 100 \times 100$ lattice points in order to manage the computational overhead to a reasonable level. At the initially very low capillary pressure, the invading front overcomes the barrier pressure only at some preferential locations depending upon the pore size along with the emergence of droplets owing to strong hydrophobicity. As the capillary pressure increases, several liquid water fronts start to penetrate

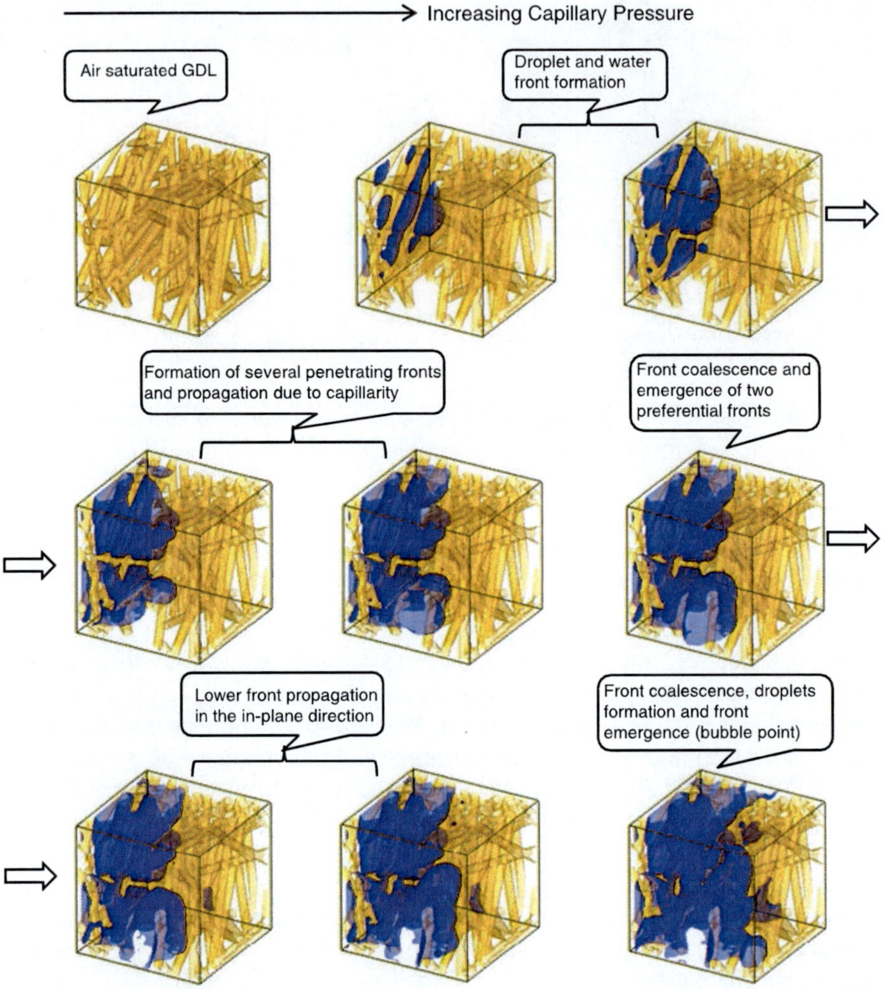

Fig. 3.8 Advancing liquid water front with increasing capillary pressure through the initially air-saturated reconstructed GDL microstructure from the primary drainage simulation using LBM (reproduced from [75] with permission from Elsevier)

into the air-occupied domain. Further increase in capillary pressure exhibits growth of droplets at two invasion fronts, followed by the coalescence of the drops and collapse into a single front. This newly formed front then invades into the less tortuous in-plane direction. Additionally, emergence of tiny droplets and subsequent growth can be observed in the constricted pores in the vicinity of the inlet region primarily due to strong wall adhesion forces from interactions with highly hydrophobic fibers with the increasing capillary pressure. One of the several invading fronts finally reaches the air reservoir, physically the GDL–channel interface, at a preferential location corresponding to the capillary pressure and is also referred to

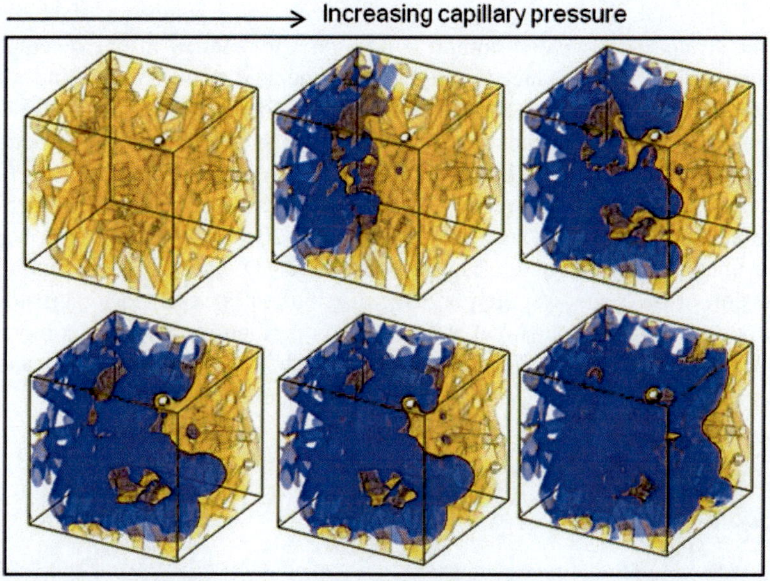

Fig. 3.9 Advancing liquid water front with increasing capillary pressure through the initially air-saturated reconstructed GDL microstructure with different fibers alignment

as the bubble point. It is worth mentioning that the LB simulation is indeed able to capture the intricate liquid water dynamics including droplet interactions, flooding front formation, and propagation through the hydrophobic fibrous GDL structure.

Figure 3.9 shows the two-phase behavior in a reconstructed GDL microstructure (porosity ~72%), where the fibers show preferential orientation in the through-plane direction. In the LBM simulation, a static contact angle of 140° is assumed. It can be observed that due to the preferential orientation of the fibers, there are several invading liquid water fronts progressing through the structure. Further, owing to the microstructure and sufficient hydrophobicity, these fronts tend to merge and build up the saturation level significantly before reaching the bubble point. This study emphasizes the influence of the microstructure on the flooding behavior in the low capillary number regime characteristics of the PEFC transport.

3.2.3 Effect of Compression

Cell clamping pressure is an important factor having strong influence on the GDL pore morphology and hence the underlying two-phase behavior [77]. In this section, we briefly discuss a reduced order compression model and two-phase transport using LBM in the GDL microstructure.

Detailed modeling of a porous material under compression is a challenging task of applied structural mechanics. The reduced compression model employed in the current study is based on the unidirectional morphological displacement of solid voxels in the GDL structure under load and with the assumption of negligible transverse strain. The reduced compression model is detailed in our recent work [79]. However, with the reduced compression model, it is difficult to find a relation between the compression ratio and the external load. The compression ratio is defined as the ratio of the thickness of compressed sample to that of the uncompressed sample. Nevertheless, this approach leads to reliable 3D morphology of the non-woven GDL structures under compression. Figure 3.10a shows compressed, reconstructed non-woven GDL microstructures with 20 and 40% compression along with the uncompressed structure and representative 2D cross sections. Due to the compression of the structure, the pore size distribution is expected to shift toward smaller pores. Figure 3.10b shows the pore size distributions of the uncompressed and 50% compressed GDL microstructures. It can be observed that the mean pore

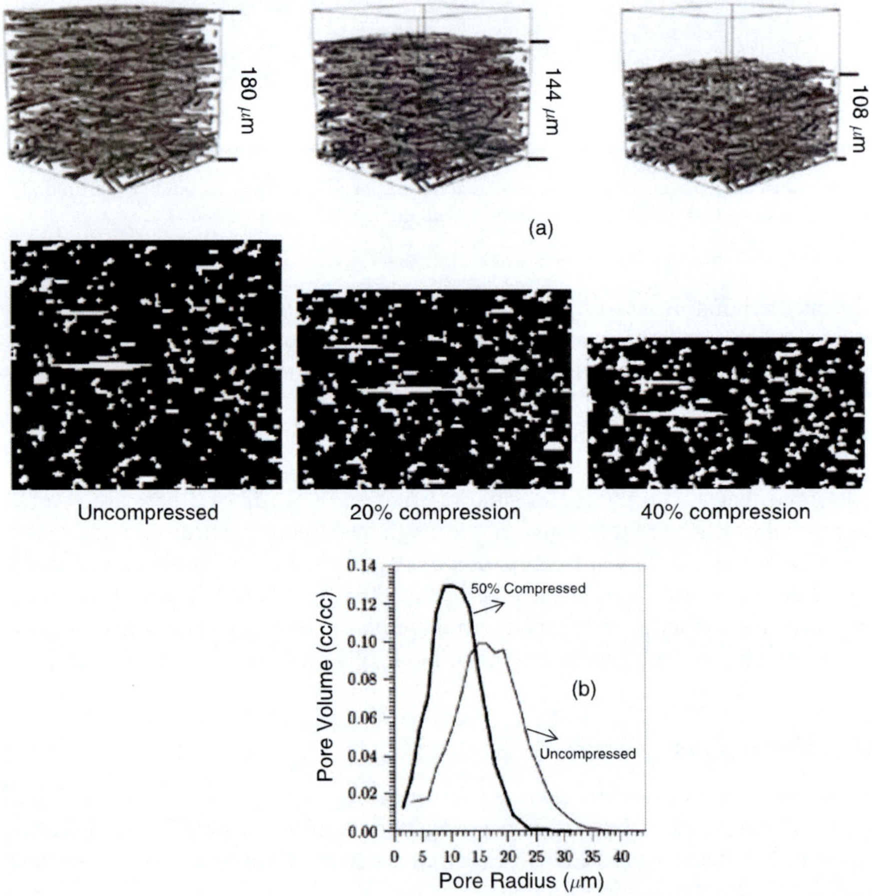

Fig. 3.10 Representative uncompressed and compressed non-woven GDL structures (**a**) along with pore size distribution (**b**) (reproduced from [79] with permission from the Electrochemical Society)

size shifts from around 17 to approximately 10 μm under 50% compression, while the width of the pore size distribution shows negligible variation.

Figure 3.11 shows the liquid water distribution as well as the invasion pattern from the primary drainage simulation with increasing capillary pressure in the initially air-saturated 30% compressed carbon paper GDL microstructure characterized by hydrophobic wetting characteristics with a static contact angle of 140°, similar to its uncompressed counterpart described earlier. The reduced compression model was used to generate the compressed structure. Similar to the uncompressed GDL, at very low initial capillary pressures, the invading front overcomes the barrier pressure only at some preferential locations depending upon the pore size. As the capillary pressure increases, several liquid water fronts start to penetrate into the air-occupied domain. Owing to the compression-induced microstructural change, further increase in capillary pressure exhibits coalescence of drops and concurrent merging of the fronts into a single front. This newly formed front propagates in the through-plane direction with liquid water buildup and subsequent action of capillarity. However, no preferential front migration toward the in-plane direction, as opposed to that in the uncompressed structure, is observed, which can be attributed to the increased tortuosity owing to the compression of the GDL. Emergence of droplets and subsequent

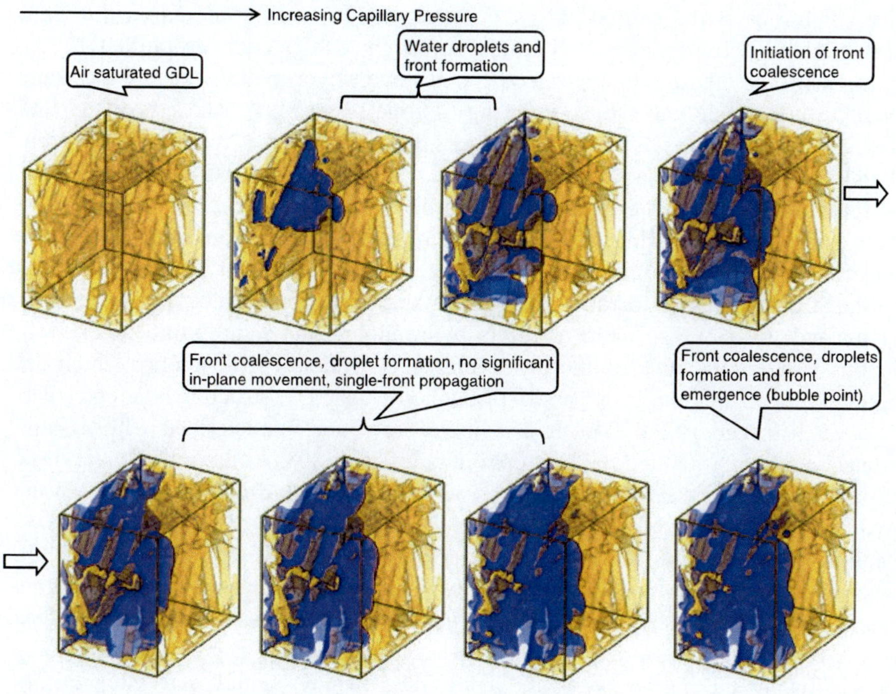

Fig. 3.11 Advancing liquid water front with increasing capillary pressure through the initially air-saturated reconstructed GDL microstructure with 30% compression from the primary drainage simulation using LBM

growth can also be observed in the constricted pores primarily due to strong wall adhesion forces from interactions with highly hydrophobic fibers with the increasing capillary pressure. Finally, the predominant front reaches the air reservoir, physically the GDL–channel interface, at a preferential location and the corresponding capillary pressure is referred to as the bubble point. Concurrent to the bubble point, the microstructural change also reflects in the movement of water in the in-plane direction which could rather be attributed to the liquid water buildup at the corresponding saturation level and differs significantly from the two-phase transport in the uncompressed GDL shown in Fig. 3.8. It is important to note that the mesoscopic LB simulations provide fundamental insight into the pore-scale liquid water transport in uncompressed and compressed GDL structures and would likely enable novel GDL microstructure design for flooding mitigation.

3.2.4 Effect of Durability

The beginning-of-life GDL exhibits hydrophobic characteristics, which facilities liquid water transport and hence reduces flooding. Experimental data, however, suggest that the GDL loses hydrophobicity over prolonged PEFC operation and becomes prone to enhanced flooding. After long-term exposure to the strong oxidative conditions at the cathode of an operating PEFC, carbon atoms on the GDL surfaces oxidize to form carboxyl groups or phenols, which are hydrophilic [80, 81]. As a result, the GDL carbon surface becomes more hydrophilic, causing a gradual increase in cathode water uptake during lifetime tests. Borup and coworkers [80, 81] have reported the GDL durability issues due to the loss of hydrophobicity from their experimental investigations.

Figure 3.12 shows the invasion pattern of liquid water from the displacement simulation with increasing capillary pressure in the initially air-saturated carbon paper GDL microstructure characterized by *mixed wettability* [77]. As mentioned earlier, after prolonged operation in a PEFC, the carbon fibers tend to lose hydrophobicity and the GDL structure exhibits hydrophobic and hydrophilic pores [80]. In this simulation, 50% of the pore volume is rendered hydrophilic, which are assumed to be randomly dispersed throughout the GDL structure, thereby characterizing an aged GDL. The hydrophobic pores are characterized with a static contact angle of 140° and the hydrophilic pores with 80°. At the initially very low capillary pressure, the invading liquid water exhibits both droplet formation in the hydrophobic pores and film formation due to the hydrophilic pores. With increasing capillary pressure, liquid water films tend to merge and assist in front movement. The front propagation is dominated by the film formation and subsequent merging phenomena. The underlying anisotropy in the GDL microstructure fails to assist in the branching and in-plane movement as in the case of a purely hydrophobic GDL. Toward the bubble point, the aged GDL displays liquid water slug formation instead of fingers and evidently leads to higher saturation level and enhanced flooding.

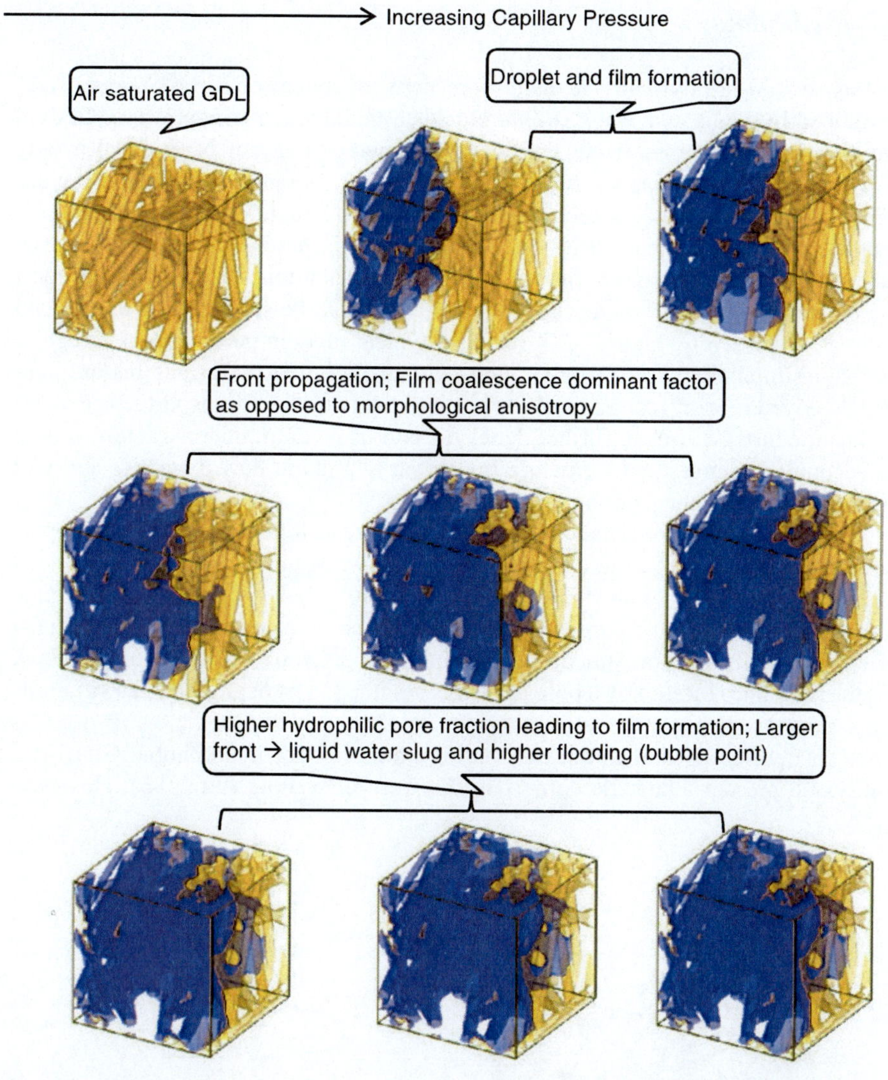

Fig. 3.12 Advancing liquid water front with increasing capillary pressure through the mixed-wet GDL microstructure with 50% hydrophilic pores from the displacement simulation using LBM

4 Two-Phase Pore Network Model

In pore network (PN) modeling, the porous medium is represented by a network of wide pores connected by narrow regions called throats and the flow is solved on this network with the relevant physics taken into account.

4.1 Methodology

In this section, a physically realistic pore network model of carbon paper GDL, developed by Sinha and Wang [82] to elucidate liquid water transport in a GDL at the pore level, is summarized. Following the methodology of Nam and Kaviany [57] in which carbon paper is regarded to consist of randomly stacked regular fiber screens, a 3D random tetragonal pore network structure with pores cubic in shape and throats of square cross section is generated. In this study, pore and throat radii are assumed to have a cut-off lognormal distribution. Pore and throat radii are defined as the radii of the largest sphere that can be inscribed in a pore and a throat, respectively. Figure 3.13 depicts a typical random pore network structure used for simplistic investigations. Such random pore network structures are usually deployed to gain fundamental insight into the two-phase transport in a generic porous medium. However, further details of porous medium microstructure should be accounted to investigate structure impact on two-phase flow dynamics and will be discussed in a later section.

The main assumptions made in the PN model are as follows: (1) wetting properties are assumed to be constant in the network; (2) while the radius of a throat serves to define its hydraulic conductance, the volume contributed by the throats is assumed to be small relative to the pore volumes; (3) only one fluid can reside in a throat; (4) flow within a throat is assumed to be laminar and given by Hagen–Poiseuille law; (5) the resistance offered by a pore to flow is assumed to be negligible; and (6) fluids are assumed to be incompressible.

In an invaded pore or throat the wetting phase (air for hydrophobic GDL) can always be present along the corners in the form of wetting films [83]. However,

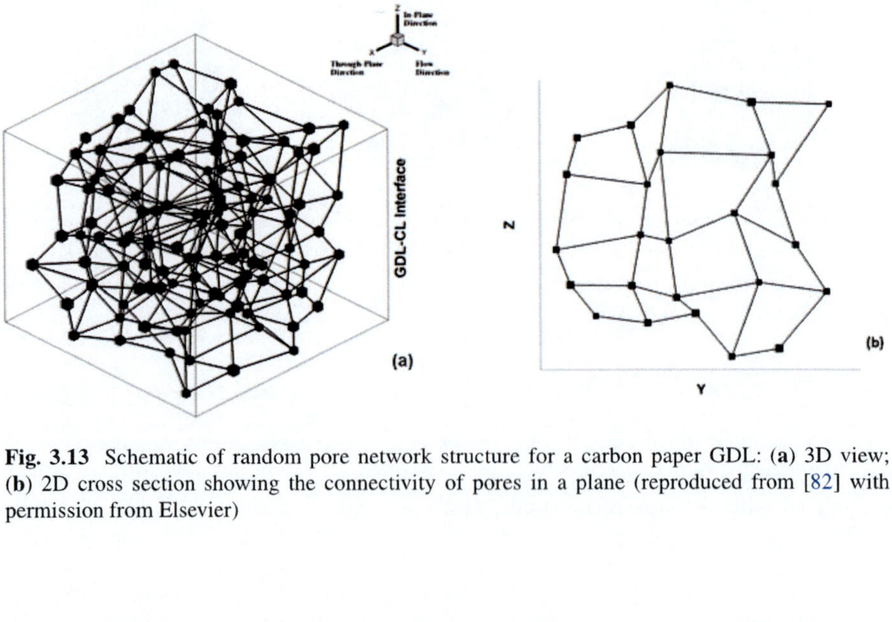

Fig. 3.13 Schematic of random pore network structure for a carbon paper GDL: (**a**) 3D view; (**b**) 2D cross section showing the connectivity of pores in a plane (reproduced from [82] with permission from Elsevier)

formation of wetting films along the corners is governed by the Concus and Finn [84] condition:

$$\theta + \gamma < 90^\circ \tag{3.37}$$

where θ and γ are contact angle between wetting phase and solid matrix and half corner angle of a pore or a throat, respectively. Thus, a contact angle between liquid water (non-wetting phase for hydrophobic GDL) and carbon fibers of 110° and square cross section of pores and throats suppress the existence of wetting films along the corners. It should be mentioned that a small fraction of pores, having half corner angle less than 20°, in the actual pore spaces of carbon paper may allow simultaneous occupancy of a throat with both fluids.

For simplicity, Hagen–Poiseuille law originally derived for circular tubes is used to represent the flow through a throat, although Patzek et al. [85] analytically derived a generalized Poiseuille law to represent flow in a throat of square cross section and showed that conductivity of a square throat is 20% lower. No substantial differences are expected as the pressure drop across a throat does not govern the transport at small capillary numbers typically encountered in a PEFC operation. More details of the PN model and its numerical algorithms can be found in Sinha and Wang [82].

The liquid water transport in an initially dry GDL and in contact with a liquid reservoir is investigated with a constant injection rate boundary condition at the inlet face. The constant injection rate of liquid water is equivalent to 2.0 A/cm^2 current density assuming that all the water produced is in liquid form. A constant pressure boundary condition is imposed on the outlet face with no coverage by liquid water assumed under high gas flow in the gas channel. No-flow boundary condition is imposed on all the other faces.

4.2 Two-Phase Transport in a Hydrophobic and Mixed-Wettability Porous Medium

Numerical visualization of liquid water front movement within the network during drainage process is depicted in Fig. 3.14, where irregular fractal patterns typical of invasion percolation can be observed. It is clear from Fig. 3.14 that liquid water moves in the GDL through several continuous clusters. As liquid water invades into the GDL, the liquid water front encounters multiple dead ends. The pressure difference across a gas–liquid interface must be larger than the capillary pressure at the interface for liquid water to invade further. Dead ends to front propagation appear when a liquid water front reaches a very narrow region with a very large entry capillary pressure. When a liquid water front invades such a narrow region, the liquid water pressure increase at the inlet face makes the water fronts unstable at several locations and liquid water invades further into the GDL there. This mechanism can be more clearly explained with Fig. 3.15, where only "open clusters" during the liquid water transport in the GDL are displayed. The open clusters are defined as the liquid water clusters having non-zero flow rate at any instant of time. Figure 3.15a

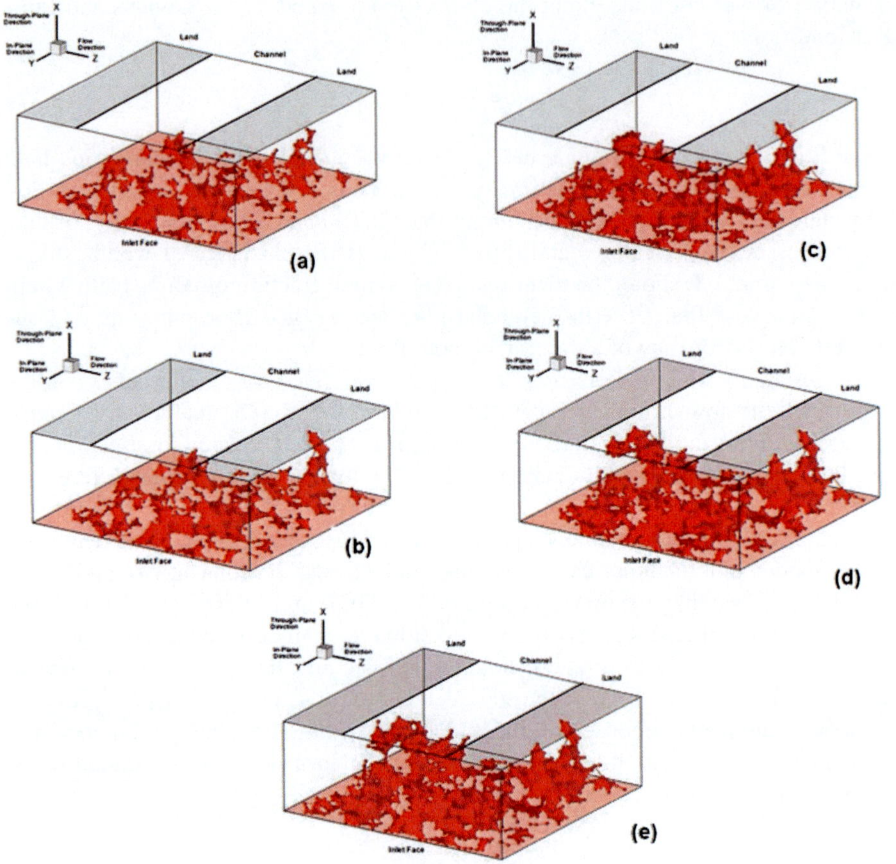

Fig. 3.14 Liquid water front movement in GDL (**a**)–(**d**) at intermediate states, showing the evolution of capillary fingers, and (**e**) at steady state (reproduced from [82] with permission from Elsevier)

shows the open clusters at an intermediate time step but as liquid water invades further into the GDL, these liquid water clusters encounter dead ends and liquid water starts flowing through other clusters in order to maintain constant flow rate at the inlet face, as shown in Fig. 3.15b, c. As mentioned earlier, capillary number for liquid water movement in a GDL under realistic operating conditions is $\sim 10^{-8}$. Thus, capillary forces control the transport of liquid water in GDL and liquid water follows a path of least resistance through the GDL.

Figure 3.16a shows the evolution of the cross-sectional averaged liquid water saturation profile along the GDL thickness during water invasion. The saturation profiles shown in Fig. 3.16a are characteristic of fractal fingering flow. The occurrence of dead ends to multiple water clusters, originating from the inlet face, and the advancement of a single cluster following the path of least resistance, at low

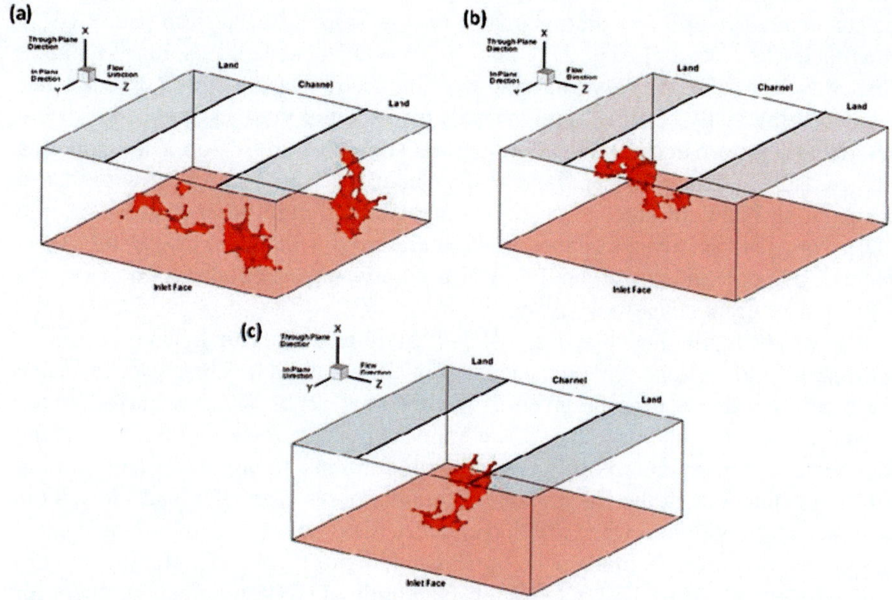

Fig. 3.15 Open liquid water clusters during liquid water transport in GDL at (**a**) initial state, (**b**) intermediate state, and (**c**) steady state (reproduced from [82] with permission from Elsevier)

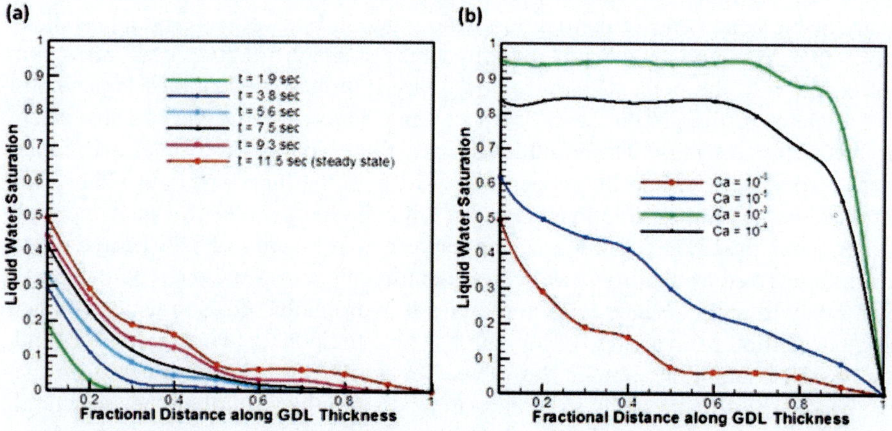

Fig. 3.16 (**a**) Evolution of liquid water saturation profiles along the GDL thickness; (**b**) steady-state saturation profiles as function of *Ca* (reproduced from [82] with permission from Elsevier)

Ca, ensue a convex shape of the steady-state saturation profile. In comparison, two-phase PEFC macroscopic models, published widely in the literature, invariably yield concave-shape saturation profiles characteristic of stable displacement. Figure 3.16b shows the steady-state liquid water saturation profiles for capillary number varying from 10^{-3} to 10^{-8} (by increasing the flow rate). As can be seen, the characteristics of the saturation profile change from a fractal form ($Ca = 10^{-8}$) to a stable flow form ($Ca = 10^{-4}$ and 10^{-3}) with increase in capillary number. This crossover from capillary fingering to stable front invasion during drainage has been experimentally observed and investigated in detail in other applications [86–89]. The crossover in the liquid flow regime shown in Fig. 3.16b also underpins the necessity to match the capillary number in any ex situ experiment with that in an operating PEFC in order to properly characterize liquid water distribution in a GDL. This does not always happen in the literature; for instance, the ex situ experiments by Djilali and coworkers [90, 91] used a water injection rate equivalent to 112 A/cm^2 current density.

It should be highlighted that despite PTFE (poly-tetra-fluoro-ethylene) treatment, hydrophilic and hydrophobic pores co-exist in GDL materials. The most exhaustive treatment, to date, is perhaps given by Weber et al. [92], who have taken mixed wettability of GDL into account via a composite contact angle as a function of the fraction of hydrophilic pores, f. They computed the maximum power and limiting current as functions of the fraction of hydrophilic pores, f, in GDL and showed the existence of an optimum value of f that entails a maximum value to limiting current and maximum power. A fundamental understanding of the liquid water transport in a mixed-wet GDL is necessary to establish an optimal PTFE treatment protocol for GDLs.

The afore-described PN model is further modified to examine the effect of mixed wettability on liquid water transport in a GDL. Since an accurate procedure to quantify the contact angle distribution inside the GDL is yet to be established, as a first approximation a uniformly random contact angle distribution is assumed, but correlated with the pore–throat sizes (i.e., larger throats are assigned larger contact angles). In the present work, contact angle is assumed to vary in the range of 100°–120° for the inlet and outlet throats, those connected to inlet and outlet faces respectively, and in the range of 65°–120° in the inner layers of GDL. The rationale behind the assumption of contact angle higher than 90° for inlet and outlet throats is based on the non-zero capillary entry pressure in PTFE-treated GDL materials. Mixed wettability of a GDL is quantified by f, the fraction of pore–throats that are hydrophilic. Hence $f = 0$ represents a hydrophobic GDL in which contact angle is distributed in a range of 90°–120°. The other pore network parameters and the model assumptions remain the same. The following results are based on the average of eight realizations of contact angle distribution with the other structural parameters kept the same. For simplicity, no masking of the outlet face with land is implemented in the following calculations. Readers can refer to Sinha and Wang [93] and references therein for more details of the mixed-wet PN model algorithms.

Figure 3.17 depicts the steady-state saturation profiles along the GDL thickness as a function of the hydrophilic fraction in a GDL averaged over eight realizations of contact angle distribution. The suppression of finger-like morphology in a mixed-wet GDL renders a change in saturation profile shape from concave, typical of fractal fingering, to convex, typical of stable front, with increase in hydrophilic fraction. The crossover from concave- to convex-shaped saturation profile, as clearly depicted in Fig. 3.17, lends support to the applicability of two-phase Darcy's law

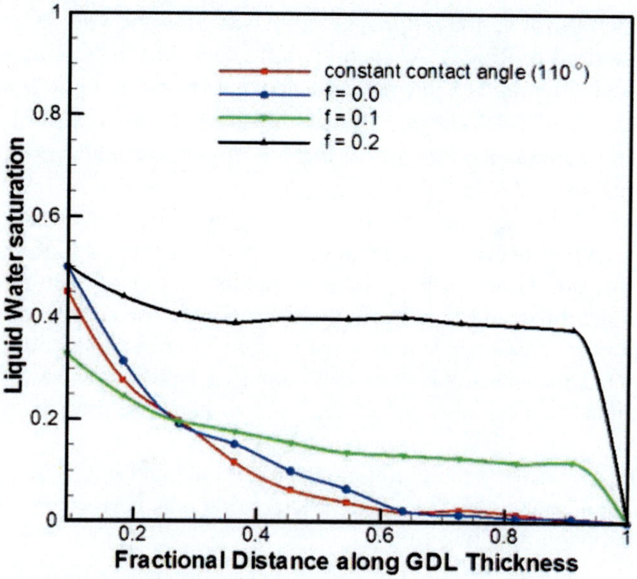

Fig. 3.17 Liquid water saturation profiles along GDL thickness as a function of the hydrophilic fraction *f* (reproduced from [93] with permission from Elsevier)

treatment to address liquid water transport in a mixed-wet GDL with $f \geq 0.2$. In recent years, Ferer and coworkers [88, 89], among others, have conducted detailed investigations identifying the physics and parameters governing the crossover from capillary fingering to stable displacement regime. The present analysis shows that crossover from capillary fingering to stable displacement can occur due to contact angle variation even at very low capillary number.

4.3 Computation of Material-Specific Capillary Pressure and Relative Permeability Relations

The construction of realistic porous medium pore morphology is the essential prerequisite for unveiling the influence of an underlying structure on the two-phase behavior. This can be achieved either by 3D volume imaging or by constructing a digital microstructure based on stochastic models. Noninvasive experimental techniques, such as X-ray and magnetic resonance computed microtomography, are the popular methods for 3D imaging of pore structures. Another alternative is reconstruction of a microstructure using stochastic simulation techniques. The low cost and high speed of data generation makes stochastic generation methods the preferred choice over experimental imaging techniques. Moreover, computer methods can not only construct topologically equivalent pore network (TEPN) structure of existing porous medium but also virtually design new structures by altering various microstructural properties. The latter advantage is particularly useful in the search for optimal materials for various applications such as improved GDL pore structures for future generations of PEFCs. As such, we have developed a systematic approach to design a virtual GDL and test its performance, consisting of the following steps:

(1) A stochastic modeling method [79] is used to generate high-resolution pore-scale 3D images of GDL.
(2) An extensive imaging analysis [94] is employed to extract the computer-generated microstructure in the form of a network of pores and throats. During this process, properties such as volume, radius, and cross-sectional shape of each pore and throat in the network are determined. The coordination number, i.e., the number of independent throats linked to a pore, is assigned accordingly.
(3) The extracted pore network is used as input to a pore network flow simulator to compute macroscopic properties, such as capillary pressure and relative permeability.
(4) Further, these macroscopic correlations can be plugged into a two-phase continuum model, such as (M2) model [95], to predict the performance of a virtual GDL in an operational fuel cell.

In this section, computer models of non-woven carbon fiber paper and woven carbon cloth GDLs are reconstructed using a stochastic generation method with structural inputs obtained from the literature and/or manufacturers. The input parameters, i.e., fiber diameter, fiber orientation, and porosity, are chosen such that absolute permeability values in both in-plane and through-plane directions match closely with those measured experimentally in the literature.

Figures 3.18 and 3.19 display SEM-like images of carbon paper and carbon cloth generated by computers. Through extensive imaging analysis of computer-constructed microstructure models, extracted pore networks are stored in data files and then used as a direct input to a two-phase flow simulator. It should be

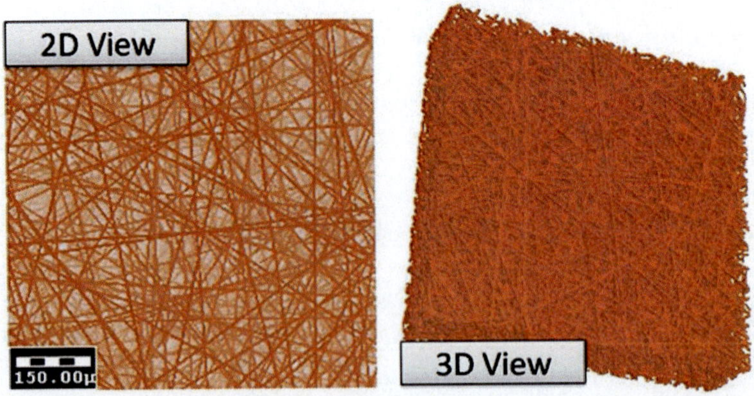

Fig. 3.18 Carbon paper microstructure generated by stochastic method (reproduced from [100] with permission from Elsevier)

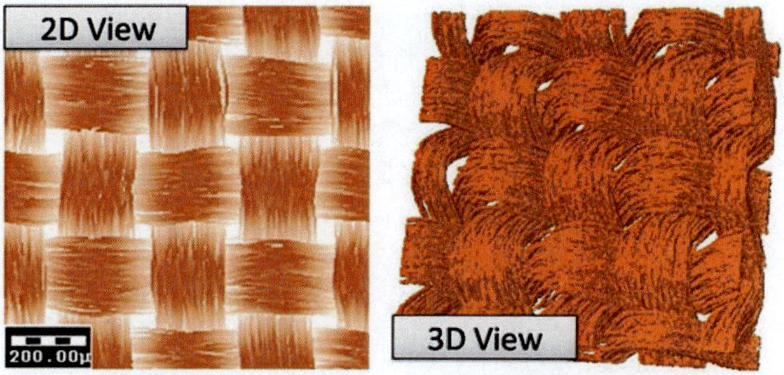

Fig. 3.19 Carbon cloth microstructure generated by stochastic method

noted that pores and throats are irregular in shape in our extracted pore networks. However, for clarity of presentation pores and throats are depicted in Figs. 3.20 and 3.21 by spheres and cylinders, respectively, with radii of their inscribed spheres. The carbon paper network represents a domain of 192×750×750 μm, containing 2537 pores and 16,501 throats. The carbon cloth network represents a domain of 360×1000×1000 μm, containing 1067 pores and 5200 throats. The pore size distributions of the extracted pore networks are shown in Fig. 3.22. A vast difference in the pore size distributions of carbon paper and carbon cloth can be clearly seen. For carbon paper, the distribution is unimodal with a peak pore size of ~20 μm. For carbon cloth, a bimodal pore size distribution spanning over a few orders of magnitude is observed. The primary mode associated with large pores (i.e., inter-thread pores) among the fiber bundles occurs between 31 and 145 μm with peak value at 100 μm. The secondary mode associated with small pores (i.e., intra-thread pores) between individual fibers appears between 1 and 31 μm with the peak value at 20

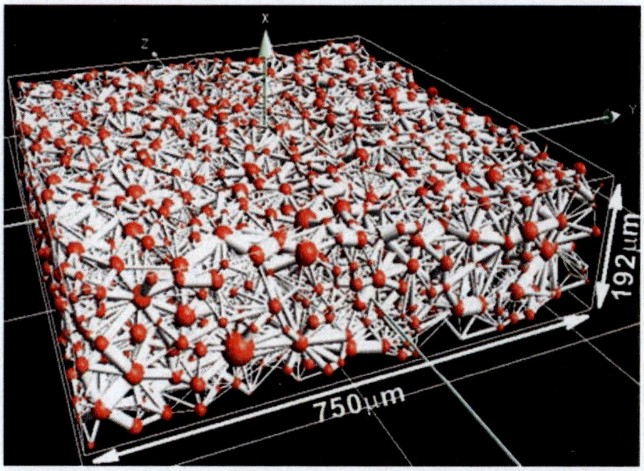

Fig. 3.20 Topologically equivalent pore network of carbon paper where pores are depicted by *grey (red online) spheres* and throats by *white cylinders* (reproduced from [100] with permission from Elsevier)

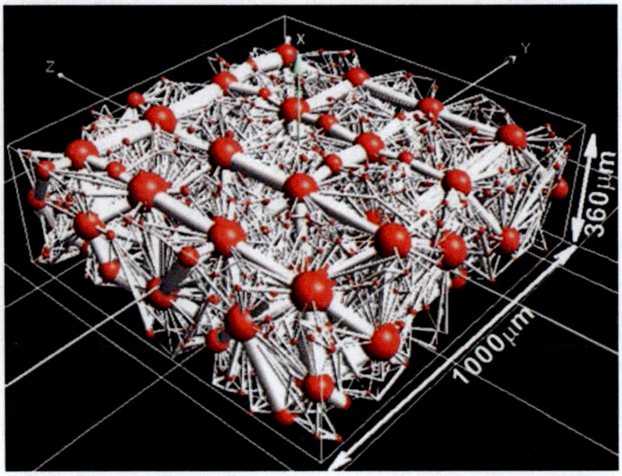

Fig. 3.21 Topologically equivalent pore network of carbon cloth where pores are depicted by *grey (red online) spheres* and throats by *white cylinders* (reproduced from [100] with permission from Elsevier)

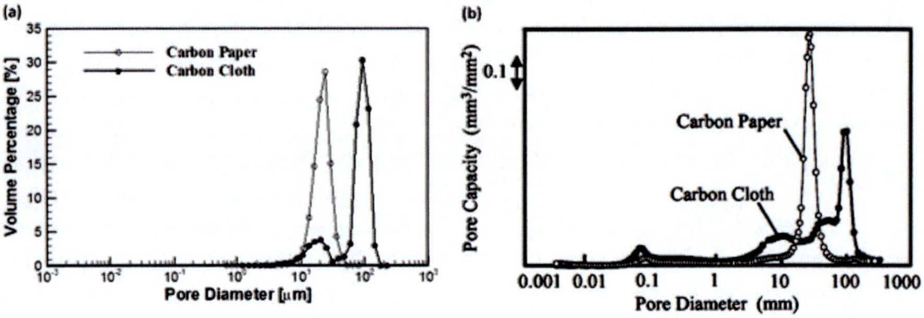

Fig. 3.22 Pore size distributions of carbon paper and carbon cloth from (**a**) topologically equivalent pore networks and (**b**) measurements [96]. Note: the unit of pore diameter should be μm instead of mm in (**b**) (reproduced from [100] with permission from Elsevier)

μm. The distribution patterns and peak values of pore diameters of the extracted pore networks are in accordance with the measured data, as shown in Fig. 3.22b [96], confirming the validity of the generated TEPN structures. Figure 3.23 contains images of throat distribution for carbon paper and carbon cloth, viewed from the through-plane direction. Together with 3D views, it shows that carbon paper and carbon cloth have distinctive fibrous structures. The microstructure of carbon paper is shown to be high topology layers formed by randomly arranged fibers along the in-plane direction. The throats are densely arranged in a highly random order. Its coordination number varies from 2 to 51, with an average value of 13. On the other hand, the carbon cloth shows the well-organized plain-weave structures. The throats are coarsely distributed in a highly ordered and structured manner. Its coordination number varies from 2 to 102, with an average value of 10. Both

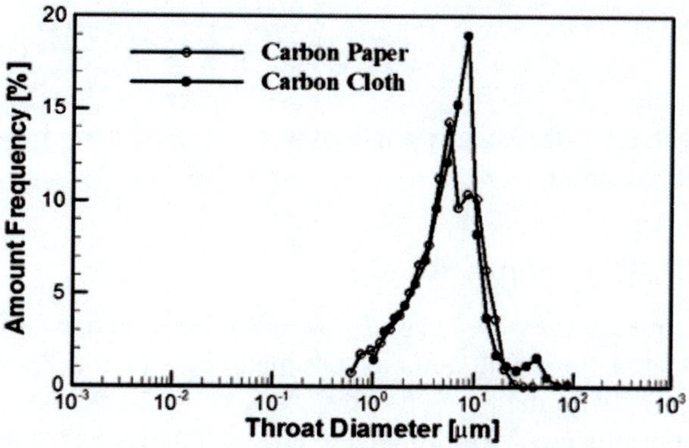

Fig. 3.23 Throat diameter distributions of topologically equivalent pore networks for carbon paper and carbon cloth (reproduced from [100] with permission from Elsevier)

carbon paper and carbon cloth have very complex microstructures, such as wide distributions of pore and throat size and high coordination numbers, implying that the ability of cubic or randomly generated pore networks to incorporate structural attributes of real GDLs is quite limited. It is extremely difficult, if not impossible, to generate random structures with high porosity and anisotropy as well as pore size distribution with large span and average coordination number higher than 6. In summary, because TEPN has an accurate description of size and position of pores and throats and interconnectedness of pores and throats and eliminates all the aforementioned limitations of the random pore networks, it becomes finally possible to predict structure-specific macroscopic transport properties pertinent to liquid water transport in fuel cell GDLs.

A quasi-static algorithm [97], applicable to flow through porous media at an infinitesimal flow rate ($Ca < 10^{-6}$) where the viscous pressure drop across the network is negligible and capillary forces completely control the fluid configuration, is deployed in the current example. For a typical fuel cell application, the capillary number is of the order of 10^{-8}. Thus the quasi-static algorithm is used here to compute the constitutive relations: capillary pressure and relative permeability as functions of liquid water saturation. In real porous media, pores and throats have complex and irregular shapes. Therefore, in the TEPN model a dimensionless shape factor is used to quantify pores and throats of arbitrary cross section [98].

$$G = \frac{A}{P^2} \tag{3.38}$$

where A is the cross-sectional area and P the corresponding perimeter. In the present work, cross sections of most throats are of triangular shape in the TEPN of carbon paper and carbon cloth. For a triangular throat connected with pores i and j, the entry pressure, P^c_{ij} is given by [99]

$$P_{ij}^{c} = \sigma\cos\theta\frac{1+2\sqrt{\pi G_{ij}}}{r_{ij}} \tag{3.39}$$

where θ is the contact angle of wetting phase and r_{ij} the minimum inscribed radius of the throat.

4.3.1 Constitutive Relations

Capillary pressure curve $P_c(S)$ and relative permeability curve $K_r(S)$ are two key relations entering into a continuum model for PEFCs. Both relations depend strongly on the pore-level distributions of the phases as well as the complexity of pore structures. The volume-averaged liquid water saturation is defined as

$$S = \frac{1}{V}\sum_{i\in \mathrm{RV}} V_i S_i \tag{3.40}$$

where V_i and S_i represent volume and liquid water saturation of pore i, respectively, and RV the representative volume. If RV represents the whole modeling sample, S will be the total average saturation, which serves as a key parameter of $P_c(S)$ and $K_r(S)$ curves. Otherwise, if the representative volume contains only a portion of pores, S is regarded as the local saturation, which is used in plotting saturation profiles. Notice from the flow assumptions that each pore can only be occupied by one fluid. Thus S_i is an integer that equals either 0 or 1.

To compute the capillary pressure curve, it is assumed that air pressure throughout the network is uniform. Initially, all the pores and throats are completely filled with air and the inlet throats are connected to a reservoir of liquid water. For liquid water to invade into a throat, the pressure difference across the liquid–gas meniscus must exceed the throat entry capillary pressure as calculated from (3.39). During each step of quasi-static displacement, a search is performed over all the interfacial positions to determine the minimum capillary pressure that will allow water to invade further into the GDL. After increasing the capillary pressure to this critical value, liquid water is allowed to invade the connecting pore and any subsequent throats and pores that can be invaded at this capillary pressure. At this time, total volume-averaged saturation is updated based on the liquid water distribution in each pore. Once no further invasion can occur, the capillary pressure is increased by a next minimum increment.

To compute the absolute permeability and relative permeability, Hagen–Poiseuille flow through throats is assumed. The volume flow rate of phase α between two neighboring pores i and j can thus be expressed as

$$Q_{ij}^{\alpha} = g_{ij}^{\alpha}(P_i - P_j) \tag{3.41}$$

where P_i and P_j are the pressures in the pores and g_{ij}^{α} is the hydraulic conductance of phases α in the throat that connects pore i and pore j. For triangular throats, the cross-sectional areas are open for both wetting and non-wetting phases. The area occupied by wetting and non-wetting phases is given by [99]

$$A_{ij}^{\mathrm{w}} = \frac{r_{ij}^2}{4G_{ij}}\left[1 - \left(\frac{\sigma}{r_{ij}P_{ij}^{\mathrm{c}}}\right)^2 (1 - 4\pi G_{ij})\right] \tag{3.42}$$

$$A^{\mathrm{nw}} = \frac{r_{ij}^2}{4G_{ij}}\left(\frac{\sigma}{r_{ij}P_{ij}^{\mathrm{c}}}\right)^2 (1 - 4\pi G_{ij}) \tag{3.43}$$

where nw and w stand for the non-wetting and wetting phases, respectively. The flow conductance is modeled by the mean hydraulic radius, R_{ij}^{α}, and the equivalent volume radius, r_{ij}^{α} [99]:

$$R_{ij}^{\alpha} = 0.5(r_{ij} + r_{ij}^{\alpha}) \tag{3.44}$$

$$r_{ij}^{\alpha} = \sqrt{\frac{A_{ij}^{\alpha}}{\pi}} \tag{3.45}$$

The conductance of the non-wetting phase through a throat between pore i and pore j is given by Poiseuille's law:

$$g_{ij}^{\mathrm{nw}} = \frac{R_{ij}^{\mathrm{nw}2} A_{ij}^{\mathrm{nw}}}{8\mu^{\mathrm{nw}} l_{ij}} \tag{3.46}$$

where μ is the viscosity and l_{ij} the length of the throat. For the wetting phase, it follows that [99]

$$g_{ij}^{\mathrm{w}} = \frac{r_{ij}^{\mathrm{w}2} A_{ij}^{\mathrm{w}}}{8\beta\mu^{\mathrm{w}} l_{ij}} \tag{3.47}$$

where β is a dimensionless flow resistance factor which accounts for the reduced conductance of the wetting phase close to the pore wall. $\beta = 2.5$ is used for a no stress interfacial boundary condition [99]. Since the fluids are assumed to be incompressible, volume conservation applies at each pore:

$$\sum_j Q_{ij}^{\alpha} = 0 \tag{3.48}$$

where j runs over all the pores connected to pore i. To compute absolute permeability, the network is forced to be fully saturated with a single fluid (liquid or air). By specifying a pressure drop across the GDL, (3.48) is solved and the volume flow rate Q at the inlet or outlet surface is calculated. The absolute permeability K is determined from Darcy's law:

$$K = \frac{Q\mu^{\alpha}L}{A_{\text{s}}\Delta P^{\alpha}} \tag{3.49}$$

where A_{s} is the area of inlet or outlet surface and L the thickness along the direction where pressure drop occurs. Similarly, relative permeability is calculated by specifying a pressure drop across the GDL for phase α, but at various saturation levels of the invasion process. Calculate the volume flow rate of phase α, Q^{α} at the inlet or outlet boundary. The relative permeability of a phase α, k_{r}^{α}, is given by

$$k_{\text{r}}^{\alpha} = \frac{Q^{\alpha}\mu^{\alpha}L}{KA_{\text{s}}\Delta P^{\alpha}} \tag{3.50}$$

Combining (3.48) and (3.49) gives

$$k_{\text{r}}^{\alpha} = \frac{Q^{\alpha}}{Q} \tag{3.51}$$

Thus k_{r}^{α} is the ratio of flow rates of phase α at the same pressure drop, but with two different phase conditions. The numerator is for two-phase flow while the denominator is for single phase flow only.

For the present simulations, the inlet surface of the GDL is assumed to be in contact with a reservoir of the non-wetting fluid (i.e., liquid water), whereas the outlet surface is connected to a reservoir of the wetting fluid (air). Constant pressure boundary conditions are imposed on the inlet and outlet surfaces. All other surfaces are subjected to no-flow boundary conditions.

Throughout this chapter, x denotes the through-plane direction and y and z the in-plane directions. For current configurations, x, y, and z are the principal axes. For the flow of a viscous fluid through an anisotropic homogeneous porous medium, it is assumed that Darcy's law must be satisfied for each phase:

$$\begin{pmatrix} u_x^{\alpha} \\ u_y^{\alpha} \\ u_z^{\alpha} \end{pmatrix} = -\frac{1}{\mu^{\alpha}}\begin{bmatrix} k_{r,x}^{\alpha}K_x & & \\ & k_{r,y}^{\alpha}K_y & \\ & & k_{r,z}^{\alpha}K_z \end{bmatrix} \nabla P^{\alpha} \tag{3.52}$$

For typical fuel cell applications where the GDL is very thin, it is reasonable to assume a constant gas-phase pressure. Thus the following expression can be derived for liquid water:

$$\begin{pmatrix} u_x^{\alpha} \\ u_y^{\alpha} \\ u_z^{\alpha} \end{pmatrix} = -\frac{1}{\mu^{\alpha}}\begin{bmatrix} k_{\text{r},x}^{\alpha}K_x & & \\ & k_{\text{r},y}^{\alpha}K_y & \\ & & k_{\text{r},z}^{\alpha}K_z \end{bmatrix} \frac{dP_{\text{c}}}{ds}\nabla s \tag{3.53}$$

Equation (3.53) implies that the liquid water flux is proportional to the liquid relative permeability, the absolute permeability, and the slope of capillary pressure curve. All these three parameters are strongly tied to the GDL microstructure. With the detailed microstructure described by TEPN, two groups of simulations, "through-plane flow" and "in-plane flow," are carried out to investigate the impact of different pore structures of carbon paper and carbon cloth on their macroscopic transport properties.

Details of the calculated values of through-plane and in-plane absolute permeabilities based on the extracted pore networks of carbon paper and cloth can be found in [100]. For carbon paper, the through-plane absolute permeability is 6.31×10^{-12} m^2, whereas the large throats in the bimodal distribution for carbon cloth increase its value to 70.9×10^{-12} m^2.

For comparison, capillary pressure curves (primary drainage curves) of both carbon paper and carbon cloth are plotted in Fig. 3.24. Capillary pressure plays an important role in the description of water flow in GDLs. For a given capillary pressure the amount of liquid water entering inside GDLs depends on the slope of capillary pressure curve. Liquid water being the non-wetting phase preferentially invades larger throats in the pore structure. Hence, characteristics of capillary pressure curve are strongly determined by both throat and pore size distributions. The liquid water invasion starts with larger throats and pores and then moves toward smaller ones in the pore and throat size distributions. Entry pressure at zero saturation is associated with the upper limit of the throat size distribution and determined by one of the largest throats in the sample. As shown in Fig. 3.23, the maximum throat diameter of carbon cloth (89.2 μm) is much larger than that of carbon paper (40.6 μm). Consequently the predicted entry pressure of carbon cloth (0.92 kPa) is significantly lower than that of carbon paper (2.15 kPa).

The slope of the capillary pressure curve is also determined by pore and throat size distributions and how these two distributions are associated. The capillary pressure curve of carbon paper has a flat slope between saturation of 0.05 and 0.8. At saturation of 0.05 and 0.8 the capillary pressures are 4.5 and 6.0 kPa, corresponding to 16.5 and 12.4 μm diameter throats, respectively. This means that the dominant

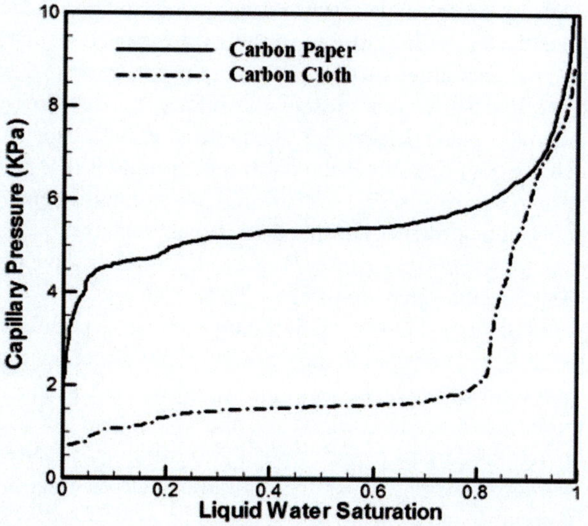

Fig. 3.24 Capillary pressure curves of carbon paper and carbon cloth predicted by TEPN (reproduced from [100] with permission from Elsevier)

throat diameter is in a very narrow range between 12.4 and 16.5 μm. About 75% pore volume is connected by dominant throats. The small decrease in throat diameter from 16.5 to 12.4 μm leads to the small increase in capillary pressure, thus the flat slope of capillary pressure curve between saturation 0.05 and 0.8. In the region beyond the saturation of 0.8, capillary pressure for carbon paper increases rapidly as liquid water invades further into smaller pores.

With a bimodal throat and pore size distribution, capillary pressure curve of carbon cloth differs significantly from that of carbon paper. The capillary pressure curve exhibits a bimodal ramp increase: a flat slope and slow increase of capillary pressure between the saturation of 0 and 0.8 and a steep slope and fast rise beyond that. At the saturation of 0.8 the capillary pressure is 2.0 kPa, corresponding to 37.5 μm diameter throats. This means that the dominant throat diameter is in a wide range between 37.5 and 81.5 μm. The flat slope region is associated with the primary mode in the pore size distribution of carbon cloth. The pore volume in this mode represents about 80% of the total volume of all pores. The sharp sloped region is associated with the secondary mode. Abrupt slope change around saturation 0.8 indicates that the peaks in the bimodal pore and throat distributions are well separated.

In Fig. 3.25, the predicted through-plane and in-plane relative permeabilities are compared for both carbon paper and carbon cloth, respectively. Relative permeability is determined by the conductance matrix of phase α flowing in throats. As such, it is influenced by the throat length, throat size distribution, and connectivity of neighboring throats. At the same liquid saturation, the connectivity of throats flown by a specific phase is strongly direction dependent, caused by the anisotropy of the structures of carbon paper and carbon cloth. This leads to a significant difference between the in-plane and through-plane relative permeability curves. Figure 3.26 displays the liquid-saturated pores at the first breakthrough. All throats are depicted by straight lines. All pores are plotted proportional to the real dimensions of the GDL structure as the inscribed spheres of the pore spaces for simplicity. The liquid pore cluster in blue color represents the connected cluster with a liquid breakthrough point at the outlet surface. Others in green color are all dead-end liquid pore clusters. Blue surfaces represent the inlet plane or reservoir. Gray surfaces represent the outlet plane. Figure 3.27 shows that carbon paper has much larger pore number density than that of carbon cloth owing to their distinctive pore structure and thickness. As shown in Fig. 3.25, liquid water cannot penetrate into the GDLs to form a continuous flow path until a critical water saturation is attained where water breaks through the outlet surface. Hence, the liquid relative permeability is zero below the breakthrough saturation in Fig. 3.25.

As displayed in Fig. 3.25, the liquid relative permeability in the in-plane direction is much smaller than in the through-plane direction. For carbon paper, below liquid saturation of 0.72, the in-plane liquid relative permeability is very small (<0.04), thereby exacerbating flooding under the land region. Hence, for high-humidity operation, carbon paper is not the best choice. However, under the low-humidity operation, carbon paper has the advantage of retaining product water and hence improving membrane hydration.

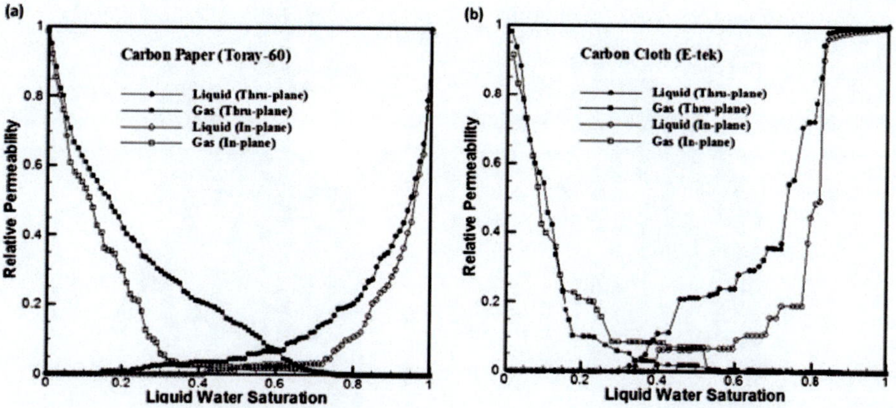

Fig. 3.25 Predicted relative permeability curves of (**a**) carbon paper and (**b**) carbon cloth (reproduced from [100] with permission from Elsevier)

Between the breakthrough saturation and saturation of 0.8, the liquid relative permeability curves for carbon cloth are not smooth, exhibiting step-wise variations. This is reflective of the unique topological structure of carbon cloth. In the saturation range, the liquid relative permeability is associated with the primary mode in the bimodal pore size distribution. Although merely 8% in number, pores in this mode are very large and occupy 80% of the total volume of all pores. Liquid volume saturation can jump by a few percent even when only one large pore is invaded. Flat regions of the curves indicate that a large change in liquid saturation hardly impacts the relative permeability. This happens when invasion proceeds in the dead-end clusters consisting of large pores. Since those clusters have no connection with the outlet, there is no contribution to the liquid permeability despite the liquid volume saturation increasing appreciably. Abrupt change regions of the curves, where a slight increase in liquid saturation causes rapid rise in relative permeability, happen when a dead-end cluster with large pores finally breaks through or connects with an open cluster. This dramatically changes the total conductance of the entire network, with a little change in the saturation. Beyond the saturation of 0.8, both through-plane and in-plane liquid relative permeability curves are seen to be very flat. This is because major liquid flow paths are already established throughout the network at the high saturation of 0.8. Adding minor liquid flow paths by invading smaller throats in the secondary mode hardly increases the total conductance of the network. Thus, beyond saturation of 0.8 the liquid relative permeability stays very close to unity.

The absolute permeability of carbon cloth is almost 15 times that of carbon paper. At the same saturation, both through-plane and in-plane relative permeabilities of carbon cloth are much higher than those of carbon paper. As indicated by (3.51), the absolute and relative permeabilities, together with the slope of capillary pressure–saturation curve, determine the water flux across the GDL. Hence, carbon cloth has a much higher capability to transport liquid water than carbon paper, leading to less

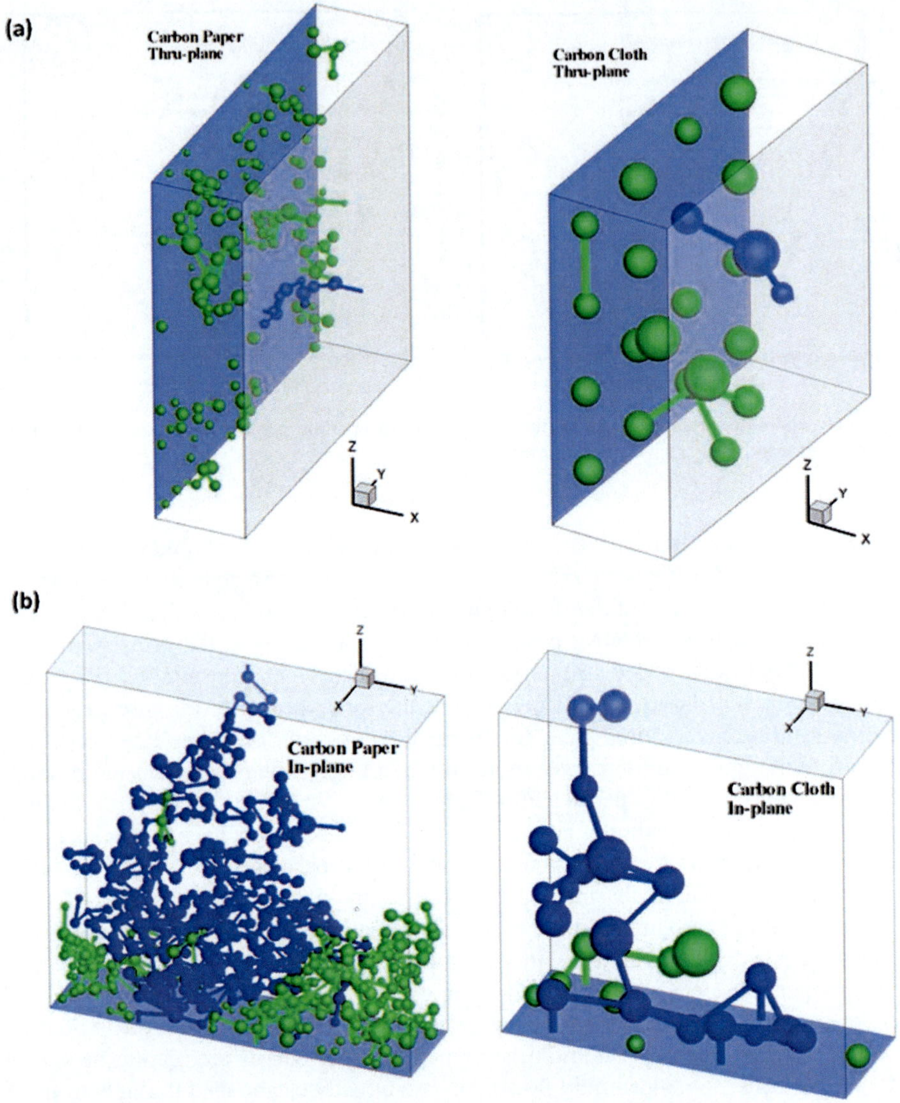

Fig. 3.26 Liquid-saturated pores at the first breakthrough for (**a**) through-plane flow and (**b**) in-plane flow (reproduced from [100] with permission from Elsevier)

retention of liquid water in carbon cloth and hence making it more suitable for high-humidity operation. For the same reason, carbon cloth GDL is inferior to carbon paper under low-humidity operation.

The gas relative permeability is calculated under the assumption that liquid-filled pores offer no conductivity to gas flow. For carbon paper, the predicted gas relative

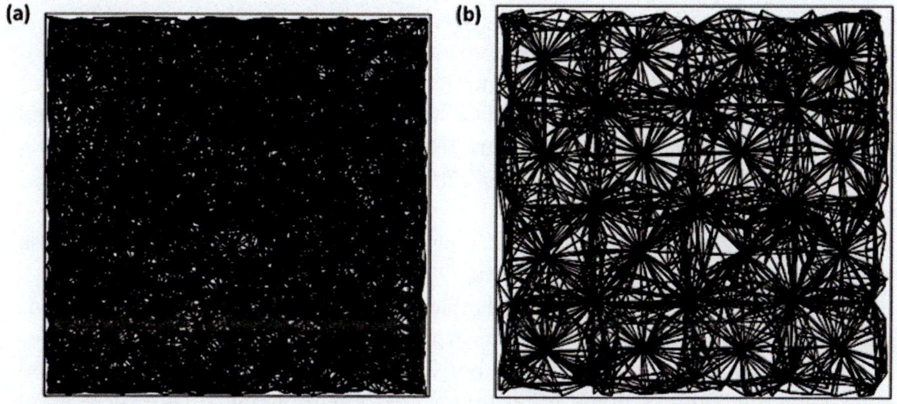

Fig. 3.27 Topologically equivalent pore networks viewed along the through-plane direction: (**a**) carbon paper and (**b**) carbon cloth (reproduced from [100] with permission from Elsevier)

permeability in the in-plane direction is much smaller than that in the through-plane direction, which could cause severe mass transport limitation under the land region. For carbon cloth, once the liquid saturation exceeds 0.55, the through-plane and in-plane gas relative permeabilities reduce to nearly 0. This means that gas transport can be effectively blocked by filling a few large pores along the main flow path. For carbon paper, the saturation needs to reach 0.7 to block the gas transport.

It should be noted that the predicted correlations are based on a single computer-reconstructed sample of carbon paper and carbon cloth. They are not as accurate and smooth as statistically averaged relative permeability, for which more stochastic generated samples are required. However, the predicted results still reveal the dramatic impact of the GDL microstructure on their macroscopic properties, and the relative permeability correlations presented herein exhibit new characteristics that have not been captured by random pore networks.

4.4 Drying of Porous Medium

Product water becomes frost or ice upon startup when PEFC internal temperature is below freezing point of water. If the local pore volume of the cathode catalyst layer (CCL) is insufficient to contain all of the accumulated water before the operating temperature of the cell rises above freezing, the solid ice may plug the CCL and stop the electrochemical reaction by starving the reactant gases. This problem, known to the automotive industry as "cold start," is important not only due to customers' demand for quick startup but also because PEFC durability is greatly affected by cold start cycles. In recent years, various investigations have probed fundamental mechanisms of cold start [101–106]. It is recognized that drier membrane prior to cold start increases membrane water uptake and thereby prevents CCL plugging by ice before the cell temperature rises above freezing. Commonly, gas purge – a procedure where an inert gas is fed into the gas channels for a short amount of time – is performed for control and minimization of residual water in a PEFC prior to shutdown. The dry gas evaporates liquid water present in PEFC porous layers and eventually dries out the membrane. Thus, the fundamental understanding of drying front evolution and its dependence on GDL microstructure, wettability, as well as purge conditions is critical.

For fundamental understanding of drying front evolution and its dependence on porous medium microstructure, a 2D random tetragonal pore network with pores and throats of square cross sections is employed. Various features and assumptions made in this model are the same as described in the previous section.

A dynamic pore network algorithm, derived from Yiotis et al. [107], is deployed to study the dynamics of liquid water removal in a hydrophobic GDL. The present model accounts for viscous and capillary transport of liquid water in addition to evaporation of liquid water and gas-phase transport. In drying, three types of pore

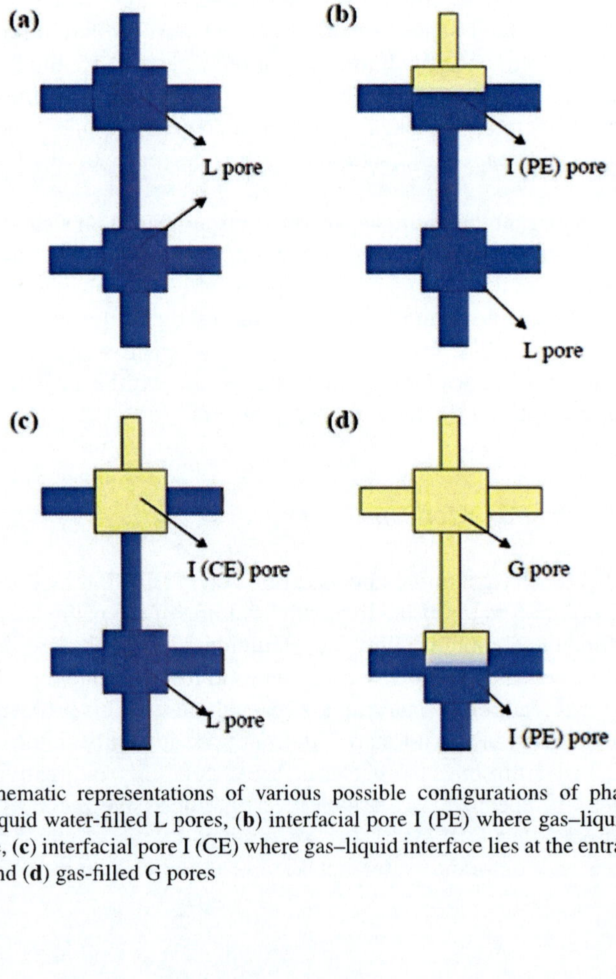

Fig. 3.28 Schematic representations of various possible configurations of phase distribution showing (**a**) liquid water-filled L pores, (**b**) interfacial pore I (PE) where gas–liquid interface lies inside the pore, (**c**) interfacial pore I (CE) where gas–liquid interface lies at the entrance of the pore I (CE) pore, and (**d**) gas-filled G pores

bodies can be identified: gas-saturated pores, denoted by G, liquid water-saturated pores, denoted by L, and pores at gas–liquid interface, denoted by I. The latter may be further sub-divided into completely empty (CE) and partially empty (PE) pores as shown schematically in Fig. 3.28. Liquid water evaporates from the gas–liquid interface at a rate governed by the mass transport in gas phase. The evaporation flux at I-type pores can be given by

$$Q_{ij}^{\text{evap}} = D\frac{C_i - C_j}{l(i,j)} \tag{3.54}$$

where D is the water vapor diffusivity in gas phase and C_i and C_j the vapor-phase concentration in a pore with non-zero liquid water saturation, which is equal to saturated water vapor concentration, and in a gas-filled pore, respectively. Water vapor concentration in a gas-filled pore, C_i, can be obtained by solving the mass conservation:

$$\sum_n DA_{\text{th},\,ij}^2 \frac{C_i - C_j}{l(i,j)} = 0 \tag{3.55}$$

where $A_{\text{th},ij}$ denotes the cross-sectional area of the connecting throat, C_j the water vapor concentration in neighboring pore, and n the total number of neighboring pores. Since purge gas is a wetting phase for a hydrophobic GDL, invasion of purge gas into a liquid water-filled pore during gas purge is equivalent to imbibition process. The following condition must be satisfied for purge gas invasion:

$$P_{\text{g}} - P_{\text{l}} > |P_{\text{c}}| \tag{3.56}$$

where P_{g} and P_{l} denote the gas and the liquid water pressure, respectively, and P_{c} represents the capillary pressure at the interface. Lenormand and Zarcone [108] among others have shown that pore filling by a wetting phase is a complex phenomenon during imbibition and is governed by the largest radius of curvature that can be achieved. This depends on the number of adjacent throats filled with non-wetting phase (liquid water for hydrophobic GDL). A pore with coordination number z can thus be filled by z–1 possible events, I_1 to I_{z-1}, each occurring at different capillary pressures, as shown schematically in Fig. 3.29 [109]. Various parametric models were presented to account for the above-mentioned cooperative pore body filling mechanism. In the present work, a parametric model proposed by Hughes and Blunt [110] is implemented to compute capillary pressure, $P_{\text{c},n}$, for an I_n mechanism:

$$P_{\text{c},n} = \frac{C_{In}\sigma\cos\theta}{r_{\text{p}}};\; C_{I1} = 2 \quad \text{and} \quad C_{I1} > C_{I2} > \cdots > C_{In-1} \tag{3.57}$$

where r_{p} is the pore radius and C_{In} the input parameters. As shown from (3.57), for I_1 mechanism, capillary pressure is maximum that makes I_1 the most favored event. With increase in the non-wetting phase-filled neighboring throats, capillary

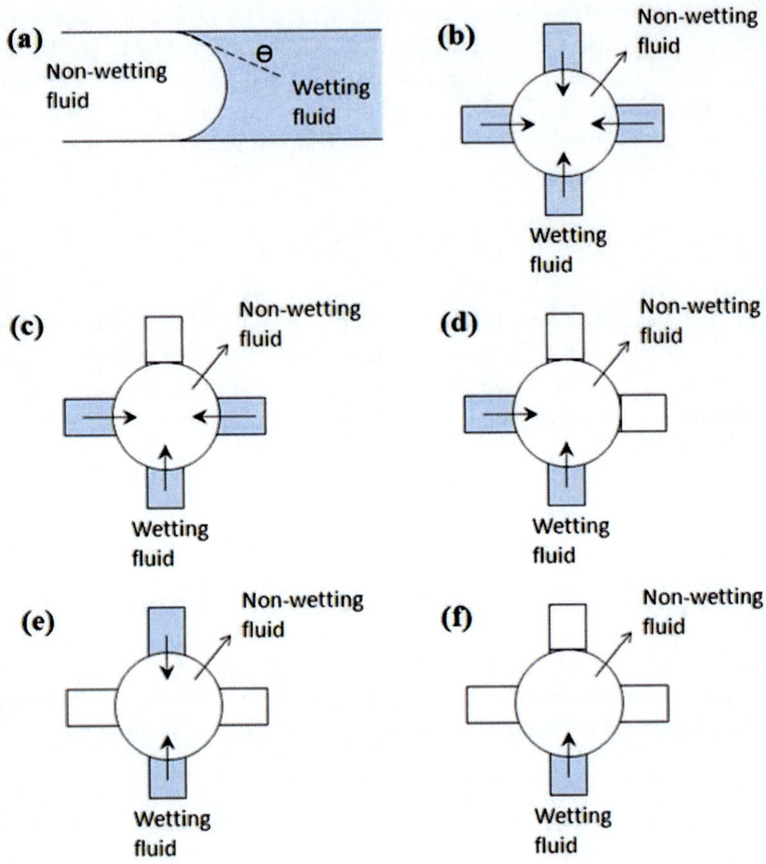

Fig. 3.29 Schematic representation of (**a**) throat filling and pore filling during imbibition (after [109]), (**b**) I_0 mechanism, not possible for incompressible fluids, (**c**) I_1 mechanism, (**d**) and (**e**) I_2 mechanism, and (**f**) I_3 mechanism

pressure decreases making the pore invasion less likely. During imbibition, another displacement mechanism called snap-off may occur due to wetting films swelling to an extent that the interface becomes unstable. Formation of wetting films along the corners is governed by the Concus and Finn [84] condition, given by (3.37). Thus, a contact angle between liquid water and carbon fibers varying in the range of 60°–130° and square cross section of pores and throats, as considered in the present work, suppress the existence of wetting films along the corners and hence the possibility of snap-off displacement mechanism. It should be mentioned that a small fraction of pores in the actual pore spaces of carbon paper may allow the formation of wetting films. Incorporation of wetting film flow, in accordance with the Concus and Finn [84] condition, and snap-off mechanisms in a topologically equivalent pore network structure of carbon paper is envisioned as a future extension of the present PN model.

Each phase, then, must obey volume conservation within each pore body:

$$V_i \frac{\partial S_i^{\alpha}}{\partial t} + \sum_{j \in N_i} Q_{ij}^{\alpha} = 0, \qquad i = 1, 2, \ldots, N \tag{3.58}$$

where S_i^{α} is the local saturation of phase α in pore i, N the total number of pores, V_i the volume of pore i, N_i the number of pores connected to pore i, and Q_{ij}^{α} the flow rate of phase α between pore i and pore j. The flow rate of any phase through a throat depends on the fluid configuration in the pores connected to that throat. The volumetric flow rate across a single phase throat, i.e., a throat connecting two gas-filled or liquid water-filled pores, is given by

$$Q_{ij}^{\alpha} = g_{ij}^{\alpha} \left(P_i - P_j\right) \quad \text{where} \quad g_{ij}^{\alpha} = \frac{2\, r_{\text{th}}(i,j)^4}{\pi\, \mu_{\alpha}\, l\,(i,j)}; \ \alpha = \text{liquid water, gas} \tag{3.59}$$

whereas in the interfacial pores purge gas can flow only if the pressure difference across the interface is higher than the capillary barrier pressure provided by the interface. If the aforementioned condition is not fulfilled, the throat and adjoining pore become "capillary blocked." However, as long as evaporation continues there is always a net liquid water flow toward the interface and can be accounted for as follows:

$$Q_{ij}^{\text{w}} = \begin{cases} g_{\text{w}} \left(P_i - P_j - P_{\text{c}}\right) + D \frac{M_{\text{w}} A_{\text{th},ij}^2}{\rho_{\text{l}}} \frac{(C_i - C_j)}{l(i,j)} & \text{if } \left(P_i - P_j - P_{\text{c}}\right) > 0 \\ D \frac{M_{\text{w}} A_{\text{th},ij}^2}{\rho_{\text{l}}} \frac{(C_i - C_j)}{l(i,j)} & \text{otherwise} \end{cases} \tag{3.60}$$

where C_i and C_j denote the water vapor concentration in pore i and j, respectively, M_{w} the molecular weight of water, and ρ_{l} the density of liquid water. The first term on the right-hand side of (3.60) accounts for the hydraulic flow of purge gas and the second term for the evaporative flux toward gas pores. It should be mentioned that the volumetric flow rate of liquid water, as represented by (3.58) and (3.59), also accounts for the evaporation-induced viscous flow of liquid water toward gas–liquid water interface, also referred to as capillary-pumping phenomena. Once the pressure field is obtained, time steps are chosen so that only one pore completely dries out during any time step. Using the value of current time step, liquid water saturation is updated in all the pores. Water vapor concentration is accordingly updated for the current phase distribution in the network. Using the extended Hoshen–Kopelman algorithm [111], various liquid water clusters are identified. All the interfacial pores are then tested for the instability of liquid water–air interface using the most recent pressure field. If the pressure drop across an interface is larger than the corresponding capillary barrier pressure, the interface is called "unstable," and purge gas invades the interfacial liquid water-filled pore. If there is no such interfacial pore in a specific liquid water cluster, the pore with least capillary barrier pressure is made "open" for purge gas flow. It should be noted that at any time step purge gas should invade the pores with least capillary barrier pressure in all the liquid water clusters unlike imbibition process where only one pore is invaded by wetting phase at any time step. Throat conductivities are updated accordingly. This algorithm is run till GDL dries out completely.

The above-described model is deployed for a GDL initially saturated with liquid water and the outlet face in contact with partially humidified purge gas. A constant pressure boundary condition with pressure equal to 1 atm is imposed on the outlet face, whereas GDL membrane interface is subjected to wall boundary condition. All the lateral faces are subjected to periodic boundary condition to eliminate the network size effect.

Physical insight can be gained from numerical visualization of liquid water front movement within the network during GDL drying, as depicted in Fig. 3.30. It is found that the drying front recedes in the GDL as soon as drying starts which is in contrast to the drying characteristics of a completely saturated hydrophilic porous medium [108, 112]. Drying front movement is controlled by vapor-phase transport ahead of the drying front and liquid water transport behind it. In a hydrophobic medium capillary transport of liquid water behind the drying front is substantially weaker than in a hydrophilic medium, resulting in an instantaneous receding of the

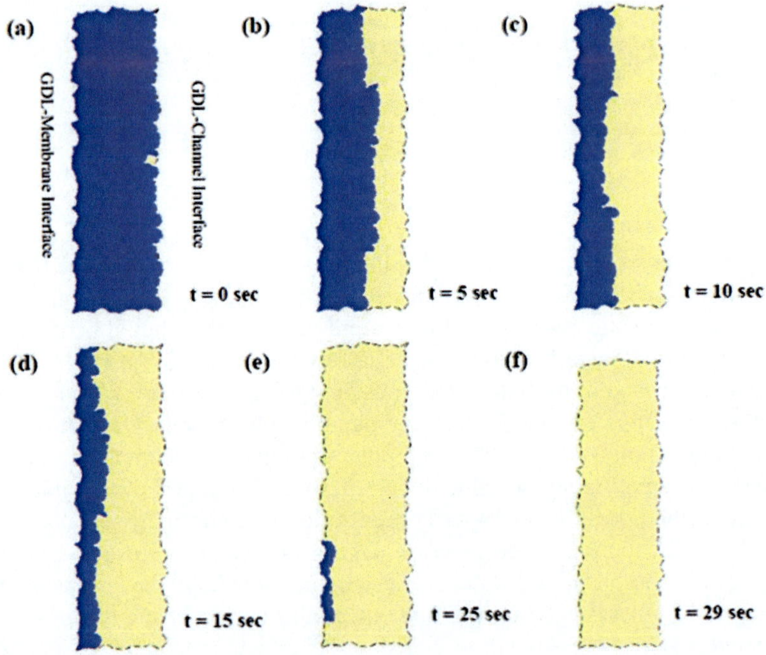

Fig. 3.30 Evolution of drying front with purge time in a uniformly hydrophobic GDL completely saturated with liquid water. Purge gas RH is 40% and temperature is 55°C

drying front in the hydrophobic GDL. As drying proceeds, the drying front evolves inside the GDL with a compact morphology as shown in Fig. 3.30. Since purge gas is a wetting phase for the hydrophobic GDL, it will invade through the pores that provide smallest radius of curvature. As explained in the previous section, invasion of a wetting phase in a pore depends not only on its radius but also on the saturation distribution in the neighboring pores. Once liquid water is removed from a pore, the probability of purge gas invasion in a neighboring pore in the same plane is substantially higher than in a pore in an adjacent plane, yielding a compact morphology for the drying front. Figure 3.31 shows the evolution of cross-sectional averaged saturation profiles along the GDL thickness. As can be seen, there is no change in liquid water saturation at the GDL–membrane interface while the layers close to the GDL–channel interface are being dried out. The corresponding saturation profiles, as shown in Fig. 3.31, are typical of a smooth drying front propagation. The saturation profiles predicted by the present PN model are also found to be consistent with the recent experimental investigation of drying in a hydrophobic porous medium [113].

For typical fuel cell applications, GDL is treated with PTFE to induce and/or enhance hydrophobicity. However, possible anomaly during PTFE treatment, surface defects, and aging can substantially alter the wetting characteristics. To elucidate the effect of GDL wettability distribution on drying characteristics, the present PN model is further extended to incorporate GDL contact angle variation from 92° to 130°. Since an accurate procedure to quantify the contact angle

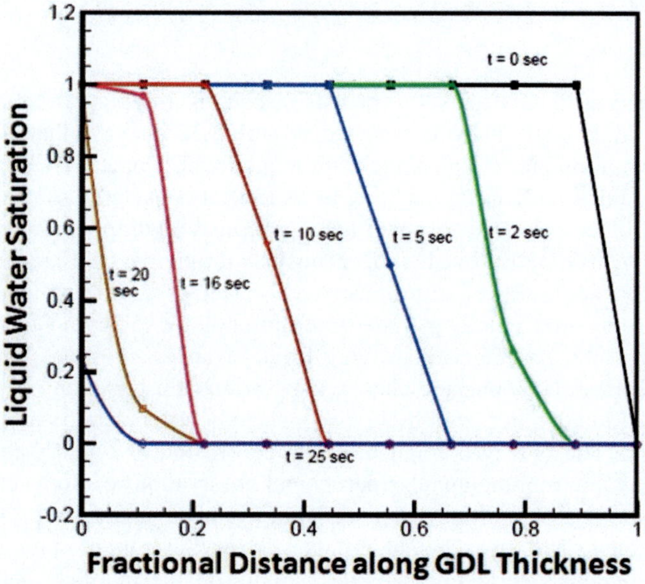

Fig. 3.31 Evolution of cross-sectional averaged saturation profile along GDL thickness direction as a function of purge time

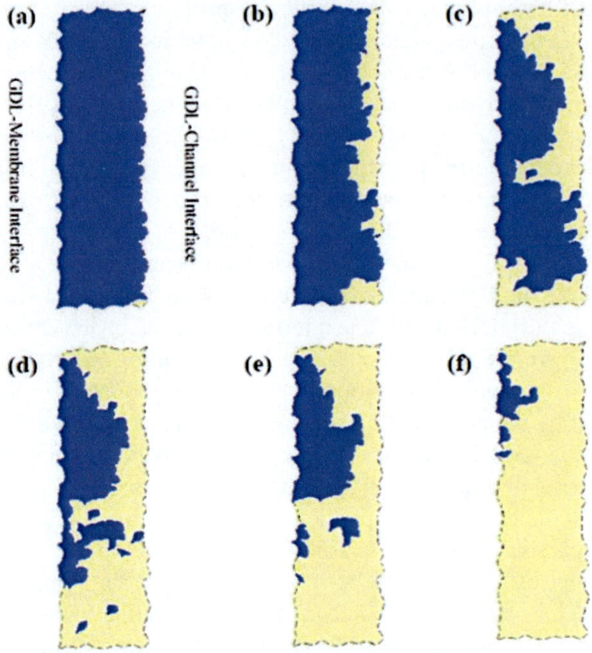

Fig. 3.32 Evolution of drying front with purge time in a hydrophobic GDL, with varying contact angle in the range of 92°–130°, completely saturated with liquid water. Purge gas RH is 40% and temperature is 55°C

distribution inside the GDL is yet to be established, as a first approximation, random contact angle distribution is assumed. Figure 3.32 displays the evolution of drying front with time for a typical realization of a hydrophobic GDL with varying contact angle. For simplicity of analysis, lands are not considered. As can be seen, the evolution of drying front in a non-homogeneously hydrophobic GDL is substantially different from that in a homogeneously hydrophobic GDL, represented by a single contact angle. Pores with higher contact angle provide preferential sites for purge gas invasion entailing finger-like morphology to drying front, as displayed in Fig. 3.32. As finger-like drying front advances, it erodes liquid water, initially present as one connected cluster, into various clusters many of which are disconnected. Liquid water from these disconnected clusters can be removed only by evaporation. The observed fragmentation and erosion of liquid water is similar to the X-ray microtomographic experimental observations of Sinha et al. [114]. Figure 3.33 shows the resulting cross-sectional averaged saturation profiles along the GDL thickness. The advancement of finger-like drying front morphology toward the GDL–membrane interface creates desaturation at the interface before liquid water is completely removed from the GDL, as shown in Fig. 3.31. Comparison of Figs. 3.33 and 3.31 clearly shows the differences in the evolution of saturation profiles for a non-homogeneously hydrophobic and a homogeneously hydrophobic

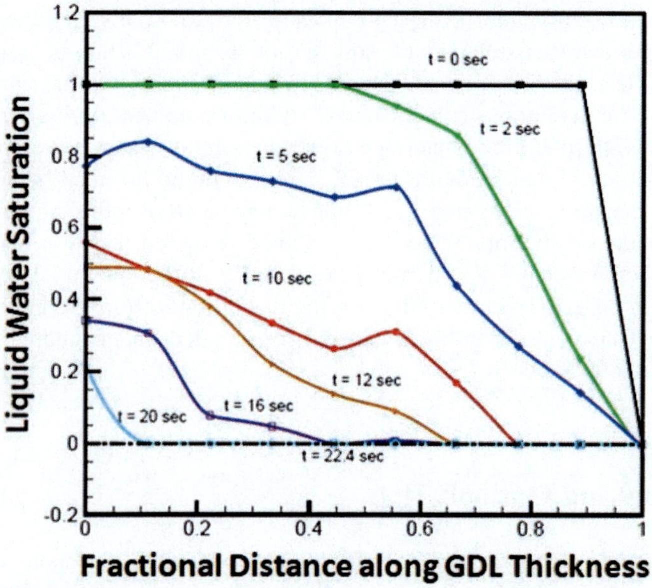

Fig. 3.33 Evolution of cross-sectional averaged saturation profile along GDL thickness direction as a function of purge time for a hydrophobic GDL with varying contact angle in the range of 92°–130°

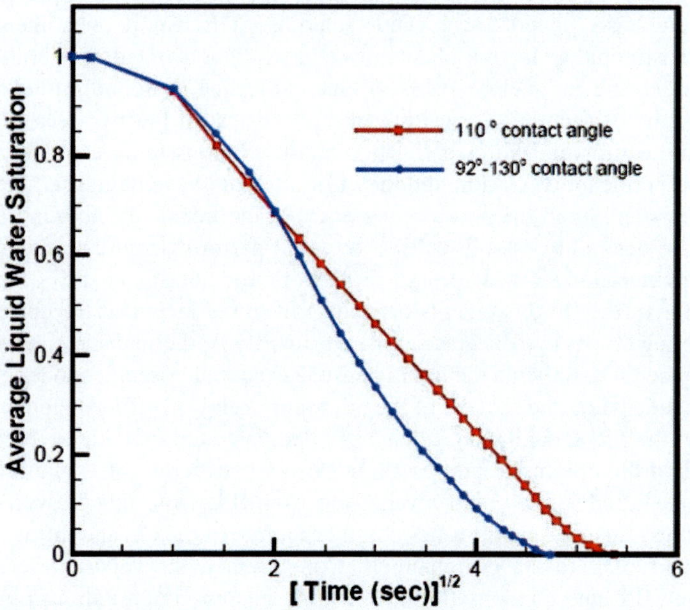

Fig. 3.34 Temporal variation of liquid water saturation during drying for GDLs with 110° uniform contact angle and varying contact angle in the range of 92°–130°

GDL. Figure 3.34 compares the drying curves for the two GDLs. Significant departures from the linear variation with the square root of time for a non-homogeneously hydrophobic GDL are observed and can be attributed to the evolution of finger-like morphology. The evolution of finger-like morphology renders drying front propagation to the GDL–membrane interface even when liquid water is present in GDL and thus can incur higher membrane HFR, a representation of membrane dryness, during GDL drying as compared to a compact drying front morphology observed in a homogeneously hydrophobic GDL. It should be noted that in a realistic fuel cell drying phenomena, GDL is partially flooded. Therefore, the initial liquid water morphology in GDL may further affect drying front propagation. Additionally, drying of catalyst layer needs to be considered for detailed understanding of drying phenomena in a fuel cell.

5 Summary and Outlook

Pore-scale modeling can drastically enhance the fundamental understanding of the intricate structure–wettability–transport interactions in porous media. In this chapter, we have specifically discussed the pore network (PN) and lattice Boltzmann (LB) modeling techniques in the context of two-phase transport in the PEFC porous structures. The multi-functionality of porous layers in the PEFC demands detailed understanding of various transport phenomena occurring at various scales in the electrode and the gas diffusion layer. In recent years, there has been increased emphasis on pore-scale understanding of transport phenomena in the PEFC microstructures. In this chapter, we have demonstrated that both the PN and LB models are uniquely capable of elucidating the profound influence of the underlying microstructure and wetting characteristics on the two-phase transport behavior and estimating the essential two-phase relations (e.g., capillary pressure and relative permeability). Although the efforts to date have delineated key fundamental understanding in the porous components, significant advances in pore-scale modeling are desired to establish PEFC material performance relation and to provide a computer-aided virtual design framework for optimizing PEFC materials. The need for further improvements originates from the facts that the current pore-scale modeling efforts in PEFC research are primarily isothermal and consider water invasion in the GDL only via the liquid phase. Recent macroscopic modeling efforts focused on understanding the role of the microporous layer (MPL) on the two-phase transport in the PEFC sandwich. It has been demonstrated that due to the presence of highly hydrophobic microporous layer (MPL) between the CL and GDL, it is unlikely that liquid water could invade into the MPL from the CL via capillary transport [115]. Instead, liquid water could evaporate in the CL and diffuse through the MPL and GDL to the gas channels. Concurrently, the vapor-phase transport through the MPL and condensation in the GDL become important, as highlighted by the macroscopic studies by Burlatsky et al. [116] and Caulk and Baker [117]. These investigations, however, did not elucidate the ensuing liquid water transport from the condensation front to gas channels. It is conjectured that there is a strong need to develop pore-scale modeling formalisms which can take into account the aforementioned water transport physics in the PEFC research and hence warrants the need for detailed consideration of heat and vapor transport in the pore-scale models.

Acknowledgments PPM acknowledges the strategic research funding in energy storage and conversion from Oak Ridge National Laboratory, managed by UT-Battelle LLC for the U.S. Department of Energy. The authors acknowledge Elsevier and the Electrochemical Society for the figures reproduced in this chapter from the referenced publications of their respective journals.

References

1. S. Pacala and R. Socolow, *Science*, **305**(5686), 968, 2004.
2. C.-Y. Wang, *Chem. Rev.*, **104**, 4727, 2004.
3. P. Baveye and G. Sposito, *Water Resour. Res.*, **20**, 521, 1984.
4. J. Bear, *Dynamics of Fluids in Porous Media*, Dover, New York, NY, 1972.
5. M. A. Celia, H. Rajaram, and L. A. Ferrand, *Adv. Water Res.*, **16**, 81, 1993.
6. C. T. Miller, G. Christakos, P. T. Imhoff, J. F. McBride, J. A. Pedit, and J. A. Trangenstein, *Adv. Water Res.*, **21**, 77, 1998.
7. S. Whitaker, *Transp. Porous Media*, **1**, 25, 1986.
8. F. A. L. Dullien, *Porous Media: Fluid Transport and Pore Structure*, Academic, San Diego, CA, 1992.
9. H. Rumpf and A. R. Gupte, *Chemie In-genieur Technik*, **43**, 367, 1971.
10. P. C. Carman, *Trans. Inst. Chem. Eng. (London)*, **15**, 150, 1937.
11. P. Z. Carman, *Flow of Gases Through Porous Media*, Butterworths, London, 1956.
12. C. Pan, Ph.D. Dissertation, University of North Carolina, Chapel Hill, NC, 2003.
13. I. Fatt, *Trans. AIME*, **207**, 144, 1956.
14. I. Fatt, *Trans. AIME*, **207**, 160, 1956.
15. I. Fatt, *Trans. AIME*, **207**, 164, 1956.
16. L. A. Ferrand and M. A. Celia, *Water Resour. Res.*, **28**, 859, 1992.
17. M. I. Lowry and C. T. Miller, *Water Resour. Res.*, **31**, 455, 1995.
18. P. C. Reeves and M. A. Celia, *Water Resour. Res.*, **32**, 2345, 1996.
19. G. G. Pereira, W. V. Pinczewski, D. Y. C. Chan, L. Paterson, and P. E. Oren, *Transp. Porous Media*, **24**, 167, 1996.
20. G. G. Pereira, *Phys. Rev. E*, **59**, 4229, 1999.
21. J. S. Andrae, Jr., D. A. Street, Y. Shibusa, S. Havlin, and H. E. Stanley, *Phys. Rev. E*, **55**, 772, 1997.
22. L. Li, C. A. Peters, and M. A. Celia, *Adv. Water Resour.*, **29**, 1351, 2006.
23. D. Kavetski, C. A. Peters, M. A. Celia, and B. Lindquist, *Geochim. Cosmochim. Acta*, **71**, A470, 2007.
24. X. Li and Y. C. Yortsos, *AIChE J.*, **41**, 214, 1995.
25. Y. L. Bray and M. Prat, *Int. J. Heat Mass Transfer*, **42**, 4207, 1999.
26. A. G. Yortsos, A. G. Boudouvis, A. K. Stubos, I. N. Tsimpanogiannis, and Y. C. Yortsos, *Phys. Rev. E*, **68**, 037303, 2003.
27. Q. Kang, I. N. Tsimpanogiannis, D. Zhang, and P. C. Lichtner, *Fuel Process. Technol.*, **86**, 1647, 2005.
28. K. E. Thompson, *AIChE J.*, **48**, 1369, 2002.
29. D. A. Wolf-Gladrow. *Lattice Gas Cellular Automata and Lattice Boltzmann Models: An Introduction*, Springer, Heidelberg, 2000.
30. F. H. Harlow and J. E. Welch, *Phys. Fluids*, **8**, 2182, 1965.

31. B. J. Daly, *Phys. Fluids*, **12**, 1340, 1969.
32. C. W. Hirt and B. D. Nichols, *J. Comput. Phys.*, **39**, 201, 1981.
33. J. A. Sethian, *Level Set Methods: Evolving Interfaces in Geometry, Fluid Mechanics, Computer Vision, and Materials Science*, Cambridge University Press, 1996.
34. S. O. Unverdi, G. Tryggvason, *J. Comput. Phys.*, **100**, 25, 1992.
35. G. Tryggvason, B. Bunner, A. Esmaeeli, D. Juric, N. Al-Rawahi, W. Tauber, J. Han, S. Nas, and Y. J. Jan, *J. Comput. Phys.*, **162**, 708, 2001.
36. G. A. Bird, *Molecular Gas Dynamics and the Direct Simulation of Gas Flows*, Clarendon, Oxford, 1994.
37. D. C. Rapaport, *The Art of Molecular Dynamics Simulation*, Cambridge University Press, 1995.
38. S. Chen and G. Doolen, *Annu. Rev. Fluid Mech.*, **30**, 329, 1998.
39. A. K. Gunstensen, D. H. Rothman, S. Zaleski, and G. Zanetti, *Phys. Rev. A*, **43**, 4320, 1991.
40. X. Shan and H. Chen, *Phys. Rev. E*, **47**, 1815, 1993.
41. X. Shan and H. Chen, *Phys. Rev. E*, **49**, 2941, 1994.
42. M. Swift, W. Osborn, and J. Yeomans, *Phys. Rev. Lett.*, **75**, 830, 1995.
43. M. Swift, S. Orlandini, W. Osborn, and J. Yeomans, *Phys. Rev. E*, **54**, 5041, 1996.
44. R. Nourgaliev, T. Dinh, T. Theofanous, and D. Joesph, *Int. J. Multiphase Flow*, **29**, 117, 2003.
45. X. He, S. Chen, and R. Zhang, *J. Comput. Phys.*, **152**, 642, 1999.
46. X. He, R. Zhang, S. Chen, and G. D. Doolen, *Phys. Fluids*, **11**, 1143, 1999.
47. Q. Kang, D. Zhang, and S. Chen, *Phys. Fluids*, **14**, 3203, 2002.
48. Q. Kang, D. Zhang, and S. Chen, *Adv. Water Resour.*, **27**, 13, 2004.
49. Q. Kang, D. Zhang, and S. Chen, *J. Fluid Mech.*, **545**, 41, 2005.
50. C. Pan, M. Hilpert, and C. T. Miller, *Water Resour. Res.*, **40**, 1, 2004.
51. H. Li, C. Pan, and C. T. Miller, *Phys. Rev. E*, **72**, 026705, 2005.
52. P. Yuan and L. Schaefer, *Phys. Fluids*, **18**, 042101, 2006.
53. S. Gottesfeld and T. A. Zawodzinski, in *Advances in Electrochemical Science and Engineering*, C. Tobias, Editor, Wiley, New York, NY, **5**, 195, 1997.
54. C. Y. Wang, in *Handbook of Fuel Cells – Fundamentals, Technology and Applications*, W. Vielstich, A. Lamm, and H. A. Gasteiger, Editors, Wiley, Chichester, **3**, 337, 2003.
55. A. Z. Weber and J. Newman, *Chem. Rev.*, **104**, 4679, 2004.
56. U. Pasaogullari and C. Y. Wang, *J. Electrochem. Soc.*, **151**, 399, 2004.
57. J. Nam and M. Kaviany, *Int. J. Heat Mass Transfer*, **46**, 4595, 2003.
58. K. S. Udell, *Int. J. Heat Mass Transfer*, **28**, 485, 1985.
59. M. C. Leverett, *AIME Trans.*, **142**, 152, 1941.
60. J. T. Gostick, M. W. Fowler, M. A. Ioannidis, M. D. Pritzker, Y. M. Volfkovich, and A. Sakars, *J. Power Sources*, **156**, 375, 2006.
61. J. D. Fairweather, P. Cheung, J. St-Pierre, and D. T. Schwartz, *Electrochem. Commun.*, **9**, 2340, 2007.
62. T. V. Nguyen, G. Lin, H. Ohn, and X. Wang, *Electrochem. Solid-State Lett.*, **11**, B127, 2008.
63. K. G. Gallagher, R. M. Darling, T. W. Patterson, and M. L. Perry, *J. Electrochem. Soc.*, **155**, B1225, 2008.
64. I. S. Hussaini and C. Y. Wang, *J. Power Sources*, **195**, 3830, 2010.
65. M. F. Mathias, J. Roth, J. Fleming, and W. Lehnert, in *Handbook of Fuel Cells – Fundamentals, Technology and Applications*, W. Lietsich, A. Lamm, and H. A. Gasteiger, Editors, Ch. 42, Wiley, Chicester, **3**, 517, 2003.
66. P. Bhatnagar, E. Gross, and M. Krook, *Phys. Rev.*, **94**, 511, 1954.
67. H. Chen, S. Chen, and W. H. Matthaeus, *Phys. Rev. A*, **45**, R5339, 1992.
68. N. Martys and H. Chen, *Phys. Rev. E*, **53**, 743, 1996.
69. X. Shan and G. D. Doolen, *Phys. Rev. E*, **54**, 3614, 1996.
70. X. Shan and G. D. Doolen, *J. Stat. Phys.*, **81**, 379, 1995.
71. S. Hou, X. Shan, Q. Zou, G. Doolen, and W. Soll, *J. Comput. Phys.*, **138**, 695, 1997.

72. Z. L. Yang, T. N. Dinh, R. R. Nourgaliev, and B. R. Sehgal, *Int. J. Heat Mass Transfer*, **44**, 195, 2001.
73. P. P. Mukherjee, Ph.D. Dissertation, Pennsylvania State University, 2007.
74. D. Tiab and E. Donaldson, *Petrophysics: Theory and Practice of Measuring Reservoir Rock and Transport Properties*, Gulf Publishing Company, Houston, TX, 1996.
75. P. P. Mukherjee, C. Y. Wang, and Q. Kang, *Electrochim. Acta*, **54**, 6861, 2009.
76. P. P. Mukherjee, P. K. Sinha, and C. Y. Wang, *J. Mater. Chem.*, **17**, 3089, 2007.
77. P. P. Mukherjee, Q. Kang, and C. Y. Wang, *Energy Environ. Sci. – R. Soc. Chem.*, DOI: 10.1039/B926077C, 2010.
78. K. Schladitz, S. Peters, D. Reinel-Bitzer, A. Wiegmann, and J. Ohser, *Comput. Mater. Sci.*, **38**, 56, 2006.
79. V. P. Schulz, J. Becker, A. Wiegmann, P. P. Mukherjee, and C. Y. Wang, *J. Electrochem. Soc.*, **154**, B419, 2007.
80. R. L. Borup et al., *Chem. Rev.*, **107**, 3904, 2008.
81. D. L. Wood III and R. L. Borup, in *Polymer Electrolyte Fuel Cell Durability*, F. N. Buchi, M. Inaba, and T. J. Schmidt, Editors, Springer, New York, NY, 159, 2009.
82. P. K. Sinha and C. Y. Wang, *Electrochim. Acta*, **52**, 7936, 2007.
83. M.A. Ioannidis and I. Chatzis, *Chem. Eng. Sci.*, **48**, 951, 1993.
84. P. Concus and R. Finn, *Appl. Math. Sci.*, **63**, 292, 1969.
85. T. W. Patzek and D. B. Silin, *J. Colloid Interface Sci.*, **236**, 295, 2001.
86. R. Lenormand, E. Touboul, and C. Zarcone, *J. Fluid Mech.*, **189**, 165, 1998.
87. R. Lenormand, *J. Phys.: Condens. Matter*, **2**, SA 79, 1990.
88. M. Ferer, G. S. Bromhal, and D. H. Smith, *Physica A*, **319**, 11, 2003.
89. M. Ferer, G. S. Bromhal, and D. H. Smith, *Phys. Rev. E*, **67**, 051601, 2003.
90. S. Litster, D. Sinton, and N. Djilali, *J. Power Sources*, **154**, 95, 2006.
91. A. Bazylak, D. Sinton, Z. S. Liu, and N. Djilali, *J. Power Sources*, **163**, 784, 2007.
92. A. Z. Weber, R.M. Darling, and J. Newman, *J. Electrochem. Soc.*, **151**, 1715, 2004.
93. P. K. Sinha and C. Y. Wang, *Chem. Eng. Sci.*, **63**, 1081, 2008.
94. H. Dong, M. J. Blunt, *Phys. Rev. E*, **80**, 036307, 2009.
95. G. Luo, H. Ju, and C. Y. Wang, *J. Electrochem. Soc.*, **154**, B316, 2007.
96. K. Yoshizawa, K. Ikezoe, Y. Tasaki, D. Kramer, E. H. Lehmann, and G. G. Schererb, *J. Electrochem. Soc.*, **155**, B223, 2008.
97. S. Bryant and M. J. Blunt, *Phys. Rev. A*, **46**, 1992, 2004.
98. G. Mason and N. R. Morrow, *J. Colloid Interface Sci.*, **141**, 262, 1991.
99. L. S. Nilsen, P. E. Oren, and S. Bakke, *SPE J.*, **2**, 136, 1997.
100. G. Luo, Y. Ji, C. Y. Wang, and P. K. Sinha, *Electrochim. Acta*, **55**, 5332, 2010.
101. M. Oszcipok, D. Reimann, U. Kronenwett, M. Kreideweis, and M. Zedda, *J. Power Sources*, **145**, 407, 2005.
102. L. Mao and C. Y. Wang, *J. Electrochem. Soc.*, **154**, B139, 2007.
103. L. Mao, C. Y. Wang, and Y. Tabuchi, *J. Electrochem. Soc.*, **154**, B341, 2007.
104. K. Tajiri, Y. Tabuchi, and C. Y. Wang, *J. Electrochem. Soc.*, **154**, B147, 2007.
105. S. Ge and C. Y. Wang, *Electrochem. Solid-State Lett.*, **9**, A499, 2006.
106. S. Ge and C. Y. Wang, *Electrochim. Acta*, **52**, 4825, 2007.
107. A. G. Yiotis, A. K. Stubos, A. G. Boudouvis, and Y. C. Yortsos, *Adv. Water Resour.*, **24**, 439, 2001.
108. R. Lenormand and C. Zarcone, *SPE J.*, 13264, 1984.
109. K. Mogensen and E. H. Stenby, *Transp. Porous Media*, **32**, 299, 1998.
110. R. G. Hughes and M. J. Blunt, *Transp. Porous Media*, **40**, 295, 2000.
111. A. Al-Futaisi and T. W. Patzek, *Physica A*, **321**, 665, 2003.
112. M. Prat, *Int. J. Multiphase Flow*, **19**, 691, 1993.
113. N. Shahidzadeh-Bonn, A. Azouni, and P. Coussot, *J. Phys.: Condens. Matter*, **19**, 112101, 2007.
114. P. K. Sinha, P. Halleck, and C. Y. Wang, *Electrochem. Solid-State Lett.*, **9**, A344, 2006.

115. J. P. Owejan, J. E. Owejan, W. Gu, T. A. Trabold, T. W. Tighe, and M. Mathias, *J. Electrochem. Soc.*, **157**, B1456, 2010.
116. S. F. Burlatsky, V. V. Atrazhev, M. Gummalla, D. A. Condit, and F. Liu, *J. Power Sources*, **190**, 485, 2009.
117. D. A. Caulk and D. R. Baker, *J. Electrochem. Soc.*, **157**, B1237, 2010.

Chapter 4
Micro/Nano Mechanics of Contact of Solids

M.S. Bobji, U.B. Jayadeep, and K. Anantheshwara

Abstract Solid objects in contact with each other are omnipresent in technological applications. Due to the presence of unavoidable surface irregularities in the form of asperities, the contact is established at many locations in the form of islands, whose sizes vary from atomic scale to the size of the contacting objects. Understanding the mechanics of these contacts is important in predicting the friction and wear of the contacting surfaces that move relative to each other. Since the individual islands are below few microns for most of the current engineering products the mechanics at micro/nano-scales dominate the resultant macroscopic friction.

Some of the issues in the mechanics of contact at small scales are explored in this chapter through examples. First is an experimental study on sliding of single asperity contact carried out in a high-resolution transmission electron microscope. The importance of the surface forces that are typically neglected in macroscopic analysis is clearly brought out in this experiment. The second example is a case study on introducing the van der Waals force contribution in an elastic finite element framework.

Keywords Contact mechanics · Tribology · Single asperity experiments · van der Waals force · Adhesion · Nanomechanics

1 Introduction

When two solids come close to each other, contact is established at different locations depending on the geometry of the solids. For example, an ideal sphere will establish contact with an ideal flat surface at a point, while a cylinder will establish contact over a line. In real life, things are different because the solid bodies deform and the geometry is not perfect. Deformable bodies establish contact over an area rather than a point or a line. Sphere will have a circular contact area with a flat surface, while a cylinder will have a rectangular contact area [1]. Assuming that the bodies deform only elastically, the contact problem can be posed within

M.S. Bobji (✉)
Department of Mechanical Engineering, Indian Institute of Science, Bangalore 560 012, India
e-mail: bobji@mecheng.iisc.ernet.in

S. Chakraborty (ed.), *Mechanics Over Micro and Nano Scales*,
DOI 10.1007/978-1-4419-9601-5_4, © Springer Science+Business Media, LLC 2011

the ambit of the theory of elasticity, and can be solved analytically for the simple geometries [1, 2].

Apart from the macroscopic form-error in the geometry, the microscopic geometry local to the contact is defined by surface roughness [3]. Thus, when a real spherical ball touches a flat surface, contact is established over a large number of tiny areas. Analysis of the rough surface contact is complicated and often need recourse to elaborate numerical methods. Here, when a rough surface is mentioned, reference is not made to a surface like that of a rock. Rather, "rough surfaces" are used to emphasize cases where smooth surface approximations cannot be used to yield any useful practical knowledge. Any engineering surface we encounter in practice, even if it is mirror finished, is rough at the smaller scales [3, 4].

Contacts are of practical importance to human kind from time immemorial. Any instrument, device, or appliance built by humans will invariably have contacts. Some contacts are static throughout the lifetime of the component like, for example, contacts between the gold interconnects and the semiconductor devices on an integrated electronic chip, while others are dynamic with a relative motion between two contacting surfaces like, for example, the contact between the tire of an automobile and the road. Designing the contacts properly will increase the lifetime and the reliability of the devices.

A variety of phenomena take place in any practical contact. These phenomena that affect the way in which the contacts behave are quiet complex. Understanding all those phenomena to the extent that they give a predictive capability has been a long-standing endeavor of tribologists. Typically, all these phenomena are lumped together in either coefficient of friction [5] or coefficient of restitution. These coefficients help in a practical way in designing components through experimental observations and empirical relations. However, when it comes to new situations where experimentation is impossible or difficult these coefficients do not offer any insights. For example, better understanding of friction between two tectonic plates will help us to predict earthquakes. In microelectromechanical devices, lack of understanding of friction at small scales leads to a lot of rejection during manufacturing and reliability issues during operation.

In this chapter we deal with the mechanics of contacts at micro- and nanoscale. Starting with the definition of contact at atomic scales, some issues both in modeling and experimental observation are elaborated with examples. The chapter is aimed at giving a short peek to a reader with general mechanical engineering background into the issues involved in contact at small scales.

2 An Overview

The nature of the real contact is very complicated. Figure 4.1 shows the way in which we try to understand this nature and apply them. We understand nature, in general, through two ways – experimental observations and abstractions. The knowledge gained is converted to approximate models. These models give us a

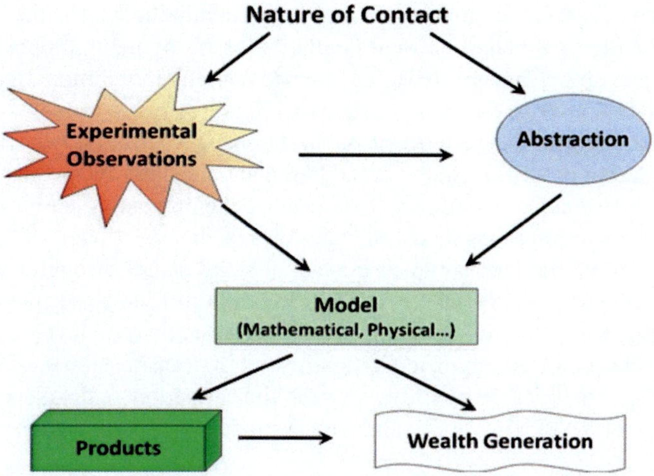

Fig. 4.1 Exploration of contact mechanics at micro/nanoscale – an overview

short-term predictive capability that helps us to apply the knowledge to practical situations. This capability helps us in developing products that will create wealth for betterment of the human life.

Experimental observations in contact mechanics are difficult because the contact is inherently hidden. The events that are taking place at the contact range from atomic dimension to the size of the contact which could be of few millimeters in typical engineering applications. The timescales involved will range from femtoseconds for atomic processes like bond formation to seconds. Any experimental observation will have a finite bandwidth and hence will be looking at filtered data. Typically we tune the experiment to be sensitive to one or few particular events, and this in turn will result in obscuring other events.

Abstractions help us to understand the nature clearly without having to be cluttered with the noisy data. Various theories developed help us to reduce the need to revert to expensive experimentation repeatedly. The complexities of practical situations often make it impossible to apply these theories to provide any solution within reasonable effort. Often, an approximate engineering solution is sufficient to give enough predictive capability to be able to design and use devices. We revert to models to help us in obtaining the solutions with necessary accuracy. The abstraction and the experimental observations help us in making the assumptions to simplify the analysis.

Often, products and solutions are developed even before the complete understanding of the issues involved through models and abstractions. For example, we have developed and have been using various knots to tie things together even before the concept of friction has been understood in a formal framework. One of the phenomena that made this possible is the fact that the coefficient of friction between two dry surfaces varies only between 0.3 and 0.6 for a wide range of materials and surface conditions. What is more intriguing is that while the coefficient of friction varies in such a narrow band, the wear coefficients vary by many orders of magnitude. In practice, this translates to design engineers giving more importance to wear reduction rather than friction.

The ultimate aim is the betterment of the human life through wealth creation. Here the definition of wealth can be broad based and need not mean only the materialistic wealth. Typically wealth creation needs development of the products that can tap into the seemingly inexhaustible nature's resources.

Here we explore the mechanics of contact at small scales through experiments as well as through mathematical modeling. In Section 4, an experiment to actually see what is happening at the contact interface at nanometric level is described. This experiment is carried out inside a high-resolution transmission electron microscope and the dynamics is recorded in real time. Section 5 details one of the possible ways of developing the mathematical models based on such experimental observations.

3 Contact of Solids

Solid is one of the states of matter that is characterized by resistance to deformation. Like other states of matter, solid is made up of atoms or molecules. The main distinguishing factor of solids is that the distance between the atoms or molecules is very small compared to that in liquid and gases. Also the relative position changes very slowly with time. Atoms (or molecules) of the solid can be said to be in force equilibrium with each other over the engineering timescales of milliseconds or seconds. To disturb this equilibrium external energy needs to be supplied. This manifests itself as the rigidity of the solid.

Even though it seems very simple to determine when two solid bodies have come into contact with each other, it is very difficult to precisely define it. For example, let us take two perfectly spherical, rigid, solid bodies and bring them close to each other from infinity. At non-zero separations, they experience no force and at the separation equal to the sum of their radii contact is established. If we try to reduce the separation further down, the bodies will experience infinite force and this behavior is referred to as hard-wall repulsion.

First reality check, comes from the fact none of the objects are rigid, and hence they ought to deform under the infinite forces. Assuming continuum and elasticity, the contact between spheres can be analytically solved [1]. Further complex models have been developed for various other behaviors of the material [1] and have been successfully applied to predict the contact behavior and in solving practical contact problems at macroscopic scales.

Next, the objects are not smooth. The practical surfaces deviate "locally" from the desired geometry and these deviations are termed differently as form error, waviness, and roughness depending on the definition of how local is "local." Rough surface contacts have been studied assuming that the surface height distribution follows a Gaussian distribution. The fact that surface roughness may not be Gaussian but may follow a fractal power law has been realized long ago [6]. There have been some attempts to bring the fractal concepts in contact mechanics. The bottom line is that roughness brings in a certain uncertainty in the contact behavior especially at smaller length scales, smaller timescales, and lower contact loads.

As one looks closer and closer at the contact, it becomes evident that the atomistic nature of the material starts affecting the contact behavior. The presence of roughness would mean that there will be many small contact islands within the macroscopic contacts region. The contact behavior is thus influenced by the atomic interactions taking place at small scales. One of the main implications for this is to introduce the van der Waals attractive force between the surfaces. Several attempts have been made in incorporating this in the continuum contact mechanics framework.

Friction is an average manifestation of many individual microscopic interactions. Depending on the system, the dominant mechanisms and the scale of these interactions will vary. For example, in metallic contacts, adhesion and plasticity have been used to explain the friction phenomena [1, 2]. The asperity level individual interaction will depend on the local geometry and the material behavior. The scale of the asperity will also determine the mechanism of the energy dissipation of friction. In this chapter, we are concerned with the experimental/numerical study of single asperity sliding with minimal or no wear.

In spite of 300 years of attention, the reason why we know so little about either energy dissipation or friction process is that "we cannot see what is taking place at the interface during sliding" [7]. Atomic force microscopes have been used to perform contact measurements and characterize contact conditions. However, they had not been able to "see" the interface. One recent development, which exploits the ability of the transmission electron microscope to see through the material to overcome this obstruction, is described in the following section.

4 Visualizing Single Asperity Sliding[1]

4.1 Introduction

When two surfaces are slid against each other, contact is established at many locations. The tribological behavior is the total sum of all interactions taking place at these locations. To understand these individual interactions, it is typical to design and carry out the single asperity experiments. The contact interaction is simulated with a sphere on flat or sphere on sphere configurations. Since the material property depends critically on the deforming volume, it is important that in the single asperity experiments, the length - and timescales and varied over a wide range from

[1]This section is adapted from Anantheshwara, K. and Bobji, M.S.: In situ transmission electron microscope study of single asperity sliding contacts. Tribol. Int., 43(5–6), 1099–1103 (2009).

very small to the engineering scales. In this section an experiment carried out with a nanometer scale asperity is elaborated.

4.2 Experimental Details

The asperities are simulated with the help of two sharp tips. The sliding experiments between these two tips are carried out inside a transmission electron microscope (TEM) to enable the visualization of the contact at very high resolution. The presence of organic contaminants in the vacuum system results in the growth of a thin carbon layer about the sharp tips. In effect, though the tips made of silicon and tungsten are used the sliding is carried out between the carbon films.

The sliding experiments were carried out with the help of specially developed mechanical probe [8]. A schematic of this probe is given in Fig. 4.2. The tungsten probe is movable relative to the silicon AFM cantilever probe aligned with the electron beam axis of the TEM. The tungsten probe is moved with the help of a three-axis coarse positioner. This positioner, based on inertial slider [8, 9, 10], is compact enough to be accommodated within a standard TEM sample holder. The compact design ensures minimal vibration and thermal drift problems apart from giving mechanical stiffness required for the nano-contact experiments. The coarse positioner is capable of bringing the tungsten probe from millimeters distance to within about 300 nm of the silicon probe.

The fine positioning and the sliding of the tungsten probe with respect to the silicon probe are carried out with the help of a four-quadrant piezo-electric tube. Figure 4.3 shows the two probes at low magnification. The alignment of the tungsten probe is carried out by looking at the Fresnel fringes at progressively higher

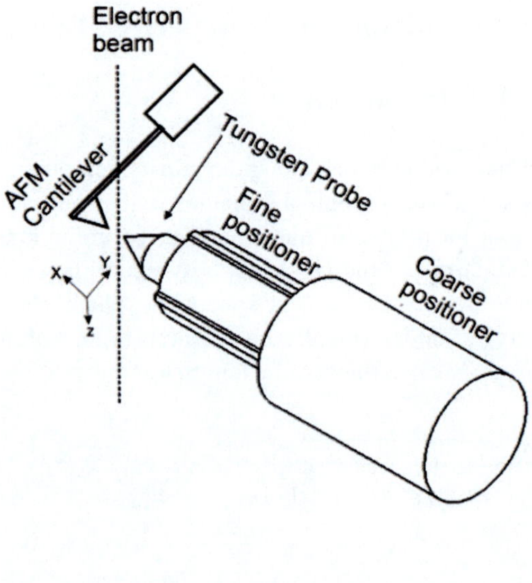

Fig. 4.2 Schematic diagram of in situ TEM holder. AFM tip is used as the sample and the tungsten probe is used to study the sliding interaction with the tip. Reprinted from [8] with permission from Elsevier

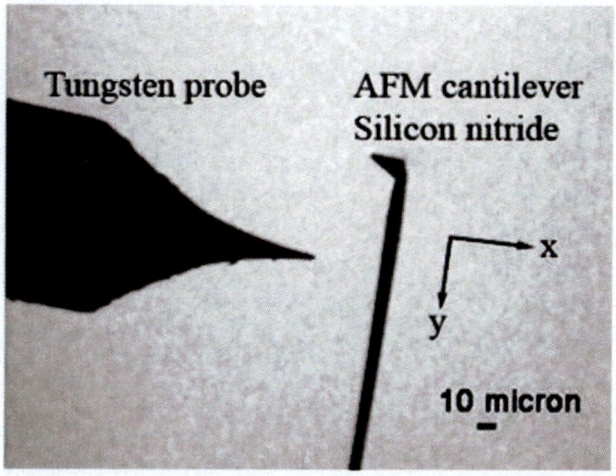

Fig. 4.3 TEM micrograph of tungsten probe and the AFM cantilever showing the orientation of the AFM tip with respect to tungsten probe. Reprinted from [8] with permission from Elsevier

magnification. Once the tungsten probe is brought in the focal plane of the electron optics, the sliding motion needs to be aligned to be in the same plane. This is ensured by applying proportionally varying voltages to the four quadrants of the piezo-electric tube.

Figure 4.4 shows the two asperities contacting each other. Since the silicon tip is mounted on a cantilever (Fig. 4.3) of stiffness 0.15 N/m, the contact force can be measured by measuring the deflection of the cantilever. For the contact shown in Fig. 4.4 the normal force is about 6.8 nN. The stiffness of the cantilever determines the maximum force that can be applied and has been chosen such that there is no appreciable plastic/permanent deformation in the carbon films. Experiments with higher support stiffness have also been carried out [10]. In such cases higher contact forces can be used to break the carbon film.

The sliding contact is simulated by moving the tungsten probe laterally (y-direction in Fig. 4.3) in the focal plane at the electron optics. The TEM images are recorded using a side-mount CCD camera, at 15 frames per second. The sliding rates are varied from 2.6 to 10.9 nm/s, and are chosen such that enough number of

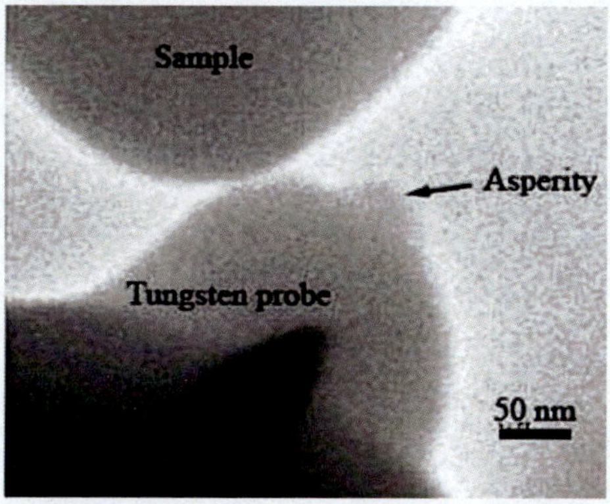

Fig. 4.4 TEM image of AFM tip and the tungsten probe in contact. A uniform layer of carbon coating is observed as a light contrast in the image. Reprinted from [8] with permission from Elsevier

frames are recorded in the video per sliding cycle. The location of the tungsten probe and the deflection of the silicon tips are determined from the individual frames of the video using cross-correlation [8].

4.3 Results

When the tungsten probe is far away from the silicon tip, say on the right-hand side of Fig. 4.4, it is not in contact with the AFM probe. As it is moved laterally (P_y) along the y-axis, it comes in contact with the silicon tip of the AFM probe and causes the cantilever to deflect along positive x-axis (t_x). This deflection is plotted in Fig. 4.5. The displacements P_y and t_x are obtained by tracking particular features in the probes through cross-correlations. The curve plotted in Fig. 4.6 is the line profile that would have been obtained from AFM in a constant height contact mode of operation. The line profile is the geometry of the sample convoluted with the geometry of the probe [11]. In Fig. 4.6 it can be seen that there is a snap-out just before the contact is broken. Attractive van der Waals interaction between the probes is the main reason for this behavior.

When two probes approach each other, there is an attractive van der Waals force between them resulting from the interaction of the dipoles present in all the atoms/molecules of the probes (see the following section for details). The AFM probe is mounted on a cantilever of stiffness 0.15 N/m that is much lower than the stiffness with which the tungsten probe is held. An instability results because of

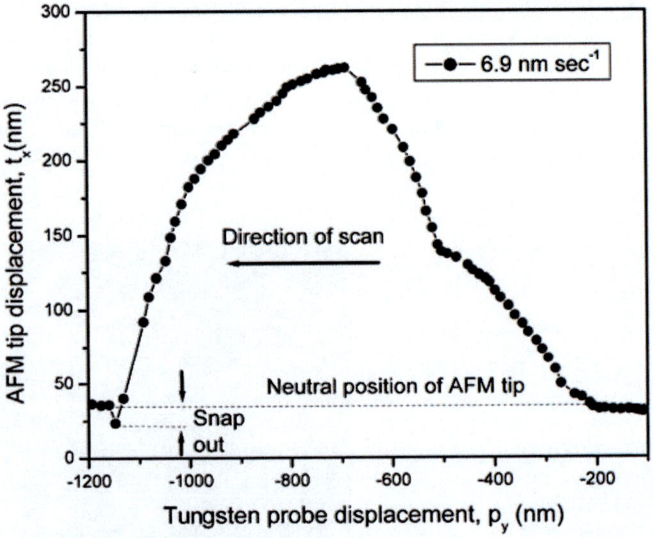

Fig. 4.5 Typical profile obtained by tracking AFM tip and the tungsten probe. Reprinted from [8] with permission from Elsevier

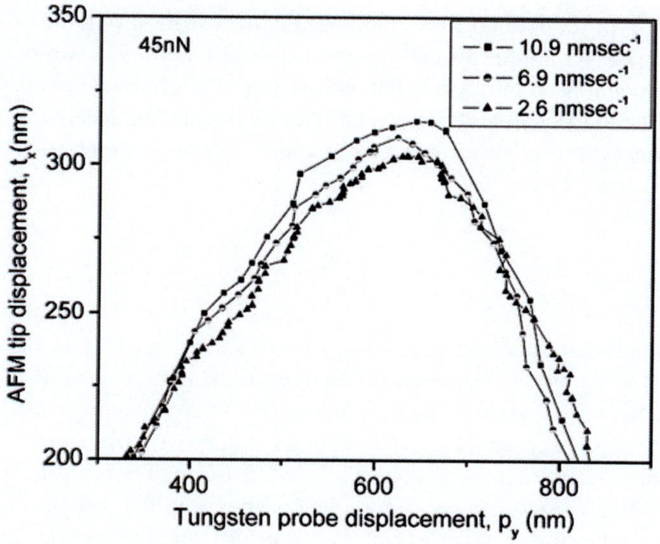

Fig. 4.6 Line profiles measured at 45 nN load showing the taller line profile with increasing speed. Reprinted from [8] with permission from Elsevier

this, leading to the snap-in (jump-to-contact) behavior during approach and a snap-out behavior during separation. In the current experiments, the snap-in behavior is barely resolved in some trials, while the snap-out is clearly seen in almost all the trials.

Table 4.1 shows the magnitude of the various snap-out events measured in these experiments. These distances when multiplied by the stiffness of the cantilever (0.15 N/m) give the attractive pull-out force between the probe and AFM tip before losing the contact. From the table, it can be seen that these pull-out forces are varying with the sliding speed as well as the maximum contact force to which the probes were subjected.

The maximum contact forces experienced by the probes can be varied by varying the interference between the probes. This is achieved by moving the tungsten probe in the positive *x*-direction (Fig. 4.3). The deformation of the probes can then

Table 4.1 Snap-out distance measured from the video at different rate slidings. Reprinted from [8] with permission from Elsevier

Experiment number	Maximum contact force (nN)	Sliding speed (nm/s)	Forward snap-out (nm)	Reverse snap-out (nm)
1	40	2.6	5.7	62.8
2	40	6.9	11.4	57.1
3	40	10.9	17.1	4
4	45	2.6	21	11.4
5	45	6.9	23	46
6	45	10.9	–	54

be measured from the line profiles obtained. Figure 4.6 shows the line profile at a maximum load of 45 nN for different sliding speeds. Only the crown of the whole profile corresponding to the asperity shown in Fig. 4.4 is shown for clarity. It can be seen that the deformation is decreasing with higher sliding speed, i.e., the AFM tip displacement is more. This implies that the material is becoming stiffer at higher strain rate.

4.4 Discussion

When two nominally rough surfaces come in contact with each other the contact is established at a number of locations depending on the normal load between them. The real area of contact which is the sum of area of all the contacting islands is a very small fraction of the apparent area [1]. As one surface starts sliding over the other, the real contact area remains constant while the number of contacting islands and their spatial distribution changes continuously. The asperities which were in contact with each other lose contact while new asperities come in contact. The current experiments simulate the behavior of one of the many contacting asperities through a contact cycle.

Figure 4.7 shows a schematic of the rough surfaces in contact. The inset shows enlarged view of two asperities that are going to interact with each other. As the surface A moves toward the left, the asperity $A1$ will come in contact with the asperity $B1$. The exact location at which the contact will occur will depend on the interference, h, between the asperities. This interference will be determined by the local geometry as well as the global contact distribution at the time and the normal load.

As the asperities come in contact with each other they will come under the influence of the attractive electrostatic and van der Waals forces. Depending on the local stiffness the asperities will jump to contact [12, 13]. We do not observe this snap-in in most of the cases in the current experiments. In one instance when the probe was moving at 2.6 nm/s in the reverse direction a snap-in was observed when the gap (s_i) between the probe and the sample was 8 nm. The amount of energy lost, E_i, due to this would depend on the average interaction force and the snap-in distance.

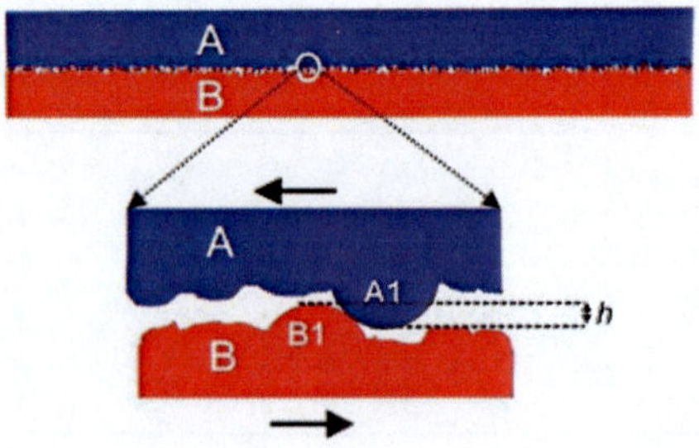

Fig. 4.7 Schematic diagram of two rough surfaces (*A* and *B*) in contact is shown. The contact of two asperities that are going to interact with each other is shown in the *inset*

Once the asperities come into contact, they can either slide over each other as observed in these experiments or they can deform plastically or otherwise dissipating energy or both. The exact nature of this interaction would be dependent on the geometry and the material properties of the asperities. Another important parameter, that will determine the interaction nature, is the stiffness with which these asperities are supported. This stiffness will depend on the contact stiffness of all the other contacting asperities. It has been assumed that the normal load between the surfaces is kept constant allowing the mean separation between the surfaces to vary. The stiffness of the cantilever in our experiment is thus indicative of the contact stiffness. The contact stiffness will vary as the surfaces slide against each other. However, it is not easy to see the variation in the stiffness during the short time when the two asperities under observation are interacting.

By varying the stiffness of the AFM cantilever it should be possible to simulate the different types of interactions experimentally. As the asperities break the contact, the adhesive forces result in a snap-out dissipating energy, E_o. The snap-out distance s_o would be predicted by the JKR and DMT theories [12]. The snap-out distance measured in our experiment is tabulated in Table 4.1. It can be seen that it depends strongly on the direction of sliding indicating that the local geometry of the asperity probably plays an important role in determining the adhesive force. The order of magnitude of the energy dissipated would be given by

$$E_o = F_{adh} s_o = k\, {s_o}^2 \tag{4.1}$$

The total energy dissipated during a single sliding experiment would be E,

$$E = E_i + F_{fric} \times s_o + E_o \tag{4.2}$$

The middle term is the work done due to the sliding friction between the probe and the sample. This energy is dissipated by two asperities sliding past each other (Fig. 4.10). If l_{asp} is the mean distance an asperity has to travel before encountering another asperity, this energy loss can manifest itself as a fraction of the macroscopic frictional force of

$$F = E/l_{asp} \tag{4.3}$$

Thus, if snap-out distance is known, then we can determine the component of the adhesive interaction between the asperities in the macroscopic friction.

5 Modeling Adhesive Interaction of Asperities

5.1 Introduction

Contact simulations at the nanoscale exhibit some remarkable differences from the macroscale contact simulations. At the macroscale, contact presents a very severe nonlinearity, wherein the contact interactions are assumed to begin at the instant when the objects come into contact. Hence, well-defined in-contact and out-of-contact configurations can be distinguished in a simulation. However, at the nanoscale, though the analyses are still highly nonlinear, we have a much gradual development of contact interactions. The contact interactions begin when there is a separation of few nanometers between the bodies, beyond which even the long-range forces like van der Waals force are negligible. As seen from the experiments at nanoscale, including the ones described in the previous section, at small separation there is usually attraction between the objects, resulting in *adhesion*, which is almost always considered as an insignificant effect in the macroscale studies. When the separation is further reduced, the attractive force increases and reaches a maximum, and then starts reducing as the repulsive force develops between the objects due to the overlapping of electronic orbits (Pauli repulsion). Though the repulsive force is extremely short ranged compared to the attractive van der Waals force, it increases at a much higher rate with further reduction in separation. Therefore, at small enough separations, usually less than a few tenths of a nanometer, net contact force changes from attractive to repulsive. It is interesting to note that, in such situations, the surface layer of atoms might be repelling each other, while atoms below the surface would be attracting the atoms in the other object. All these aspects point to the fact that in case of nanoscale contacts, it might be even meaningless to say whether the objects are in contact or out of contact. Further conceptual difficulties arise as even the surfaces are not precisely defined at the nanoscale. In spite of all these, the nanoscale contact simulations can help to quantify adhesion between the solids, thus helping us to take a step forward in understanding the fundamental mechanisms of friction and wear, in addition to better interpreting the results of experimental studies like AFM measurements.

At the nanoscale, the adhesive interactions between two spheres or between a sphere and a flat surface are the configurations, which are studied extensively. Studies on adhesive interactions could be considered to have begun with the expression for adhesive force between two rigid spheres developed by Bradley [14], and Derjaguin approximation [15] can be used to generalize this relation to a variety of geometries. JKR and DMT models were developed in the 1970s for adhesive interaction between elastic spheres. JKR model [16] represents the case of contact between large spheres made of soft materials with high surface energy, while DMT model [17] is a good approximation for small spheres of hard materials with low surface energy. Maugis Dugdale model [18] was developed for the transition regimes, which lie between these two extremes. The regimes of suitability of these different models can be obtained from the "adhesion map" developed by Johnson and Greenwood [19]. However, the applicability of these analytical techniques is limited to simplified geometries, and they are not able to account for the effect of deformation on the adhesive interactions.

Assuming the continuum hypothesis to be valid, we can perform analysis of contact at nanoscale using finite element method (FEM). In order to capture the adhesive interactions, van der Waals force needs to be incorporated in such studies. Van der Waals force, being extremely short ranged compared to the dimensions of even microscale solids, is usually approximated as a surface force. However, the analytical integration, employing the Derjaguin approximation [15], used in calculating the magnitude surface forces, becomes more and more inaccurate when the size of the object is progressively reduced to the nanoscales. Hence, finite element analysis (FEA) of adhesive interactions between nanoscale solids should be carried out by including van der Waals force as a volume force (body force) [20, 21, 22]. By this approach we can analyze interacting bodies of more complicated geometries, and more accurate results can be obtained for nanoscale objects.

5.2 Nanoscale Contact Simulation – A Case Study[2]

In this section, we present a case study in which a single nanoscale asperity is brought close to a rigid half-space, with the objective of analyzing the adhesion force, deformations, and stresses, and the adhesion-induced instability or *jump-to-contact* instability. Since, the scope of this study is limited to the adhesive interactions before the contact is established, only the van der Waals force is incorporated in the analysis, i.e., the repulsive force is not considered. The asperity is modeled as a smooth elastic hemisphere, and the effect of the rigid half-space is incorporated in the finite element formulation by summing up the contribution due to the half-space [15]. Adhesion force (per unit volume) experienced by an elemental volume at a distance d from a half-space is given by

$$f_{\mathrm{adh}} = -\frac{A}{2\pi d^4} \tag{4.4}$$

where A is the Hamaker constant. This adhesion force is incorporated into finite element framework as a body force (force distributed over the volume of the body). Equilibrium is reached when the elastic restoring force in the deformed configuration becomes equal to the force due to adhesion. However, if the gradient of the adhesion force with separation is greater than the stiffness, then adhesion force cannot be balanced by the elastic restoring force [12]. The resulting instability causes a progressive increase in deformation, till the asperity jumps into contact with the surface of the half-space. To capture this adhesion-induced instability (jump-to-contact) through finite element simulations, we adopt an iteration technique based on "residual stress update algorithm." The general expression for elemental force vector ($\{r_e\}$) at any iteration in the absence of initial strain and surface tractions is given by

$$\{r_e\} = \int [N]^T \{F\}\, dV - \int [B]^T \{\sigma_0\}\, dV \tag{4.5}$$

[2]This section is adapted from Bobji et al. [22].

where $[N]$ is the shape function matrix, $\{F\}$ is the body force vector, $[B]$ is the strain-displacement matrix, and $\{\sigma_0\}$ is the residual stress. The residual stress is initialized to zero for the first iteration, i.e., $\{\sigma_0\} = \{0\}$ for all points. In subsequent iterations, we analyze the deformed configuration (updated using the deformation up to the previous iteration) for the incremental change in deformation and adhesion force. Since we solve for incremental deformations, global equilibrium equation has the general form

$$[K]\{\delta\}^m = \{\Delta R\} \tag{4.6}$$

where the superscript m indicates the iteration number. The stiffness matrix $[K]$ for each iteration is evaluated at the corresponding deformed configuration, $\{\delta\}^m$ is the incremental displacement vector, and $\{\Delta R\}$ is the incremental load vector. The total displacement at each node after any iteration step is the sum of the incremental displacements of that node up to that iteration. The results are said to be converged when displacement for any iteration falls below a predefined tolerance limit. It is found that the analysis diverges when the adhesion-induced instability sets in, and the maximum initial separation at which this occurs is termed as the *instability separation*. If the initial separation is larger than this instability separation, a converged solution is obtained after a few iterations.

The schematic diagram of the asperity used in this study is given in Fig. 4.8. The asperity is modeled using four-noded axisymmetric elements. As indicated in Fig. 4.8, the top nodes of the model are constrained in the axial direction, and the nodes along the axis are constrained in radial direction. The body force at any point in the hemisphere is calculated based on the axial distance of that point from the half-space. The analyses are performed for hemispheres of various radii and Young's moduli and by varying the initial separation (h) between the tip of the asperity and half-space. Starting at a large initial separation, a series of analyses are carried out

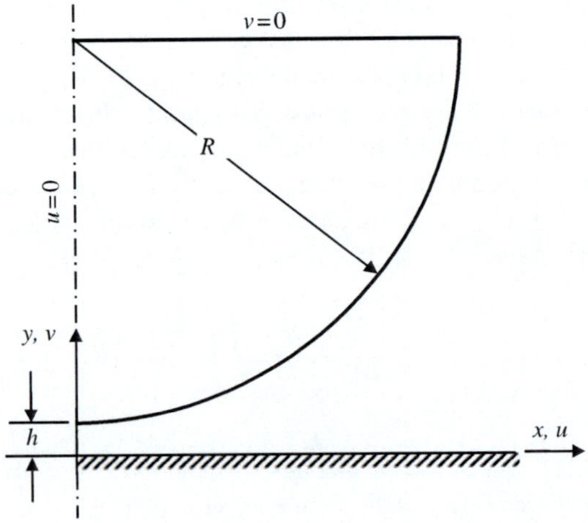

Fig. 4.8 Schematic diagram of the geometry of the asperity model used for analysis. Reproduced from [22] with kind permission from Springer Science + Business Media

for each hemisphere model by progressively reducing the separation. Finally, when the instability separation is reached, the analysis shows divergence.

5.2.1 Deformation, Adhesive Force, and Instability of Asperity

In the finite element simulations, a value of 5×10^{-19} J has been used for Hamaker constant which is a typical value for metals. Young's modulus is varied from 70 GPa (aluminum) to 400 GPa (tungsten), and Poisson's ratio used is 0.3. Asperity radii ranging from 10 to 200 nm is used in the model with the aim of capturing the behavior of the typical scanning probes used in experimental studies.

Figure 4.9 shows the variation of total adhesive force with separation for a hemisphere of 25 nm radius near a rigid half-space. The total adhesive force (F) is obtained by summing up the reaction forces acting at the axial support locations corresponding to each initial separation (h). The force values obtained from finite element simulation is compared with the analytical expressions for rigid sphere [12]

$$F = \frac{AR}{6h^2} \tag{4.7}$$

and the approximate analytical expression for elastic spheres by Vinogradova and Feuillebois [23]:

$$F \approx \frac{AR}{6h^2} + \frac{2A\sqrt{R}}{15E^* h^{7/2}} \tag{4.8}$$

This expression had been obtained by them as a first-order correction to (4.7) using the deformation resulting from the rigid sphere adhesion force.

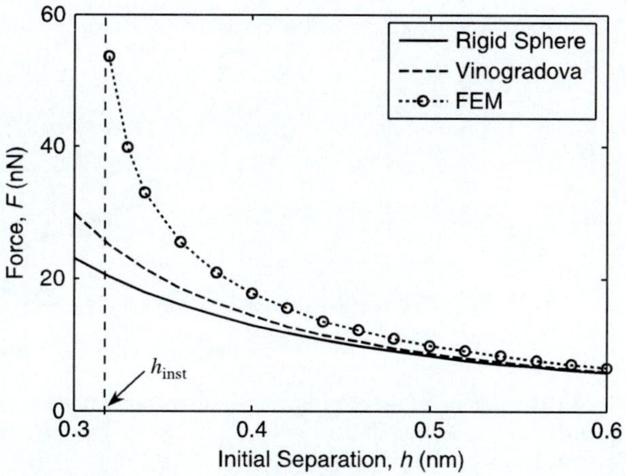

Fig. 4.9 Variation of total adhesive force with separation for 25 nm sphere. The instability separation from FE analysis (0.31 nm) is indicated by the *vertical dotted line*. Reproduced from [22] with kind permission from Springer Science + Business Media

As mentioned earlier, the simulation is carried out starting at a large initial separation and progressively reducing it till the instability separation, h_{inst}, is reached. Since the simulations give equilibrium adhesive force, corresponding to a final separation, they are much higher than those given by either (4.7) or (4.8), when the initial separation is close to the instability separation. At large separations, the deformation is too small to make any significant change in separation, and hence the values given by analytical expressions and FEM are similar.

Figure 4.10 gives the variation of maximum deformation in the asperity (w) at different initial separations (h). The maximum deformation occurs at the tip of the asperity (point in the asperity that is closest to the half-space), and is hence called "tip deformation." Similar to the case of force, results from the analytical expression and that from FE analysis match very closely at large initial separation (>0.4 nm), since the corresponding deformations are very small. However, closer to the instability separation, the tip deformation from FE solution is found to be much higher than that from the analytical expression due to the significant reduction in separation, which is not considered in the analytical expression.

Figure 4.11 gives the comparison of the instability separation (initial separation at which jump to contact occurs) as obtained from finite element analyses, with the expressions obtained by Pethica and Sutton [12] and Attard and Parker [24]. Consistent with the higher forces and the deformations, the finite element analysis gives a higher separation at which the instability occurs. From a tribology perspective, higher instability separation would mean higher predicted values of coefficient friction and higher rate of material removal by adhesive wear. The best fit curve for the finite element analysis results, when the elastic modulus is 70 GPa, gives the empirical relation for instability separation (with h_{inst} and R in nanometers) as

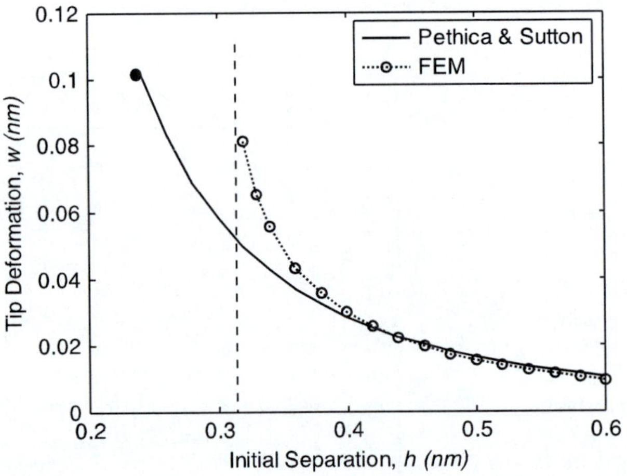

Fig. 4.10 Variation of tip deformation with separation for 25 nm sphere. Both the curves end at the instability separation of 0.24 and 0.31 nm. Reproduced from [22] with kind permission from Springer Science + Business Media

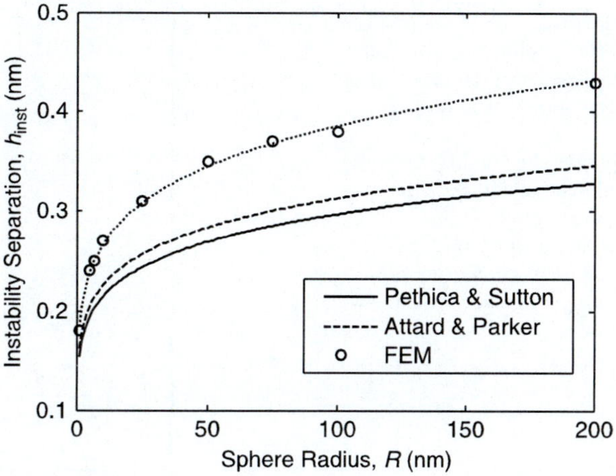

Fig. 4.11 Variation of instability separation with change in sphere radius. The instability separation obtained by finite element analysis is higher than the two approximate analytical expressions, since effects of deformation on the adhesive force are not considered in both of them. Reproduced from [22] with kind permission from Springer Science + Business Media

$$h_{inst} = 0.184 \times R^{0.161} \tag{4.9}$$

as compared to the equation by Pethica and Sutton (for the above material properties):

$$h_{inst} = 0.154 \times R^{0.143} \tag{4.10}$$

and that by Attard and Parker [24]:

$$h_{inst} = 0.162 \times R^{0.143} \tag{4.11}$$

5.2.2 Stresses and Yielding in Asperity

The stress induced in the asperities due to the attractive forces could even be as high as the breaking strength of the material [25]. For metals, the asperities would undergo plastic deformation much before the breaking stresses are reached, and the plasticity reduces the local stiffness drastically, leading to higher instability separations. Assuming that von Mises yielding criterion is applicable at this length scale, we studied the initiation of plastic deformation in an asperity.

The von Mises stress distribution for a typical case is given in Fig. 4.12. The maximum stress is occurring at approximately 1 nm below the surface of the asperity, which is different from the stress distribution if surface force approximation is used [1]. The stresses are seen to be very high, and can induce plastic deformations in many of the engineering materials.

In order to study the initiation of yielding in asperities, 25 nm spheres made of different materials are analyzed, and the corresponding variation of maximum von Mises stress is shown in Fig. 4.13. It is seen that for all the cases considered, the

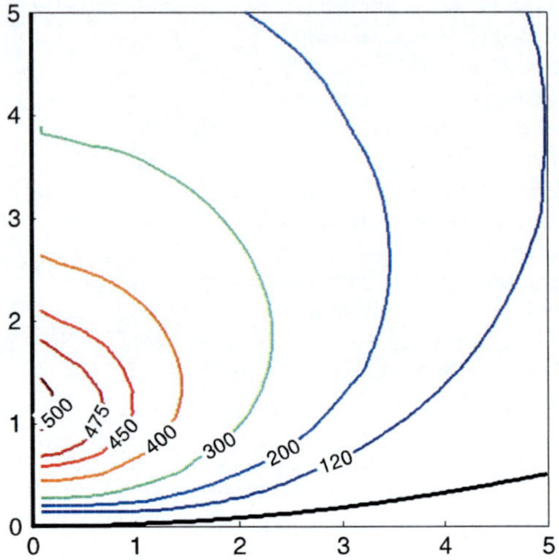

Fig. 4.12 von Mises stress (MPa) contours near the tip of 25 nm sphere (initial separation is 0.35 nm). The maximum von Mises stress is found to be 508 MPa, which is higher than the yield strength of aluminum. Reproduced from [22] with kind permission from Springer Science + Business Media

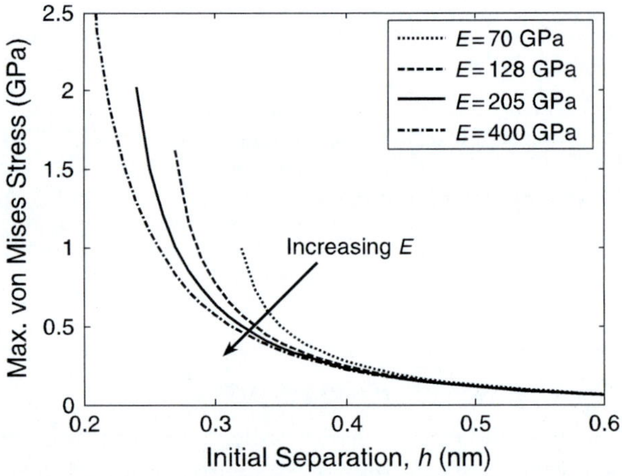

Fig. 4.13 von Mises stress variation with separation for 25 nm sphere for materials of different Young's moduli. For any given separation, the stress in the material with higher Young's modulus is less due to smaller deformations, which results in lower value of adhesion force. Reproduced from [22] with kind permission from Springer Science + Business Media

separation at which yielding occurs (yield point separation) is higher than the corresponding instability separation. For example, assuming the engineering value of yield strength, the yield point separation for copper is 0.59 nm, while the instability separation is 0.27 nm. In general, when geometry is the same, the difference between yield point separation and the instability separation will be higher for a material with higher Young's modulus or lower yield strength.

This being an elastic analysis, the predicted values of stresses are not expected to be correct after yielding has initiated; however, it is interesting to note that the stresses near the instability separation for tungsten (or in general for a material with higher Young's modulus) are much higher than that for aluminum. Hence, even with

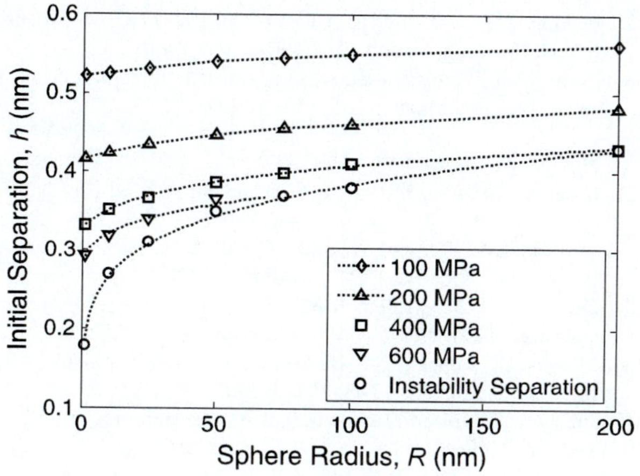

Fig. 4.14 Comparison of yield point separation curves and instability separation for 25 spheres made of materials with different yield strengths (100–600 MPa). Reproduced from [22] with kind permission from Springer Science + Business Media

the high value of yield strength, tungsten spheres are very much prone to yielding before reaching the instability.

Figure 4.14 gives the comparison of yield point separation and the instability separation for spheres of different radii and different yield strength (keeping Young's modulus to be constant at 70 GPa). The figure indicates that the plastic deformations are more severe for smaller asperities and that pure elastic jump to contact is possible in case of larger asperities, or in case of materials of high yield strength. The exact radius, at which the yield point separation and instability separation are equal, will depend on Young's modulus and yield strength of the material.

References

1. Johnson, K.L.: Contact Mechanics, Cambridge University Press, Cambridge (2003).
2. Maugis, D.: Contact, Adhesion and Rupture of Elastic Solids, Springer, Berlin (2000).
3. Thomas, T.R.: Rough Surfaces, 2nd Ed, Imperial College Press, London (1999).
4. Bobji, M.S., Venkatesh, K., and Biswas, S.K.: Roughness generated in surface grinding of metals. J. Tribol. Trans. ASME, 121(4), 746–752 (1999).
5. Cowin, S.C., and Doty, S.C.: Tissue Mechanics, Springer, New York, NY (2007).
6. Archard, J.F.: Elastic deformation and the laws of friction. Proc. R. Soc. Lond. A 243, 190–205 (1957).
7. Singer, L., and Pollock, H.M.: Fundamentals of Friction: Macroscopic and Microscopic Processes, Kluwer, The Netherlands (1992).
8. Anantheshwara, K., and Bobji, M.S.: In situ transmission electron microscope study of single asperity sliding contacts. Tribol. Int., 43(5–6), 1099–1103 (2010).
9. Bobji, M.S., Ramanujan, C.S., Pethica, J.B., and Inkson, B.J.: A miniaturized TEM nanoindenter for studying material deformation in situ. Meas. Sci. Technol., 17, 1–6 (2006).
10. Wang, J.J., Lockwood, A.J., Peng, Y., Xu, X., Bobji, M.S., and Inkson, B.J.: The formation of carbon nanostructures by in situ TEM mechanical nanoscale fatigue and fracture of carbon thin films. Nanotechnology, 20(30), 305–703 (2009).
11. Meyer, E., Hug, H.J., and Bennewitz, R.: Scanning Probe Microscopy, Springer, Berlin (2004).

12. Pethica, J.B., and Sutton, A.P.: On the stability of a tip and flat at very small separations. J. Vac. Sci. Technol. A, 6(4), 2490–2494 (1988).
13. Bobji, M.S., Pethica, J.B., and Inkson, B.J.: Indentation mechanics of Cu–Be Quantified by an in-situ TEM Mechanical probe. J. Mater. Res., 20(10), 2726–2732 (2005).
14. Bradley, R.S.: The cohesive force between solid surfaces and the surface energy of solids. Phil. Mag., 13(86), 853–862 (1932).
15. Israelachvili, J.N.: Intermolecular and Surface Forces, 2nd Ed., Academic, San Diego, CA (1992).
16. Johnson, K.L., Kendall, K., and Roberts, A.D.: Surface energy and the contact of elastic solids. Proc. R. Soc. Lond. A, 324, 301–313 (1971).
17. Derjaguin, B.V., Muller, V.M., and Toporov, Y.P.: Effect of contact deformations on the adhesion of particles. J. Colloid Interface Sci., 53(2), 314–326 (1975).
18. Maugis, D.: Adhesion of spheres: the JKR-DMT transition using a Dugdale model. J. Colloid Interface Sci., 150(1), 243–269 (1992).
19. Johnson, K.L., and Greenwood, J.A.: An adhesion map for the contact of elastic spheres. J. Colloid Interface Sci., 192, 326–333 (1997).
20. Cho, S.S., and Park, S.: Finite element modeling of adhesive contact using molecular potential. Tribol. Int., 37, 763–769 (2004).
21. Sauer, R.A., and Li, S.: An atomic interaction-based continuum model for adhesive contact mechanics. Finite Elem. Anal. Des., 43, 384–396 (2007).
22. Bobji, M.S., Xavier, S., Jayadeep, U.B., and Jog, C.S.: Adhesion-induced instabilities in asperities. Tribol. Lett., 39, 201–209 (2010).
23. Vinogradova, O.I., and Feuillebois, F.: Interaction of elastic bodies via surface forces. 1. Power-law attraction. Longmuir, 18, 5126–5132 (2002).
24. Attard, P., and Parker, J.L.: Deformation and adhesion of elastic bodies in contact. Phys. Rev. A, 46(12), 7959–7971 (1992).
25. Yao, H., Ciavarella, M., and Gao, H.: Adhesion maps of spheres corrected for strength limit. J. Colloid Interface Sci., 315, 786–790 (2007).

Chapter 5
Mechanics of Peeling of a Flexible Adherent Off a Thin Layer of Adhesive

Animangsu Ghatak

Abstract Soft, deformable pressure-sensitive adhesives are encountered in several engineering and scientific applications, in which the adhesive remains as a sandwiched layer of viscoelastic glue between two flexible or rigid adherents. How does the thin layer of glue adhere two surfaces or bodies together is a question. While chemical character of the adhesive, its rheological characteristics, interaction with the adhering surface are important for adhesion, recent developments in adhesion science suggest that the physical mechanism of adhesion too play an equally important role. For example, the adhesion strength can simply be varied by altering the thickness of the adhesive layer without effecting any chemical alteration. In this chapter we have considered the simplest situation of adhesion of a thin layer of elastic adhesive to a flexible adherent. We have described a rigorous theoretical analysis for estimating the work of adhesion on such an adhesive. Next, we have considered surface instability at the contact of the adhesive and the adherent. The adhesive layer confined between two adherents does not deform uniformly, but turns undulatory with a wavelength that scales linearly with the thickness of the adhesive. We have shown that the instability appears at a threshold value of a dimensionless quantity, termed as the degree of confinement of the layer. While, in all these examples, a pre-existing cusp-shaped crack propagates because of lifting of the flexible adherent, the problem of initiation of this crack has also been discussed in the context of crack propagation on incision-patterned adhesives. We have shown that on such adhesive surfaces the adhesive strength increases by an order of magnitude simply because of the patterns.

Keywords Thin adhesive film · Work of adhesion · Flexible plate · Confinement · Elastic instability · Wavelength · Crack initiation

---

A. Ghatak (✉)
Department of Chemical Engineering, Indian Institute of Technology (IIT), Kanpur 208016, India
e-mail: aghatak@iitk.ac.in

S. Chakraborty (ed.), *Mechanics Over Micro and Nano Scales*,
DOI 10.1007/978-1-4419-9601-5_5,

1 Introduction

History of adhesion is as old as that of the mankind, as there is enough evidence that the pre-historic man had indeed developed the method of attaching the stone-made blade with the shaft of his hunting tool using pitch prepared from "bark of the birch" [1, 2]. The same material was also found to be used as the coating for pottery. Centuries later, adhesive of a different kind was used to preserve the mummies in Egypt. Today adhesives are ubiquitous as we use them in different forms in variety of applications, e.g., as pressure-sensitive adhesive, solvent-based adhesive, the instant cure, conductive adhesives, and adhesive for packaging tiny components, e.g., electronic chips and sealants. With the advent of new materials and novel applications, the demands on adhesives are also growing and getting more versatile. While some applications demand strong adhesion, some need permanent bonding, some applications demand that end products remain flexible, yet others demand strong yet repeatable adhesion. Clearly, systematic study is necessary on different aspects of the adhesive to meet all these diverse demands.

How does one characterize an adhesive? Adhesives are characterized by the adhesion strength defined as the energy released when two adherents are brought in contact or the energy which is to be spent when they are separated by the application of an external force. For most applications, the magnitude of these two adhesion strengths are, however, not same, because the work of adhesion estimated during detachment far exceeds the value when the two adherents are attached together. In the language of thermodynamics, during bonding, the two surfaces are merged together to form an interface. Since the energy of the interface, γ_{12}, is smaller than the total energy of the two surfaces, γ_1 and γ_2, energy of the quantity of $\gamma_1 + \gamma_2 - \gamma_{12}$, known as the thermodynamic work of adhesion, is released [3], leading to adhesion of the two surfaces. However, during separation of these two surfaces, the loads are not applied exactly at the interface but away from it, as a result, if there exist dissipative factors involved in the process, for example, friction, viscous flow, plastic deformation, and similar other irreversible processes, energy gets dissipated with the result that the effective adhesion strength is increased. Two aspects are important for adhesion between two materials: the interaction between the surfaces and the deformation of the adherents. In general, adhesives comprise long polymeric molecules which can interact with the two adherents physically or chemically; in some cases these molecules can interdigitate into the two surfaces; in some other applications, polymeric molecules between two surfaces can entangle forming a bridge and so on. An example is the solvent-based adhesives which are applied as polymeric solution in a suitable solvent which flows into the confined gap between the adhering surfaces and into the minute grooves of the surface roughness, eventually the solvent evaporates out and the polymer is permanently set between the two adherents. In this case, adhesion occurs via chemical bonding or mechanical interlocking, so that a permanent bond is formed, which once broken cannot be joined again within a short interval of time. Consequently, the adhesive loses its ability to re-adhere to a surface.

The subject of this chapter is a different class of adhesive, known as the pressure-sensitive adhesive (PSA), which essentially comprises a layer of soft, crosslinked elastic, and viscoelastic material sandwiched between two rigid or flexible substrates. The deformability allows such adhesive to follow the roughness of the surface of the adherents on application of small pressure. Here the adhesive may or may not undergo permanent deformation depending on the prominence of the viscous component. If the adhesive material deforms permanently because of application of the load, e.g., viscous flow, it generally loses its ability to adhere after using it for even once. For example, it is a common experience that most adhesive tapes adhere strongly to a perfectly smooth and clean surface for the first time, but after it is peeled off, it hardly adheres again to the surface with the same adhesion strength. Careful examination under optical microscope reveals roughening of the adhesive surface after the first peel which prevents its subsequent re-adhesion. In some other situations, the adhesive separates cleanly from the adhering surface when peeled at small rate but leaves a residue when peeled at a faster rate. Here, the dynamic effect of viscous flow vis-a-vis the peeling rate seems to determine the location of debonding of the adhesive: at the interface of the adherents or at the bulk of the either material. Such complex behavior involving viscous flow is not observed with elastic materials. However, here too during peeling off the adhesive from the adherent, debonding can occur at the interface or at the bulk depending upon the adhesion strength at the interface and the fracture toughness of the adhesive. The adhesion strength at the interface depends upon the type of interaction present; for example, for most PSAs, adhesion primarily occurs via physical interaction, e.g., van der Waals intermolecular forces [4] of attraction. The adherents may interact via different other forces, e.g., polar acid–base interaction, hydrogen bonding, and capillary forces [3]. Whereas the van der Waals interaction results in low adhesion, interaction of other types result in stronger, even permanent bonding.

We encounter the PSAs as the glue on the back of labels stuck on all kinds of items in supermarket, on post-it, postal stamps, as the glue on adhesive tapes of different types, and so on. We see them also in host of engineering applications, e.g., as metal–polymer and polymer–polymer joints in automotive and aeronautical industry and even in biology at the feet of many living organisms, for example, many insects like grasshoppers, spiders, as lizards like gecko, and amphibians like frogs. In these variety of applications, the adherents along with the sandwiched adhesive layer are subjected to variety of forces, e.g., extension and compression, shear, bending, and torsional loads and many often mix of several of these forces. In addition, the adhesive is subjected to variety of environmental conditions, e.g., heat, humidity, and light. How the glue deforms in these different conditions, importantly, how these parameters affect the adhesion strength has been a subject of interest to adhesion scientists for a long time.

How to measure the adhesion strength of an interface is a question. In fact, for viscoelastic adhesives it is hardly possible to measure the thermodynamic work of adhesion at the interface, as the measured value includes also the energy dissipated because of viscous flow of the adhesive. Therefore, for the sake of simple analysis of the mechanism of adhesion, in this chapter we consider a thin layer of elastic adhesive sandwiched between two adherents. The adhesive interacts only via van der Waals forces which result in low adhesion strength between the surfaces. As a result during separation of the adhesive from the adherent, debonding occurs at the interface, thus precluding many complexities of the viscoelastic adhesives. Here we have first presented a rigorous method of estimating the thermodynamic work of adhesion of such an adhesive. Next, we have shown that surface of an incompressible elastic adhesive, when confined between two adherents, does not deform uniformly, but turns undulatory with a wavelength that scales linearly with the thickness of the adhesive. We have obtained the threshold degree of confinement of the layer at which the instability appears. Finally, we have examined how crack propagates on an incision-patterned adhesive.

2 Adhesion of a Flexible Adherent

2.1 Lifting Plate Experiment

Figure 5.1a depicts the geometry of the system being considered here. It consists of a thin, soft film of poly(dimethylsiloxane) (PDMS) of thickness $h = 40$–$1000\ \mu\text{m}$ and shear modulus $\mu = 0.25-2.0$ MPa strongly bonded to a rigid substrate. In contrast to most pressure-sensitive adhesives which are viscoelastic, the adhesive film considered here is elastic so that the viscous effects can be neglected. This simplification allows easy analysis of the static equilibrium of forces rather than considering a more complicated situation involving dynamic viscous effects. In order to estimate the adhesion strength of the adhesive, a flexible plate of rigidity $D = 0.02$–2.0 Nm is kept in partial contact with this film in the form of a curved elastica supported on a spacer of height Δ. The flexural rigidity of thin plates is defined as $D = \mu' t^3/6\left(1-\nu'\right)$ where μ' and ν' are, respectively, the shear modulus and Poisson's ratio of the plate and t is its thickness [5]. In another variant of the experiment, the flexible plate is lifted off at displacement-controlled mode by applying a load at its free end. While Fig. 5.1a essentially shows the experiment being carried out in rectangular geometry, similar experiment can also be carried out in cylindrical geometry, where loads are applied at the center or along the periphery of a circle. In all these cases, the contact line between the adhesive and the adherent propagates further away from the location of the spacer, i.e., the line of application of the load, finally coming to rest, as an equilibrium is established between the bending force of the plate and the adhesion force of the adhesive. The equilibrium distance of the contact line from the point of application of the load depends upon various characteristics, e.g., geometric properties like the thickness of the adhesive film; material properties like the elastic modulus of the film and the flexural rigidity of the contacting plate; and the interfacial property like the adhesion strength of the interface. A manifestation of this dependence is the observation that the contact line rests further away from the load as the flexural rigidity of the plate is increased

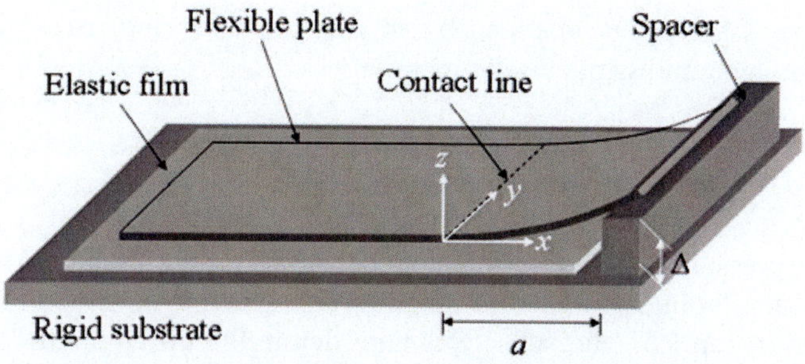

Fig. 5.1 Schematic of the experiment in which a model elastic adhesive remains bonded to a rigid substrate and a flexible plate is detached from it with the help of a spacer inserted at the opening of the crack. For a critically confined film, the contact line does not remain smooth but becomes undulatory as shown in video micrograph

or the film thickness is decreased. Similarly, the contact line remains closer to the spacer in experiments in which the interfacial adhesion strength is increased. These observations then suggest that it should be possible to manipulate the dual effects of adhesive thickness and adherent flexibility to obtain an estimate of the adhesion strength between the adhesive layer and the contactor. In fact, this experiment can be useful also for estimating the hysteresis of the adhesive if there is any, because the adhesion strength can be calculated independently for the crack opening and closing experiment. In both these cases measurement of the crack length a, defined as the distance of the line of contact between the two contactors and the spacer, can yield the independent values of the adhesion strength.

2.2 Displacement and Stress Field

Estimation of the interfacial adhesion strength, however, requires the knowledge of the functional forms of the deformations in the film and that of the flexible adherent. Figure 5.2 shows the schematic of the displacement field in which $u\,(x, z)$ and $w\,(x, z)$ are displacements along x- and z-directions, respectively. The displacements are independent of y because of the plane strain conditions along this axis. The figure shows also the characteristic lengths of the experimental geometry: film thickness h along the thickness co-ordinate and L along the lateral direction, i.e., x. The

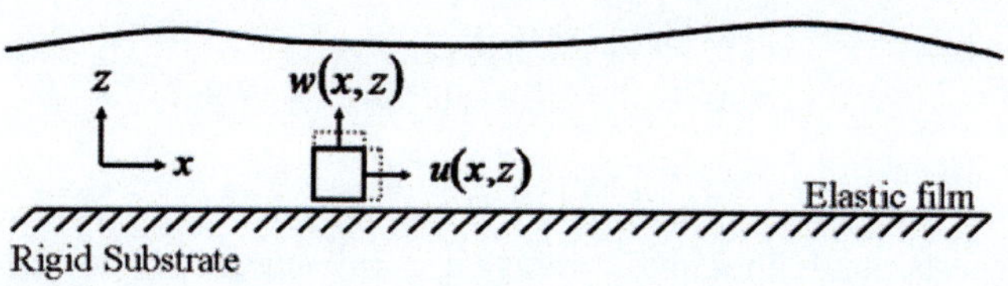

Fig. 5.2 Schematic depicts the displacement field in the elastic film

expressions for u and w are obtained by solving the following stress equilibrium relations for an incompressible elastic material:

$$\begin{aligned} u_{xx} + u_{zz} &= p_x/\mu \\ w_{xx} + w_{zz} &= p_z/\mu \end{aligned} \tag{5.1}$$

where p is the pressure field in the film. Notice that (5.1) is the simplified form of the more general three-dimensional stress equilibrium relation subjected to the plane strain approximation [5]. The subscripts here define the differentiation. Equation (5.1) along with the incompressibility relation $u_x + w_z = 0$ is solved under appropriate boundary conditions. Furthermore, for the condition that characteristic length along x far exceeds that along z, i.e., $L >> h$, (5.1) is simplified via lubrication approximation similar to that in fluidic system [6]. A simple yet rigorous treatment presented in detail in [7, 8] yields the following form of the displacement field in the elastic film:

$$\begin{aligned} u(x,z) &= \frac{6z(z-h)F'}{qh^3}\phi_1(x) \\ w(x,z) &= \frac{z^2(3h-2z)F'}{h^3}\phi_2(x), \quad x < 0 \end{aligned} \tag{5.2}$$

where

$$\phi_1(x) = e^{qx/2}\left(aqe^{qx/2} + \frac{3aq+4}{\sqrt{3}}\sin\left(\frac{\sqrt{3}qx}{2}\right) - aq\cos\left(\frac{\sqrt{3}qx}{2}\right)\right)$$

$$\phi_2(x) = e^{qx/2}\left(aqe^{qx/2} + \frac{3aq+2}{\sqrt{3}}\sin\left(\frac{\sqrt{3}qx}{2}\right) + (aq+2)\cos\left(\frac{\sqrt{3}qx}{2}\right)\right)$$

and

$$F' = 3\Delta \Big/ \left(6 + 12aq + 9(aq)^2 + 2(aq)^3\right)$$

Notice that the displacement of the flexible contacting plate in the region $x < 0$, at which it remains in contact with the adhesive film, is equal to that of the adhesive at its surface, $\psi(x < 0) = w(x, z = h)$. At $0 < x < a$, the displacement of the plate is obtained as

$$\psi(x) = F'\left(2(aq+1) + (3aq+2)qx + aq(qx)^2 - (qx)^3/3\right), \quad 0 < x < a \tag{5.3}$$

Analysis of the expressions in (5.2) and (5.3) brings out several points. For example, (5.2) suggests that both u and w vanish at $z = 0$ suggesting no-slip conditions that the film remains strongly bounded to the substrate. At the other interface, i.e., at $z = h$ too, u is calculated to be zero, which implies that here too the film is not

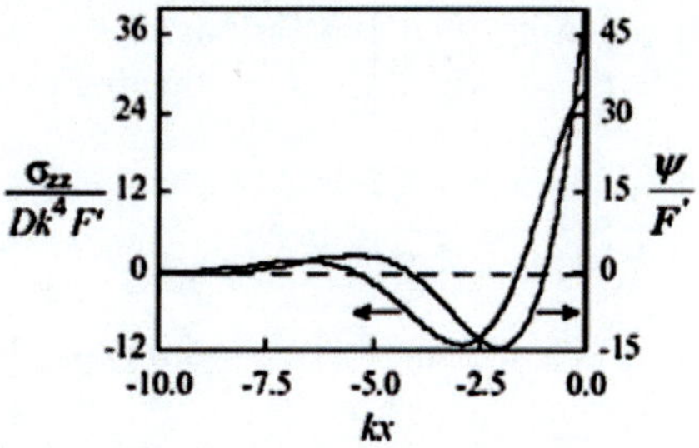

Fig. 5.3 Dimensionless normal traction and the vertical displacement of plate show oscillatory variation with respect to the dimensionless distance from the line of contact

allowed to slip on the surface of the contacting plate because of frictional resistance. A competing situation can be that the friction is negligible at the interface of the film and the flexible plate, which implies a condition of complete slippage. This situation calls for a different set of boundary conditions and consequent solutions for the displacements, which have been presented in detail elsewhere [7]. The actual condition in experiment is expected to be that of partial slippage with a non-zero finite frictional coefficient. Calculations have shown that all these cases do not result in any significant difference in the estimated values for the adhesion strength at the interface. Important point is that, in the region at which the flexible adherent remains in contact with the adhesive, i.e., at $x < 0$, the functional dependence of displacements u and w on x is estimated to be oscillatory with the film being subjected to tensile and compressive loads periodically. In Fig. 5.3, curve 2 shows this variation of $\psi = w(x, z = h)$; it is maximum at the contact line, i.e., at $x = 0$ but oscillates with exponentially decreasing amplitude away from it; finally, it decays to zero at $x \to -\infty$. The figure shows also the spatial variation in the normal stress in the film which too is oscillatory with maximum tensile stress occurring at the contact line. The periodicity of these oscillations scales with a material length scale $q^{-1} = \left(Dh^3/12\mu\right)^{1/6}$ which in essence is the ratio of deformability of the film and that of the plate and defines the distance from the contact line within which the stresses remain concentrated [7, 9]. Putting representative numbers $h = 100\,\mu\text{m}$, $D = 0.02$ Nm, $\mu = 1.0$ MPa, the wavelength of the oscillations $5q^{-1}$ is calculated as 1.72 mm, which is significantly larger than the thickness of film.

2.3 Work of Adhesion

The above expressions for displacement field in the adhesive film and that of the flexible plate can be used to determine the work of adhesion W in the light of the principle that at equilibrium the total energy of the system is minimized. This energy comprises the bending energy of the cover plate, the elastic energy in the film, and the interfacial energy of adhesion, the summation of which can be written as

$$\Pi(a) = \int_{-\infty}^{a} \frac{D}{2}\left(\frac{d^2\psi}{dx^2}\right)^2 dx + \int_{-\infty}^{0}\int_{0}^{h} \frac{\mu}{4}\left(\frac{du}{dz} + \frac{dw}{dx}\right)^2 dz dx + Wa \tag{5.4}$$

Substituting the expressions for deformations depicted in (5.2) and setting $\partial\Pi/\partial a = 0$ we obtain the following expression for the work of adhesion:

$$W = \frac{9D\Delta^2}{2a^4} g(aq)$$

$$g(aq) = 8(aq)^4 \left(12 + 46(aq) + 72(aq)^2 + 56(aq)^3 + 21(aq)^4 + 3(aq)^5\right) / 3\left(6 + 12(aq) + 9(aq)^2 + 2(aq)^3\right)^3 \quad (5.5)$$

in which the function $g(aq)$ is obtained as the correction to the classical result of Obreimoff [10] for peeling off a rigid substrate. Figure 5.4 depicts typical results from experiments in which elastic films of poly(dimethylsiloxane) (PDMS) of different shear moduli and thicknesses are used as the adhesive and glass cover slips are used as the adherent. Here (5.5) is rewritten as

$$\frac{\Delta D^{1/2} q^2}{\gamma^{1/2}} = \left(\frac{W}{12\gamma}\right)^{1/2} f_1(aq) \quad (5.6)$$

where $f_1(aq) = (8(aq)/3g(aq))^{1/2}$ and $\gamma = 20$ mJ/m^2 is the surface energy of the adhesive layer, i.e., the poly(dimethylsiloxane) (PDMS) films. Typical experimental parameters are the following: film thicknesses $h = 50-450\,\mu$m, shear modulus of the layer $\mu = 0.2-0.9$ MPa, flexural rigidity of contacting plate $D = 0.02-0.84$ Nm. The glass plate is coated with a monomolecular layer (SAM) of hexadecyltrichlorosilane (HC) molecules which minimizes hydrogen bonding or other interactions and the consequent hysteresis in adhesion [11]. The crack length a for different spacer heights Δ has been scaled following (5.6) and plotted in Fig. 5.4 which shows that all data collapse on a single straight line going through the origin yielding the slope $(W/12\gamma)^{1/2} = 0.43$. Therefore, $W = 44$ mJ/m^2. Similar values are obtained also in Johnson, Kendall, and Roberts (JKR) contact mechanics experiments [12] of the same elastomeric networks in contact with HC-coated glass substrates [13].

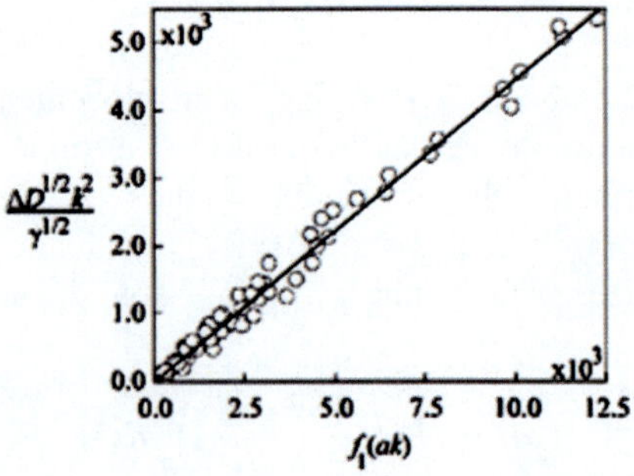

Fig. 5.4 The data from large number of experiments are scaled according to (5.6) and plotted as $\frac{(\Delta D^{1/2}k^2)}{\gamma^{1/2}}$ as a function of $f_1(aq)$. The slope of the master curve yields the work of adhesion at the interface

3 Adhesion-Induced Elastic Instability

While in discussion so far, the line of contact between the flexible plate and the elastic film is considered to remain straight, in many experiments, however, it turns wavy and undulatory. Similar wavy contact lines are observed also with viscoelastic adhesive tapes, when peeled off a solid adherent [14, 15]. In fact, during peeling off the tape from the substrate, the process of separation itself begins with the formation of these finger-like structures at the contact line [16, 17] which is followed by several other complex phenomena like formation of columns and fibrils, drainage and thinning down of fibrils, and finally collapse. Naturally, understanding the origin of these instabilities is key for achieving the desired design parameters for the adhesive. Competing hypotheses for this instability to occur are the following: (a) viscous flow of the adhesive material, (b) the elastic strain energy of the adhesive, and (c) the coupled effect of both the viscous and elastic characters of the adhesive.

Here we have explored the possibility of the elastic deformation of the adhesive being responsible for the fingering instability to occur. Therefore, the adhesive system essentially remains the same as in the previous section, i.e., a thin, crosslinked elastomeric layer of poly(dimethylsiloxane) (PDMS), without any viscous effect, which remains bonded to a substrate as depicted in Fig. 5.1. In addition, we describe several other corollary geometries as depicted in Fig. 5.5 to explore this phenomenon. Figure 5.6 depicts the collage of optical micrographs of such instability patterns which appear in different experiments of Fig. 5.5. The schematic of the experiment presented in Fig. 5.5a depicts a cylindrical geometry in which a tiny glass sphere is placed between the adhesive film which remains bonded to a rigid substrate and a flexible contacting plate. Here, the sphere is used as the spacer which does not allow the flexible contactor to come in complete contact with the adhesive, instead, results in a stress profile which varies radially but remains symmetric in the azimuthal direction. Figure 5.5b too depicts a cylindrical geometry, but a different experiment, in which the adhesive film remains bonded to a flexible backing which is then brought in contact with a rigid circular disk used as a contactor. This disk remains fixed with a rotatable table which can be driven by a motorized drive. The film remains in complete contact with the disk when it is free of any load. However, when dead weights in the shape of annular rings are placed on the glass backing, it bends such that the film tends to come out of contact from the contactor toward the central zone of the contacting area. Here too the stress field remains radially symmetric. In addition to subjecting the adherents to normal loads these experiments allow also shear load to be applied at the interface.

For example, the round table can be rotated relative to the adhesive film with the help of a motorized drive. However, since the film remains in contact with disk indenter, in order to prevent it from rotating, a barrier is used. Optical micrographs in Fig. 5.6a–d show the instability patterns as uniformly spaced waves which appear in the rectangular geometry of Fig. 5.1. These images captured for increasing flexural rigidity D of the contactor show that the characteristic distance between the fingers, the wavelength, remains nearly independent of D; however, the length of the fingers, the amplitude, increases. At first sight these patterns appear to be similar to the

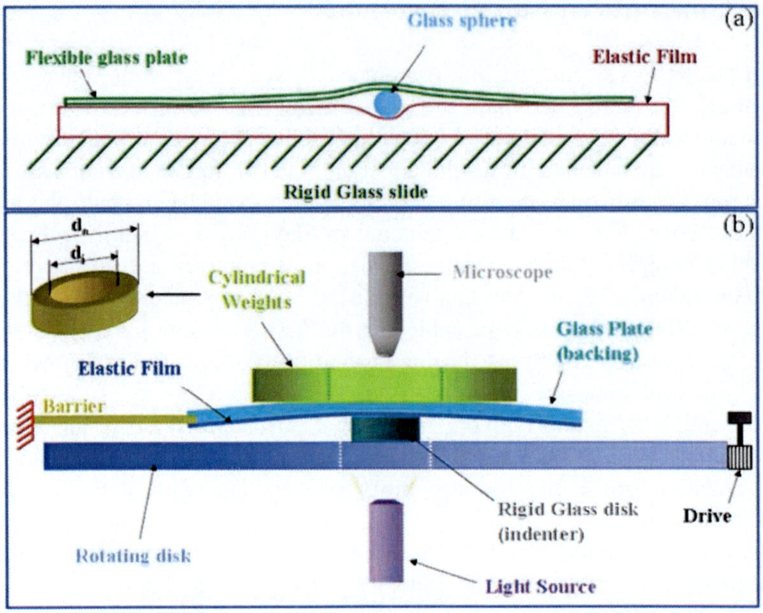

Fig. 5.5 (**a**) The figure shows the schematic of a peeling experiment cylindrical geometry. Here we use a glass sphere of diameter ~25 μm as the spacer for supporting a flexible glass cover plate. As a result the stress field remains symmetric along the θ direction but varies radially. (**b**) Here too peeling is done in the cylindrical geometry. However, here, in contrast to the experiment in (**a**), the elastic film remains bonded to a glass backing while a glass disk is attached to a round frame. Dead weights in the shape of annular rings are used for carrying out debonding and bonding of the film

well-known Saffman Taylor instability [18] which appears in confined geometry, e.g., between two parallel plates, when a liquid having larger viscosity is displaced by one with lower viscosity. The wavelength of the Saffman instability [18–20] depends not only on the gap between the plates but also on the velocity of the interface and the interfacial tension [21]. However, unlike the liquid system, here the patterns remain stable and do not vanish even when the contact line remains static. They appear spontaneously during crack opening or crack closing and even when the crack remains static, suggesting that the occurrence of these patterns does not depend on the dynamics of the system. Similar to the patterns in Fig. 5.6a–d, the images in Fig. 5.6e–h show patterns which appear when the experiments are done as in Fig. 5.5(a). Depending on the material and geometric characteristics of the adherents, the patterns appear both as fingers and also as cavitating circles. These cavitation patterns are rather similar to that found with rubber-like materials in experiment in which a thin block of rubber material confined between two rigid substrates is pulled apart by applying a normal load [22, 23]. While these cavities appear at the bulk of the rubber, in experiment of Fig. 5.5a cavities appear at the interface. What remains within these cavitating circles is a question. It is conjectured that the crosslinked PDMS material contains dissolved air which occupies the space

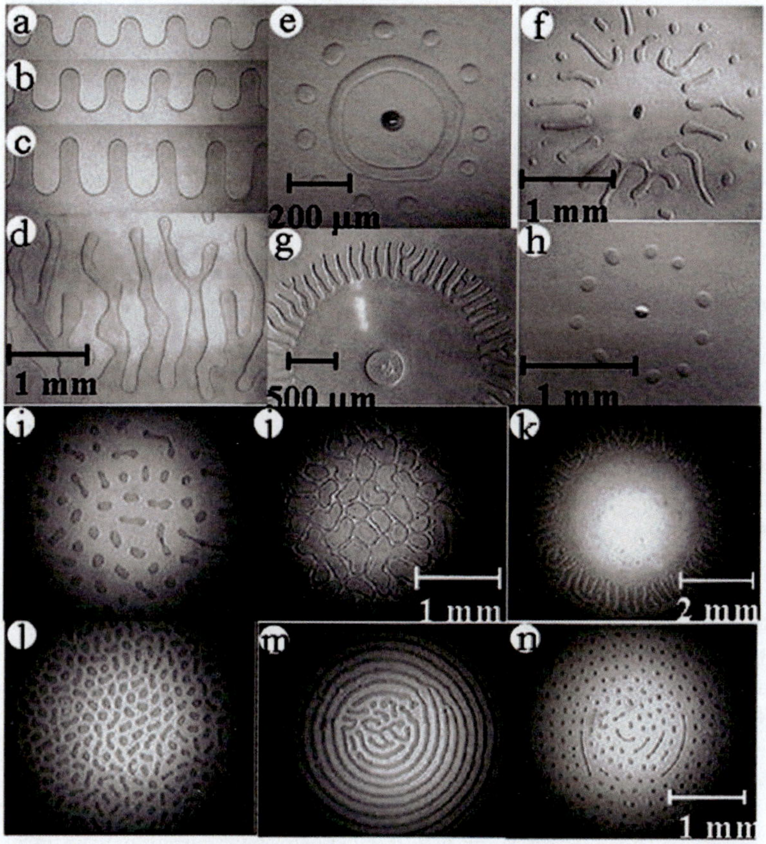

Fig. 5.6 Collage of patterns of instability at the interface of an adhesive and a flexible contacting plate. Here, the two adherents are brought in contact in different experimental geometries. For example, optical micrographs (**a–d**) are obtained in the experiment presented in Fig. 5.1, micrographs (**e–h**) are obtained in the experiment in Fig. 5.5(a), and (**i–l**) are obtained in Fig. 5.5(b)

within the cavity. These cavities do not go away even when the whole setup is placed inside a vacuum chamber. This observation suggests that the cavities do not appear because of occurrence of low-pressure region at the interface. However, the cavitation patterns do not appear randomly but keeping a constant spacing between them, which signifies that the occurrence of these patterns is possibly governed by the same physics as that of the fingers.

In particular, the instability patterns of Fig. 5.6a–h are characterized by directionality as the fingers and the cavities appear along the contact line between the adherents or along the periphery of a circle. The optical micrographs in Fig. 5.6i–n, however, show isotropic instability patterns which appear for experiments done as in Fig. 5.5b. Appearance of these patterns occurs in three different phases with progressive increase in load. Initially the cavitation patterns nucleate at lower loads, albeit not randomly, but keeping a uniform separation distance between themselves. This distance is obtained from the fast Fourier transform (FFT) analysis of the images. At the completion of the nucleation phase, the patterns grow eventually coalescing to form larger patches at which the film comes out of contact with the rigid indenter. Eventually, the film comes out of contact with the indenter throughout a circular area, with the appearance of fingering patterns at the periphery. As it so happens these isotropic instability patterns are not exactly stable because when a rotational shear is applied by rotating the indenter disk relative to the film, the isotropic cavitation patterns immediately transform into ring-like patterns. When the applied rotational shear is perfectly symmetric, the rings remain concentric; however, spiral rings emanating from the center of the contact area appear for slightly asymmetric shear load, applied at the interface. The rings remain continuous toward the central region of the contact area, however, break up into circles toward the edge. Thus the directionality of the instability patterns is restored on application of both normal and shear loads.

These instability patterns are characterized by two different length scales: the separation distance λ between the waves which scales with the thickness of the films as $\lambda \sim 4h$ [24–26] and the length of the fingers A which varies with D and μ as $A \sim (D/\mu)^{1/3}$ [8, 26, 27]. Wavelength λ depends only on the thickness of the film remaining independent of the elastic modulus of the film and the flexural rigidity of plate (Fig. 5.7). However, patterns do not appear for all combinations of geometric and material properties; experiments with varying film thicknesses and moduli and the plates with different flexural rigidities show that the contact line between the film and the plate becomes wavy when the film thickness h decreases below a critical value $h_c \sim (D/\mu)^{1/3}$ [26], implying that the threshold thickness of the film below which instability appears increases with increase in the flexural rigidity of the plate and decrease in shear modulus of the film. These observations can be written in a more compact form by defining a new parameter $\varepsilon = hq$ which is a ratio of two length scales, the thickness of the film h, and the lateral length scale, q^{-1}. In essence ε is a measure of the confinement of the film; lower the value of ε, more confined is the film and consequently more susceptible it is to undergo the instability. In fact,

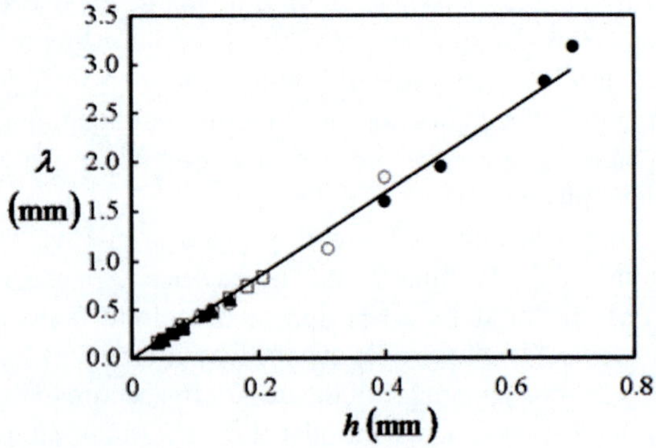

Fig. 5.7 Wavelength of the instability patterns scales linearly with thickness of the film as $\lambda = 4h$. The data remain independent of the elastic modulus of the film and the flexibility of the plate in the range of values in which experiment is done

from the experimental data stated above, the threshold value of the confinement parameter is found to be 0.35. The consequence of this confinement effect is that the adhesive stresses at the interface do not always result in uniform deformation throughout the whole area of contact, rather spatially varying deformations [22–25] attain lower energy for the system.

3.1 Excess Displacements

A perturbation analysis of the experiment brings out the dual effects of the incompressibility of the elastic film and its confinement which engenders the instability. In the perturbation analysis, the displacement field is expanded in terms of $\varepsilon = hq$ as

$$t = t_0(x, z) + \epsilon^2 t_1(x, y, z) + \cdots \tag{5.7}$$

where $t = u, v, w$, and p, respectively. Notice that in expression of (5.6), t_0 remains invariant along y; therefore, it represents the base solution of the displacement and the pressure fields which have been obtained in (5.2). The rest of the terms at the right-hand side of (5.7) account for the perturbation along y and is considered to be the excess quantities. It is apparent from the video micrographs as in Fig. 5.1b that the adhesive film and the plate remain in contact whole through the area of the finger implying that the deformation of the film is not perfectly sinusoidal, as it would mean a line contact between the plate and the film. At the simplest level, the variation of displacement and pressure field on y can be expressed as

$$\begin{aligned} t_i(x, y, z) &= t_{i0}(x, z)\sin(ky), \quad \text{where } t = u, w, \text{ and } p \\ v_i(x, y, z) &= v_{i0}(x, z)\cos(ky), \quad \text{where } i = 1, 2, 3 \end{aligned} \tag{5.8}$$

where $k = 2\pi/\lambda$ represents the wave number of surface undulations. The expression in (5.7) and (5.8) are substituted in the three-dimensional stress equilibrium relation and the incompressibility relation. The advantage of the perturbation expansion of variables as in (5.7) and (5.8) is that it allows for the terms with same order in ϵ in the resultant equation to be considered separately. The consequent relations are then solved to yield the following relations for the excess quantities [8]:

$$p_1 = 0$$
$$u_1 = 0$$

$$v_1 = \frac{c(x)}{k}\left(\left(\frac{-2kh\left(e^{kh} + e^{-kh}\right)}{e^{kh} + e^{-kh} + 2khe^{kh}}\right)\sinh(kz) + kz\left(\frac{e^{kh} + e^{-kh} - 2khe^{-kh}}{e^{kh} + e^{-kh} + 2khe^{kh}}e^{kz} - e^{-kz}\right)\right)\cos(ky)$$

$$w_1 = \frac{c(x)}{k}\left(-\frac{2e^{kh} + 2e^{-kh} + 2kh\left(e^{kh} - e^{-kh}\right)}{e^{kh} + e^{-kh} + 2khe^{kh}}\sinh(kz) + kz\left(\frac{e^{kh} + e^{-kh} - 2khe^{-kh}}{e^{kh} + e^{-kh} + 2khe^{kh}}e^{kz} + e^{-kz}\right)\right)\sin(ky) \tag{5.9}$$

Here, it is enough to consider only the lowest order terms in ϵ, since the higher order terms do not enhance accuracy significantly. Since in experiments, $\epsilon < 0.3$, the solutions for excess displacement suggest that the excess deformations in the film occur under very small excess pressure which is of the order $\epsilon^4 < 0.01$. This excess pressure, however small, varies along y implying that it still depends upon the gap between the film and the plate and is a signature of distance-dependent intermolecular forces [28, 29] present only in the immediate vicinity ($< 0.1\ \mu$m) of the contact line between the film and the plate. In other words, it suggests sharp increase in gap between the two adherents as could be observed in atomic force microscopy (AFM) images of the permanent patterns of surface undulations. Hence, it does not contribute significantly to the overall energetics. It should be noted that the instability occurs because of the deformation of adhesive film varies along y; however, the plate does not bend along y, so that the excess deformation of the plate ψ_1 and its derivatives remains uniform along this axis. The dependence of the excess quantities on x is incorporated via the coefficient $c(x)$ which is obtained as [8]

$$c(x) = \frac{c_0}{a(3A + 4a)}\left(-\left(3A^2 + 6aA + 2a^2\right)(x/A)^3 - 3\left(2A^2 + 3Aa\right)(x/A)^2 - 3\left(A^2 - 2a^2\right)(x/A) + a(3A + 4a)\right) \tag{5.10}$$

Here $c_0 = c(x = 0)$ is such that the excess energy of the system comprising the adhesive and the adherent is minimized.

3.2 Excess Energy

Total energy of the system consists of the elastic energy of the film, bending energy of the plate, and the interfacial energy; the detailed expression of it can be written as

$$\begin{aligned} \Pi &= \Pi_e + \Pi_b + \Pi_i \\ &= \frac{\mu}{4}\int_{-\infty}^{0}\int_{0}^{2\pi/k}\int_{0}^{h}\left(\left(v_z + w_y\right)^2 + \left(u_y + v_x\right)^2 + \left(u_z + w_x\right)^2\right) dzdydx \\ &+ \frac{D}{2}\int_{-\infty}^{a}\int_{0}^{2\pi/k}\left(\psi_{xx}\right)^2 dydx + W_A\left(2\pi a/k + A_{\text{finger}}\right) \end{aligned} \tag{5.11}$$

where A_{finger} is the interfacial area of contact at $-A < x < 0$. The total energy consists of two components: that corresponding to the base value of the displacement and that due to the excess quantities. In order to obtain the threshold condition for instability, it is enough to estimate the excess energy defined as $\Pi_{\mathrm{excess}} = \Pi - \Pi_0$, because the range of values of different parameters at which Π_{excess} turns negative should engender the surface instability. In order to probe this point systematically, it is best to write the excess energy in dimensionless form and in terms of dimensionless parameters; details of which can be found in [8, 27]. Here we explain the threshold condition of instability in Fig. 5.8 which depicts a typical plot of dimensionless excess energy Π as a function of the dimensionless amplitude $\xi = Aq$, for range of values of the confinement parameter, $\epsilon = 0.48 \pm 0.05$. Here Π is calculated for dimensionless crack length $aq = 25$ and dimensionless wave number $K = 2\pi/(\lambda/h) = 1.91$.

The plots show that Π varies non-monotonically with ξ, going consecutively through a maxima and a minima. The instability occurs when the minima of the excess energy turns negative. For example, we do not expect the contact line to be unstable for $\epsilon = 0.48$ as at this confinement, the excess energy Π remains positive for all values of ξ. However, the threshold condition occurs at $\epsilon = 0.365$ beyond which minima of Π does attain negative values, suggesting that the adhesive film gets more than critically confined. Notice that the occurrence of a maxima in the energy plot suggests also that only perturbations with large enough amplitude grow while others decay as observed in experiments. These observations are further illustrated in the bifurcation diagram in Fig. 5.8b which depicts the maximum and minimum values of the dimensionless amplitude of the fingers as a function of the confinement parameter ϵ. This figure too corroborates with the experimental observations that no instability is expected to occur beyond a critical value of

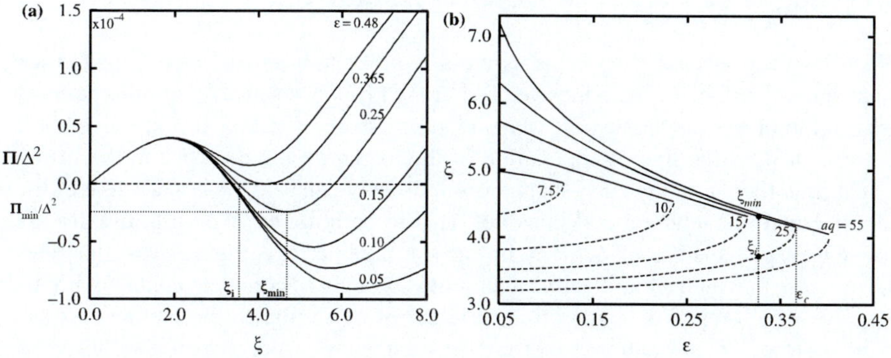

Fig. 5.8 (**a**) Dimensionless excess energy is plotted against the dimensionless amplitude of the waves for different values of the confinement parameter ϵ. The curves are obtained using representative values for the dimensionless parameters: $aq = 25$ and $K = 1.91$. (**b**) Bifurcation diagram showing variation of ξ w.r.t. ϵ for different values of aq and $K = 1.91$. The *dotted* and the *solid lines* represent, respectively, ξ_i vs. ϵ and ξ_{min} vs. ϵ. No solution for ξ exists beyond ϵ_c

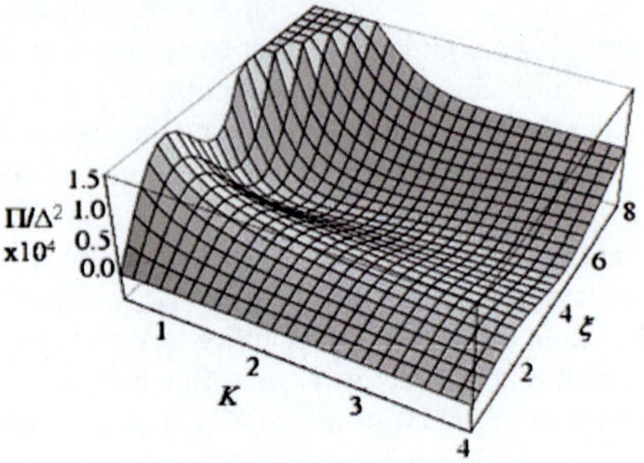

Fig. 5.9 Excess energy Π for $\epsilon = 0.2$ and $aq = 25$ is plotted w.r.t amplitude ξ and wave number K

the confinement parameter and that the amplitude of the instability waves should increase for experiments in which the film is increasingly more confined.

Figure 5.8, however, presents a simplistic picture in which the effect of confinement of the film on the amplitude of the instabilities is estimated at a specific wavelength. However, in actual experiment the wavelength and amplitude are non-linearly coupled, which is illustrated in Fig. 5.9. The data depict a typical plot of dimensionless excess energy Π as a function of the dimensionless amplitude $\xi = Aq$ and wave number kh, which shows a minima at $kh = 1.6$. This result implies that wavelength varies with thickness of the film as $\lambda = 3.93h$. These calculations bring out the important point that surface instabilities can indeed occur in thin elastic films and that the fingering instabilities observed with the viscoelastic adhesive tape owe its genesis at least to some degree to the confinement of the adhesive.

4 Peeling Off a Patterned Layer of Adhesive

What has been discussed so far is essentially propagation of a cusp-shaped crack on a smooth surface of the adhesive. The crack propagates away from the point of application of load as the flexible plate is progressively lifted off the adhesive but it traverses toward the line of application of the load with the decrease in the lifting height. In either case, the cusp-shaped crack already exists at the contact line of the adhesive and the adherent and does not have to be initiated. We will now discuss how the cusp-shaped crack actually initiates at the interface. Figure 5.1a illustrates the problem in which a thin flexible adherent is brought in complete contact with the adhesive which remains bonded to a rigid substrate similar to the previous experiments. However, in contrast to geometries depicted earlier, here the adhesive is made to have a sharp, defect-free edge by incising it with a sharp razor blade. The plate is then lifted off the adhesive to initiate the crack [30] from this edge. The difference in these two experimental geometries engenders from different conditions at the line contact between the adhesive and the flexible adherent. For the experiments described earlier, the stress field remains maximally tensile at the contact line: $\left.\frac{\partial \sigma_{zz}}{\partial x}\right|_{x=0} = 0$, because of van der Waals force of intermolecular interaction between the adhesive and the adherent. For the present experiment, the stress at the contact edge remains zero: $\sigma_{zz}|_{x=0} = 0$, as it remains open to atmosphere. As a result, the crack does not initiate from the edge of the adhesive but little away from it: $x < 0$, with the appearance of cavitation bubbles as captured by series of video micrographs in Fig. 5.10. A simple theoretical analysis similar to that presented in earlier sections shows that the stress field becomes maximally tensile at a distance $b = 0.74q^{-1}$ away from the edge ($x < 0$) but remains oscillatory with exponentially decaying amplitude further away from it. Experiments with adhesive layers of different thicknesses and shear moduli and flexibility of adherents corroborate with this scaling law with slight deviation in the value of the coefficient [30]. Here too the characteristic length scale along the lateral direction is deduced as q^{-1}.

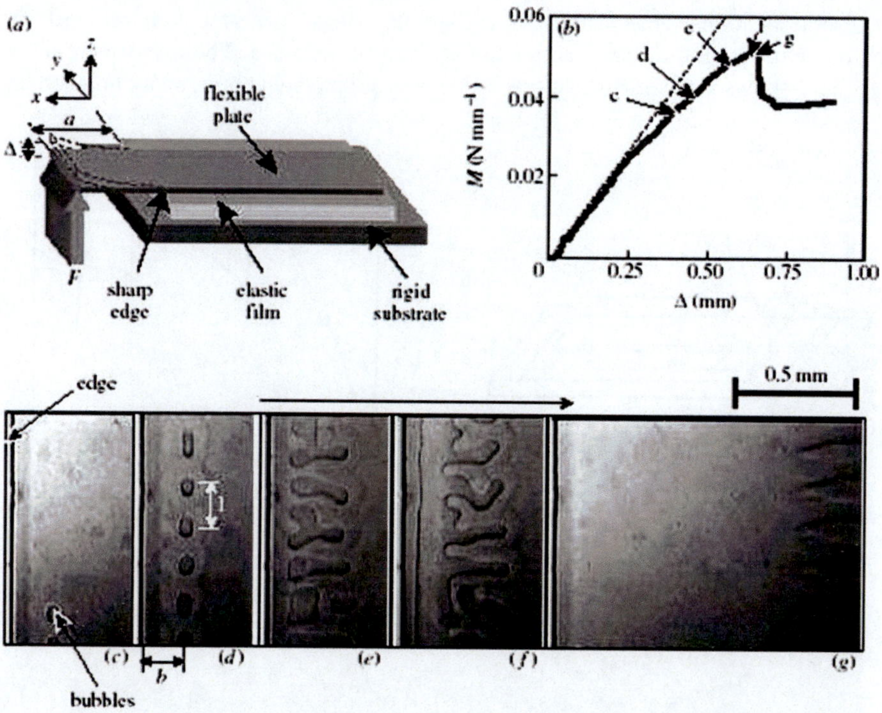

Fig. 5.10 (**a**) Schematic of an experiment in which the adhesive layer with sharp edge remains bonded to a rigid substrate. A flexible contactor is lifted off at a constant rate in order to initiate an interfacial crack. (**b**) The lift-off torque $M = Fa$ is plotted against the displacement Δ of the plate. (**c–g**) Video micrographs capture the sequence of events leading to the initiation of the crack at the interface. These data correspond to an adhesive layer of thickness $h = 40$ μm and shear modulus $\mu = 1.0$ MPa and a contactor of rigidity $D = 0.02$ Nm. The crack initiates with the appearance of cavitation bubbles at a distance away from the edge. Eventually the bubbles coalesce and merge with the edge leading to catastrophic propagation of the crack.

Fig. 5.10 shows also the corresponding torque vs. displacement plot. The torque here is obtained by multiplying the load with the distance from the edge of the adhesive, $M = F \cdot a$. This figure shows that with progressive increase in the lifting height of the plate, the load on the plate increases but the crack does not initiate till a threshold torque is reached. First, the bubbles start cavitating progressively while keeping a constant separation distance $\lambda = 4h$ between them, then the bubbles grow and coalesce, eventually, the cavitating bubbles reach the edge of the contact and the crack then propagates catastrophically away from the contact line till a new equilibrium is reached between the elastic forces in the adherents and the adhesion force at the interface. Concomitant to it, the load decreases abruptly to a significantly lower value, following which the load does not change significantly with further increase in the lifting distance. In essence, crack initiates at a larger load than that required to drive it on a smooth adhesive.

The difference between the crack initiation and propagation is further demonstrated by introducing several parallel incisions along the width of the adhesive film at several locations. Thus, a smooth, continuous adhesive is effectively replaced by the one with several discontinuities in the form of incisions. The experiment as in Fig. 5.11 shows that the surface crack does not propagate continuously on such an

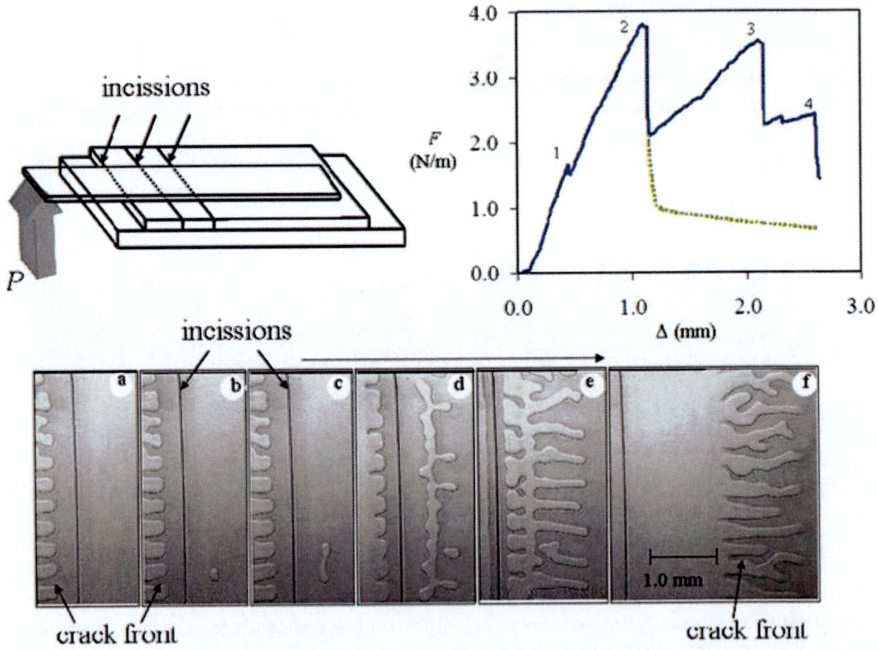

Fig. 5.11 Experiment on an elastic film patterned with several parallel incisions. On such a film, the crack does not propagate continuously but intermittently with intermediate crack arrests and initiations. The plot of lifting torque M vs. displacement Δ shows several peaks corresponding to the incisions. **(a–f)** The crack arrest and initiation phenomenon is captured by the sequence of video micrographs. An adhesive film of thickness 80 μm and shear modulus $\mu = 0.9$ MPa and a contactor of flexural rigidity $D = 0.02$ Nm is used in experiment

adhesive surface but propagates intermittently with crack arrests and initiations. The crack gets arrested at the vicinity of an incision and does not cross it, although the flexible plate is continued to be lifted with consequent increase in the load. At a threshold load, the crack initiates again but at the other side of the incision with the appearance of cavitating bubbles in a similar way as in Fig. 5.11. Eventually, the bubbles grow, coalesce, and finally reaches the incision beyond which the crack propagates again catastrophically. Thus, for adhesive films patterned with several incisions, crack arrest and initiation occur repeatedly resulting in a torque vs. displacement plot which comprises several peaks and catastrophic falls. When the incisions are very finely placed, so that the adhesive is partitioned into small islands, the effect of the individual incisions is no longer felt, instead the torque vs. displacement plot shows the combined effect of all the incisions [30, 31]. Consequently, the crack propagates uniformly on such an adhesive at a constant torque which remains larger than that required to drive the crack on a smooth adhesive. The fracture toughness G of the adhesive interface is estimated by calculating the area under the force vs. displacement plot which shows that G on patterned adhesive surfaces can be an order of magnitude larger than that on a smooth adhesive surface. In essence, in order for the crack to propagate on an island, it gets stretched. However, the elastic energy stored in one island does not get transferred to the one behind the crack because of physical discontinuity. As a result, once the crack crosses over an island, it relaxes back to zero load leading to dissipation of energy (Fig. 5.12).

This result underscores the importance of mechanics in adhesion as it shows that the adhesion strength of an interface can be manipulated by simply altering the physical geometry of the adhesive without altering the chemical character or its rheological behavior. This principle now forms the basis of a new class of PSAs for which the chemical nature of the adhesive no longer plays the primary role, rather adhesion is manipulated via physical patterning with nano- to microscopic structures.

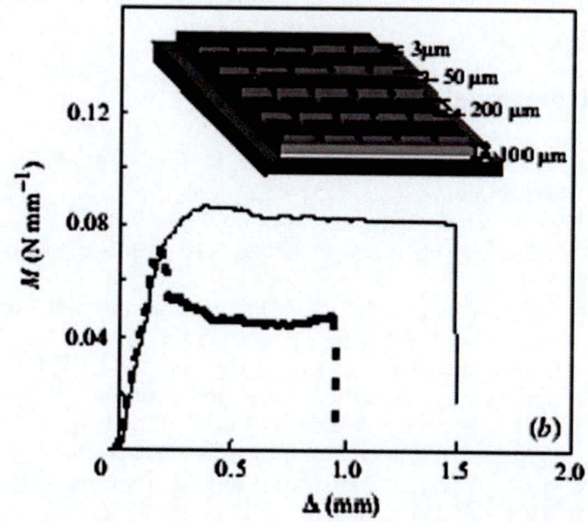

Fig. 5.12 Experiment on an adhesive film surface textured with rectangular islands shows that the lifting torque remains constant at a larger value than that for an adhesive layer with smooth surface

5 Summary

To summarize, we have described in this chapter a simple yet rigorous method for measuring the adhesion strength of a model adhesive which is elastic, incompressible, and of finite uniform thickness. This method complements to two other conventional methods of adhesion measurement: the JKR (Johnson, Kendall, and Roberts) experiment [12] and the classical peel experiment [32, 33]. Whereas the JKR method is based on a rigorous theory, it is applicable only for adhesives considered to be elastic half space; it does not account for the effect of the finite thickness of the adhesive layer. On the other hand, the peel experiment in which an adhesive bonded to a flexible backing is peeled off, a substrate is widely used because of its simplicity, but it is not very rigorous, e.g., it allows for large deformation of the adhesive and the plastic backing. As a result, it is difficult to estimate the actual contribution of the molecular interaction at the interface of the adhesive with the adherent. The lifting plate experiment presented in this chapter addresses these drawbacks of these methods. We have depicted also a new kind of instability that occurs with thin films of elastic adhesive. These instabilities are often observed as column-like structures at the vicinity of the three-phase contact line during peeling of a scotch tape. We have shown that these instabilities are signature of confinement of the adhesive film as it is the confinement coupled with the incompressibility of the adhesive that leads to the favorable energetics for the perturbations to grow. Distance-dependent surface forces are important but do not play any significant role. Finally, we have highlighted the significance of topographical patterning of adhesive surface which allows enhancement of adhesion strength of the adhesive interface via physical mechanism. It is worth noting that all our analysis presented here is restricted to incompressible elastic films. However, these experiments and the theoretical analysis prepare the ground for more analyzing more complex problems associated with viscoelastic adhesive materials.

Acknowledgment The author acknowledges financial support from the Department of Science and Technology, Government of India, and Mr. and Mrs. Gian Singh Bindra Research Fellowship.

References

1. F. Sauter, E. Hayek, W. Moche, U. Jordis, Identification of betulin in archaeological tar, Z. Naturforsch **42(11)**, 1151–1152 (1987).
2. F. Sauter, A. Graf, C. Hametner, J. Fröhlich, Studies in organic archaeometry III 1. Prehistoric adhesives: alternatives to birch bark pitch could be ruled out, ARKIVOC **2001** (v) 21–24 (2001).
3. A. W. Adamson, A. P. Gast, Physical Chemistry of Surfaces, 6th Edition, Wiley Interscience, New York, NY, (1997).
4. J. N. Israelachvili, Intermolecular and Surface Forces: With Applications to Colloidal and Biological Systems, Second Edition, Academic, New York, NY (1991).
5. L. D. Landau, E. M. Lifshitz, Theory of Elasticity, 3rd Revised Edition; Course of Theoretical Physics, Pergamon, New York, NY, Vol. VII (1986).
6. G. K. Batchelor, Introduction to Fluid Dynamics, Cambridge University Press, Cambridge, (1967).

7. A. Ghatak, L. Mahadevan, M. K. Chaudhury, Measuring the work of adhesion between a soft confined film and a flexible plate, Langmuir, **21**, 1277 (2005).
8. A. Ghatak, Confinement-induced instability of thin elastic film, Phys. Rev. E **73**, 041601 (2006).
9. D. A. Dillard, Bending of plates on thin elastomeric foundations, J. Appl. Mech. **56**, 382 (1989).
10. J. W. Obreimoff, The splitting strength of Mica, Proc. R. Soc. Lond. A. **127**, 290–297 (1930).
11. M. K. Chaudhury, G. M. Whitesides, Direct measurement of interfacial interactions between semispherical lenses and flat sheets of poly (dimethylsiloxane) and their chemical derivatives, Langmuir **7**, 1013 (1991).
12. K. L. Johnson, K. Kendall, A. D. Roberts, Surface energy and contact of elastic solids, Proc. R. Soc. Lond. A **324**, 301 (1971).
13. K. Vorvolakos, M. K. Chaudhury, The effects of molecular weight and temperature on the kinetic friction of silicone rubbers, Langmuir, **19**, 6778–6787 (2003).
14. Bi-min Z. Newby, M. K. Chaudhury, H. R. Brown, Macroscopic evidence of the effect of interfacial slippage on adhesion, Science **269**, 1407–1409 (1995).
15. Bi-min Z. Newby, M. K. Chaudhury, Effect of interfacial slippage on viscoelastic adhesion, Langmuir **13(6)**, 1805–1809 (1997).
16. R. J. Fields, M. F. Ashby, Finger-like crack growth in solids and liquids, Philos. Mag. **33**, 33 (1976).
17. Y. Urhama, Effect of peel load on Stringiness phenomena and peel speed of pressure sensitive adhesive tape, J. Adhes. **31**, 47 (1989).
18. P. G. Saffman, G. I. Taylor, The penetration of a fluid into a porous medium or Hele-Shaw cell containing a more viscous liquid, Proc. R. Soc. Lond. A **245**, 312 (1958).
19. G. M. Homsy, Viscous fingering in porous media, Annu. Rev. Fluid Mech. **19**, 271–311 (1987).
20. J. Nittmann, G. Daccord, H. E. Stanley, Fractal growth viscous fingers: quantitative characterization of a fluid instability phenomenon, Nature **314**, 141 (1985).
21. G. H. McKinley, A. Öztekin, J. Byars, R. A. Brown, Self-similar spiral instabilities in elastic flows between a cone and a plate, J. Fluid Mech. **285**, 123 (1995).
22. A. N. Gent, P. B. Lindley, Internal rupture of bonded rubber cylinders in tension, Proc. R. Soc. Lond. A, **249** (1257), 195–205 (1959).
23. A. N. Gent, D. J. Tompkins, Nucleation and growth of gas bubbles in elastomers, J. Appl. Phys. **40**, 2520 (1969).
24. A. Ghatak, M. K. Chaudhury, V. Shenoy, A. Sharma, Meniscus instability in a thin elastic film, Phys. Rev. Lett. **85**, 4329 (2000).
25. W. Mönch, S. Herminghaus, Elastic instability of rubber films between solid bodies, Europhys. Lett. **53**, 525 (2001).
26. A. Ghatak, M. K. Chaudhury, Adhesion-induced instability patterns in thin confined elastic film, Langmuir **19**, 2621 (2003).
27. A. Ghatak, M. K. Chaudhury, Critical confinement and elastic instability in thin solid films, J. Adhes. **83**, 679–704 (2007).
28. V. Shenoy, A. Sharma, Pattern formation in a thin solid film with interactions, Phys. Rev. Lett. **86**, 119 (2001).
29. J. Sarkar, V. Shenoy, A. Sharma, Patterns, forces and metastable pathways in debonding of elastic films, Phys. Rev. Lett. **93**, 018302 (2004).
30. A. Ghatak, L. Mahadevan, J. Y. Chung, M. K. Chaudhury, V. Shenoy, Peeling from a biomimetically patterned thin elastic film, Proc. R. Soc. Lond. A **460**, 2725–2735 (2004).
31. J. Y. Chung, M. K. Chaudhury, Roles of discontinuities in bio-inspired adhesive pads, J. R. Soc. Interface **2**, 55–61 (2005).
32. D. H. Kaelble, Theory and analysis of peel adhesion: bond stresses and distributions, Trans. Soc. Rheol. **4**, 45–73 (1960).
33. J. L. Gardon, Peel adhesion. II, A theoretical analysis, J. Appl. Polym. Sci. **7**, 643 (1963).

Chapter 6
Liquid Thin Film Hydrodynamics: Dewetting and Pattern Formation

Rabibrata Mukherjee

Abstract Ultra-thin polymer films become unstable due to various types of interaction forces like van der Waals interaction, steric forces, molecular level recoiling, sudden release of residual stresses or due to the presence of defects on substrate or the film, resulting in disintegration and rupture of the film, which is also associated with morphological evolution and formation of mesoscale surface features. In this chapter we introduce the concept of dewetting first, followed by a brief theoretical discussion on the conditions under which a thin liquid film can spontaneously become unstable, based on a linear stability analysis. We subsequently discuss the morphological evolution sequence under true experimental conditions. The instability-mediated structures are inherently random and isotropic, thereby having limited practical utility. We discuss how dewetting on a topographically patterned substrate might be useful in imposing long-range order to the dewetted structures. Finally, we discuss some recent developments on suppressing dewetting in unstable film by incorporation of nanoparticles or nanofillers in extremely low amount to the polymer matrix. This approach of stabilizing ultra-thin films will be extremely useful from the standpoint of coatings, which should not degrade and disintegrate with time.

1 Introduction

Stability and dynamics of thin liquid films and coatings are important in many areas like molecular electronics [1], flexible display screens [2], optical sensors [3], structural color [4], reusable super adhesives [5], super hydrophobic and self-cleaning surfaces, [6], and scaffolds for tissue and genetic engineering [7]. Even every blink of the eye involves the motion of the eyelid over a thin layer of corneal film that isolates any direct contact between the lid and the cornea. The film is known as the "tear film" and a rupture of which results in the so-called dry eye syndrome and can result in epithelial tissue damage and corneal ulceration [8]. Further, for people using contact lenses, the interaction between the tear film and the lens surface

R. Mukherjee (✉)
Department of Chemical Engineering, Indian Institute of Technology (IIT),
Kharagpur 721302, India
e-mail: rabibrata@iitkgp.ac.in

S. Chakraborty (ed.), *Mechanics Over Micro and Nano Scales*,
DOI 10.1007/978-1-4419-9601-5_6, © Springer Science+Business Media, LLC 2011

is important in ensuring prolonged contact lens tolerance and the lens is invariably coated with a wetting and cushioning solution prior to its insertion in the eye in order to enhance the preferential wetting of the tear film on the lens surface [8]. Similarly, the film integrality is of utmost importance in any coating application. However, liquid thin films often generate intriguing and complex flow regimes, which may, under some specific conditions, lead to the disintegration of the film itself, with the formation of regular or chaotic structures which may have submicron feature size and periodicity, depending on the film thickness [9–58]. The latter concept of structure formation during thin film instability is progressively finding application as a viable alternative to soft lithography for the formation of meso- and nanoscale surface patterns, particularly in soft materials like polymers and gels. It is therefore important to understand the hydrodynamics of a thin liquid film, in order to understand how or under what condition its rupture can be inhibited (coating application) or what type of structures will evolve based on its instability (patterning application).

The subject of flow of thin liquid films spreading on solid surfaces has been investigated for a long time, as Reynolds himself was one of the first ones to work in this area [59]. His work on lubrication flows eventually led to the development of the "lubrication theory," which is extensively used even today to describe flow with negligible inertial effects at low magnitude of the flow components associated with the dynamics of the flow [59]. Instability in a thin liquid film has been extensively studied in various settings under the action of various forces like gravity, centrifugation, capillarity, thermo-capillarity, and intermolecular forces on smooth/structured and impermeable/porous surfaces with and without evaporation/condensation. Most of these studies involve tracking the motion of the three-phase contact line [60].

However, this chapter is not intended to act as a full-scale review and we will keep our discussion limited only to interfacial interaction-driven instability in ultra-thin, low inertia static films, which exhibit rich dynamics and pattern formation [9–58]. While these types of instability are possible in all types of viscous liquids, recent attention is significantly on the study of polymer thin films, which are both important from the practical standpoint as well as their high viscosity and low vapor pressure allow performance of time-resolved studies. Further, the structures can be frozen or made permanent by simply quenching the unstable film below glass transition temperature, T_G (for most common homopolymers $T_G \sim 100°C$), thereby allowing the formation of permanent patterns and further detailed ex situ investigation (for example, atomic force microscopy). The stability of an ultra-thin film (thickness $\sim$100 nm or below) depends on a competition between destabilizing long-range intermolecular forces and the stabilizing surface tension and other short-range repulsive forces. This chapter is organized as follows: Section 1 covers a basic understanding of spontaneous dewetting in ultra-thin films, including identification of the necessary and sufficient conditions for film instability based on linear stability analysis (Section 2) [9–19]. In Section 3, some recent results on experimental investigation of dewetting and pattern formation in a polymer thin film is presented, including the morphological evolution at different stages of dewetting [20–43]. Section 4 provides some examples of using a topographically or a chemically patterned substrate template to align the dewetting patterns [44–56]. Section 5 presents some very recent results on working out strategies for inhibiting instability and dewetting in ultra-thin polymer films [61–75].

2 Basics of Dewetting

2.1 Complete and Partial Wetting

At this point we introduce the concept of complete and partial wetting of a solid surface by a liquid. Wettability of a chemically homogeneous, smooth, flat, rigid surface by a liquid depends on several factors, including the surface tensions of the liquid (γ_L) and the solid (γ_S) and the interfacial tension between the two (γ_{SL}). A typical configuration of a liquid drop resting on a defect-free solid surface is shown in Fig. 6.1a, where the angle θ_E refers to the intrinsic equilibrium contact angle, and close to the solid surface, can be obtained from a balance of the horizontal components of the concerned surface and interfacial tensions, which is the well-known Young's equation and is given as

$$\gamma_S = \gamma_{SL} + \gamma_L \cos\theta_E \tag{6.1}$$

It is obvious from the above equation as well as Fig. 6.1a, b that the geometry of a drop, for a given volume of liquid, on a solid surface is a function of the surface and interfacial tensions. In a scenario where other body forces like gravity is negligible as compared to the surface forces and there is no internal pressure distribution within the drop, it (the drop) would typically assume the shape of a section of a sphere [76]. Depending on the magnitude of θ_E, two distinct wetting regimes are possible: (i) complete wetting, where the liquid drop spreads fully on the solid surface ($\theta_E \sim 0°$, Fig. 6.1c), and (ii) partial wetting, where θ_E has a non-zero finite value which in other words means that the liquid forms a drop on the solid surface, instead of spreading. A further classification for partial wetting is possible, depending on the value of θ_E [76]. In cases where θ_E is acute ($0 < \theta_E < 90°$, Fig. 6.1a), the substrate surface is typically termed as lyophilic and in contrast, when θ_E is obtuse ($90° < \theta_E < 180°$, Fig. 6.1b), the surface is known to be lyophobic. Lyophobicity implies that it is thermodynamically favorable to have the bare solid surface rather than to

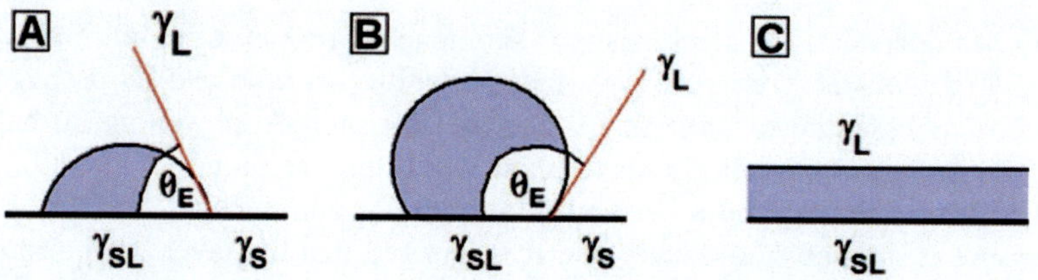

Fig. 6.1 Illustration of (**a**, **b**) partial wetting and (**c**) complete wetting of a solid surface by a liquid

have it submerged under a liquid film. In cases where the concerned liquid is water, a surface exhibiting an acute equilibrium water contact angle (WCA) is known as a hydrophilic surface and a surface with WCA in the range of 90–180° is known as a hydrophobic surface [76]. It is well known and can also be seen from (6.1) that a lower value of γ_S will result in higher value of θ_E, that is, on a lower energy surface the liquid will show less spreading vis-a-vis lower liquid–solid interfacial contact area [76]. Thus hydrophobicity is seen on surfaces that possess lower surface tension. There are theories that show that WCA on a surface can depend on the roughness of the surface also, in addition to lower surface energy alone leading to wetting regime transition to the well-known Cassie and Wenzel states, which is the basic of structural superhydrophobicity [77, 78]. However, discussion on those topics is beyond the scope of this chapter and an interested reader is encouraged to consult several excellent reviews available in the area [79, 80].

Based on the discussion thus far we now present the basic concept of dewetting. Dewetting is opposite of spreading and results in the retraction of a three-phase contact line on a solid surface. One easy and macroscopic example of dewetting is when a stationary drop of liquid, which is in equilibrium on a solid surface making an angle θ_E with it, is disturbed, which results in forcible spreading of the drop, resulting in an increased solid–liquid area of contact vis-a-vis reduced contact angle between the drop and the substrate. As θ becomes smaller than θ_E, the balance of the horizontal components of surface tension gets disturbed, and the RHS in equation (6.1), which corresponds to the force acting inward (toward the center of the drop), becomes larger. This results in a macroscopic retraction of the contact line, with an associated shape change of the drop in the form of increasing θ. Under ideal circumstances, the retraction continues till θ becomes identical to θ_E. In reality, one would often encounter a non-spherical drop shape or non-circular contact line, which is due to inhomogeneities on the substrate that act as local pinning locations. Dewetting in a thin film occurs under similar circumstances, where the imbalance in the horizontal components of surface and interfacial tension triggers the dynamics of the contact line and the final drop shape is governed by θ_E. However, dewetting results only after the film has ruptured. The subsequent section describes the condition under which a thin film can spontaneously rupture.

2.2 *Spontaneous Film Rupture Due to Interfacial Interactions: Capillary Waves*

In order to understand the mechanism of spontaneous rupture of a thin film which is associated with the growth of a surface instability in time and its propagation in space, it is essential to understand the precise condition prevailing at the surface of a thin liquid layer. In order to understand this, we draw our attention to a layer of liquid that rests on a perfectly flat and horizontal substrate. If no external pressure is applied to the layer and it is ensured that the layer has no internal or external thermal gradient, then it can be argued that the layer represents a stagnant, static pool of liquid, where all the components of velocity at each point are identically zero. This implies that the vertical component of velocity at the liquid–air interface is zero and it is therefore reasonable to conclude that the liquid–air interface is perfectly flat. In reality, the assumption of a flat liquid–air interface is perfectly valid at the macroscopic length scale. However, if the same interface is investigated at sub-10 nm resolution, due to thermal motion of the molecules (kT, k: Boltzmann constant) as well as mechanical vibrations, extremely low amplitude (approximately few nanometers) and high-frequency molecular level fluctuations are always present at the liquid–air interface. Further, each of these tiny fluctuations increases the interfacial surface area, and therefore, the growth of them are opposed by surface tension, which tries to minimize the total surface energy. In the absence of any other destabilizing force fields, these interfacial fluctuations die down with time. However, every individual fluctuation results in a localized pressure gradient between the peak (higher pressure) and the valley (lower pressure) of the fluctuation at the fluid surface. The localized pressure difference, which is also known as the Laplace pressure, is related to the surface tension as well as local radii of curvature by the well-known Young Laplace equation which is given as $\Delta p = \gamma(1/R_1 + 1/R_2)$. The gradual flattening of the interfacial perturbations due to the stabilizing influence of surface tension results in flow of liquid from the higher pressure zones (peaks) of the fluctuations to lower pressure valleys on the liquid surface. The localized flows give rise to a spectrum of capillary waves, which are always present on every liquid surface, irrespective of thickness or depth of the layer. As the amplitude of the capillary wave spectrum is extremely low their existence is of no significance at the macroscopic scale, where the gravitational force field overwhelms all other interactions. Thus, in conventional fluid dynamics, no reference is drawn about surface tension or capillary waves while discussing open channel flow [10].

2.3 Distinction Between a Stable, Unstable, and Metastable Thin Films

The scenario changes with reduced film thickness and particularly for films thinner than about 100 nm, as the two interfaces of the film, the film–substrate interface and the film–air interface, start interacting due to intermolecular van der Waals interaction. It may be noted that attractive van der Waals interaction results between two molecules even in non-polar atoms, arising out of the interactions of instantaneous multipoles. Such interactions are attractive and scale as $1/r^6$, where r is the separation distance between the molecules and the attraction persists only up to few nanometers (~10 nm). Between two surfaces, the van der Waals interaction shows a scaling of $1/r^2$, with a slower decay as compared to that between two molecules. Consequently, the interaction stretches over few tens of nanometers (arguably around 100 nm) [11].

Thus, if a liquid layer is thinner than ~100 nm, then van der Waals interaction-induced interfacial attraction is active between its two interfaces. This attraction tries to amplify the capillary waves, by increasing the amplitude of the fluctuation

[11]. As already mentioned, the growth of the capillary waves is opposed by surface tension. In an event where the attractive interfacial interaction overcomes the stabilizing effects of surface tension, the initial thermal or mechanical fluctuations at the film–air interface experience a driving force for growth. The film ruptures with formation of holes when the growing amplitude of the wave equals the film thickness. Concurrently, the two distinct interfaces of the film also merge, forming a three-phase contact line on the substrate [17]. Subsequent dynamics on the film surface is in the form of retraction of the contact line, till the dewetting liquid attains thermodynamic equilibrium on the surface. This phenomenon of spontaneous rupture and dewetting of a thin film on a solid surface is referred to as the spinodal dewetting, due to its similarity to the well-known spinodal decomposition in fluid mixtures, where the compositional fluctuations in the fluid mixture can be regarded analogous to the height fluctuations in a thin film. Surface tension plays an interesting role in spinodal dewetting: it opposes the growth of any fluctuation at the liquid–air interface thereby playing a stabilizing role in an intact film. However, as soon as the film ruptures and holes form, surface tension acts in the direction of the hole growth leading to further dewetting of the ruptured film on the substrate, undergoing a role reversal [17].

In thicker films, the interfacial attraction is absent and therefore such films do not undergo spontaneous dewetting. However, such films may still dewet, only in the event of an instability being nucleated around a defect. Thus it becomes important to distinguish between a thermodynamically stable, unstable, and a metastable film in a quantitative manner. Typically this distinction is based on the sign of the partial derivative of ϕ (effective interface potential $= \partial \Delta G/\partial h$; ΔG: excess free energy per unit area) with respect to the local film thickness ($\partial \phi/\partial h$). It is obvious that the sign of $\partial \Phi/\partial h$ depends on the functional form of ϕ. In an unstable film (curve 1, Fig. 6.2) ϕ is constituted of antagonistic interactions with different decay rates [14]. The two major types of interactions which dominate in a general dielectric thin film are the already-mentioned longer range interfacial attraction due to van der Waals interaction and short-range steric repulsion. It is important to note that within the hydrodynamic framework no study of dewetting is possible in the absence of a short-range repulsion close to the substrate due to non-physical singularity at the contact

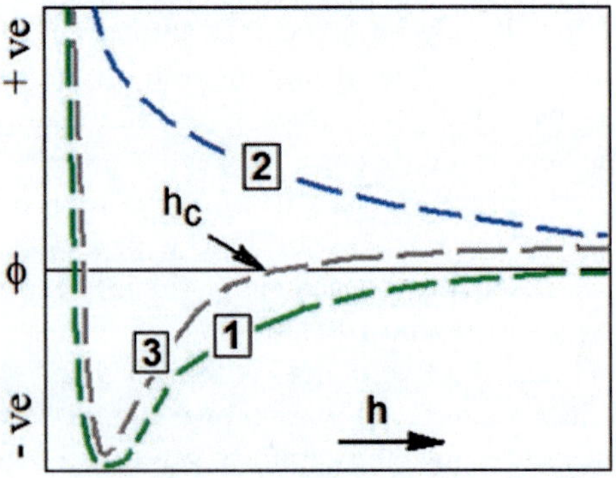

Fig. 6.2 Illustrative ϕ vs h (film thickness) dependence for a spinodally unstable (*line 1*), stable (*line 2*), and metastable (*line 3*) films, respectively

line [15–19]. A well-known representation of ϕ for an unstructured thin liquid film which captures the essential physics of a viscous thin film reasonably accurately is given as [19]

$$\phi = (A_s/6\pi h^3) - (S_P/l_P)\exp(-h/l_P) - (B/h^9) \tag{6.2}$$

The equation is well suited for simple liquids as well as amorphous polymer melts. A_s represents the effective Hamaker constant due to the van der Waals interaction; a positive or negative value of it denotes long-range interfacial attraction or repulsion [14]. S_P denotes the strength of any medium or short-range interaction. Depending on whether S_P is positive or negative, the interaction can be repulsive or attractive. l_P represents the corresponding decay length [26]. The effects of any acid–base (AB) interactions in a polar liquid like water (hydrophobic repulsion) or the entropic confinement effects due to adsorption/grafting of polymer brushes at an interface are represented by this term [14, 19]. In order to avoid the non-physical singularity at the contact line in the absence of any physical short-range repulsion, an additional term (B/h^9) is included for simulations [15, 19]. This term is akin to a contact repulsion and it plays a major role in simulations involving most polymer films, as in simple homopolymers the second term is in most cases rather insignificant. In the absence of any form of interfacial attraction, no film instability is possible (curve 2, Fig. 6.2), as a long-range repulsion tends to stabilize the film. A classic example of a stable film is a low-energy coating on a high surface energy substrate, an example of which is the oxide layer on a silicon surface (typically 1.5–2 nm thick, often referred to as the native oxide layer). In fact it is very difficult to remove the oxide layer, which is necessary for the fabrication of microelectronic chips. A film is termed metastable in which the sign of $\partial\Phi/\partial h$ changes as a function of h (curve 3, Fig. 6.2) [19, 28].

2.4 The Thin Film Equation

The linearized equation of motion for non-slipping, isothermal, single-component, Newtonian fluid film without evaporation/condensation on a physically or chemically rough substrate exhibiting long wave instability is given as by Sharma as [19]

$$3\mu h_t + [(h - af)^3\{\gamma(h_{xx} + h_{yy}) - \phi\}_x]_x + [(h - af)^3\{\gamma(h_{xx} + h_{yy}) - \phi\}_y]_y = 0 \tag{6.3}$$

where subscripts denote differentiation with respect to that particular variable; $h = h(x, y, t)$ is the local film thickness from a datum, $z = 0$; $z = af(x, y)$ quantifies a rough substrate surface; $\eta = [h(x, y, t) - af(x, y)]$ and $a = 0$ for a smooth surface, co-ordinate x is parallel to the substrate surface, t is time, μ is viscosity, and γ is the surface tension of the liquid [19]. Simplification of the x-component Navier–Stokes equation with long wave (or small slope) approximation in association with the continuity equation provides the individual components of velocity (u, w) along the

x- and z-co-ordinate axes as functions of h, Φ, and x. The small slope approximation ($\delta h/\delta x << 0$), which originates from the assumption that the wavelengths of the surface fluctuations (λ) are much larger ($\lambda >> h_0$) compared to the mean film thickness (h_0), is reasonably successful in capturing the essential physics of the system unless the film thickness is of molecular dimensions [19, 26]. The velocity components are substituted in the kinematic boundary condition (mathematical form: $\delta h/\delta t + u_s\, \delta h/\delta x = w_s$) which is akin to local mass balance at the liquid–air interface and correlates the two components of velocity resulting in (6.3). Subscript s refers to velocity at the free surface. The solution of this equation gives the mean film thickness h as a function of spatial coordinate (x) and time (t). In situations where $h_t = 0$, the steady-state film thickness is of the form $h(x)$, signifying a spatial variation of film thickness even at a steady state, which is somewhat similar to a dynamic steady state, and can be correlated to the concept of the capillary waves and their stabilization by surface tension forces [15, 18, 19].

The term ϕ_x is a summation of two different gradients [19]:

$$\phi_x = (\partial\varphi/\partial h)(\partial h/\partial x) + (\partial\varphi/\partial x)|_h \tag{6.4}$$

On a smooth surface with no heterogeneity (chemical or physical), the second term $(\partial\varphi/\partial x)|_h = 0$.

It is logical to view (6.3) as a summation of three distinct sets of interaction, which are the contributions of viscous forces, surface tension force due to the curvature at the free surface (Laplace pressure), and the excess intermolecular forces (disjoining pressure), respectively. The viscous forces do not affect the stability of the film and merely control the rate of evolution of the system. It is already discussed that the Laplace pressure term due to surface tension has a stabilizing influence in an intact film. Thus, any potential cause of destabilization invariably has its origin in the terms representing the excess intermolecular interaction (terms containing ϕ). Further, (6.3) is found to be extremely useful in investigating whether a film of uniform initial thickness h_0 remains stable with time or not [26]. The linearized equations of motion subjected to the influence of intermolecular forces (ϕ) and simplified by the lubrication theory (negligible inertial effects) give the following space periodic solutions for the film thickness as [19, 26]

$$h(x,t) = h_0 + \varepsilon \sin(kx)\exp(\omega t) \tag{6.5}$$

where

$$\omega = C[-\gamma_L k^4 - (\partial\Phi/\partial h)k^2] \tag{6.6}$$

ε is the amplitude of perturbation around mean film thickness h_0 and $\sin(kx)\exp(\omega t)$ represents the functional form of perturbation, which is periodic in space and exponential in time. Equation (6.6) provides the linear dispersion relation for the initial growth coefficient (ω) for a small amplitude ($\varepsilon << h_0$) perturbation with wave number (k). Only a positive value of the growth coefficient (ω) will lead to the growth of the initial perturbations with time and eventually destabilize the film. This means that the necessary condition for film instability is $\omega > 0$. The parameter $C = h_0^3/3\mu$ is a positive constant and is independent of k. Thus, the condition of film instability ($\omega > 0$) is possible only for $\partial\Phi/\partial h < 0$. Consequently, a film with $\partial\Phi/\partial h > 0$ for all values of h is thermodynamically stable. A film is metastable if $\partial\Phi/\partial h$ is negative for some ranges of h only. The term $\partial\Phi/\partial h$, which solely influences the possibility of spinodal dewetting, is referred to as the *spinodal parameter*. By way of dewetting, liquid moves from thinner zones of the film to thicker zones, which is a scenario similar to a negative diffusion [19, 26, 28].

Equation (6.6) further provides two critical values of ω. The first of the two can be obtained by solving for k at conditions of $\omega = 0$. Apart from the trivial solution of $k = 0$, a second solution is [26]

$$k = (-(\partial\Phi/\partial h)/\gamma)^{0.5} \tag{6.7}$$

It can be seen from (6.6) that for values of $k > (-(\partial\Phi/\partial h)/\gamma)^{0.5}$, ω becomes negative irrespective of the sign of $\partial\Phi/\partial h$. Thus, the expression obtained in (6.7) provides an upper cutoff (k_{C}) in wave number, signifying that only the fluctuations with wave number in the range $0 < k < k_{\mathrm{C}}$ can grow with time. Fluctuations with higher wave number ($k > k_{\mathrm{C}}$) fail to overcome the stabilizing influence of surface tension and eventually subside. This implies that wavelengths having periodicity shorter than $\lambda < 2\pi/k_{\mathrm{m}}$ fail to cause any destabilization even in a spinodally unstable film. This also validates the assumption about the long wave nature of instability in spinodal dewetting. The other critical wave number, k_{m}, is obtained from (6.6) by setting $\partial\omega/\partial k = 0$ and solving for k. This wave number corresponds to the condition at which the growth coefficient is maximized. The corresponding wavelength, $\lambda_{\mathrm{m}} = 2\pi/k_{\mathrm{m}}$, represents the short time initial length scale of instability. Further, λ_{m} is an important parameter as it can be directly measured in experiments and can be compared with theoretical predictions. The expressions for λ_{m}, N_{m} (number density of features), and corresponding linear timescale for the appearance of instability are given as [19, 26]

$$\lambda_{\mathrm{m}} = \left[-8\pi^2\gamma/(\partial\phi/\partial h)\right]^{1/2} \tag{6.8a}$$

$$N_{\mathrm{m}} = \lambda_{\mathrm{m}}{}^{-2} = -(\partial\phi/\partial h)/8\pi^2\gamma \tag{6.8b}$$

$$\tau = 12\,\gamma\mu\left[h^3(\partial\phi/\partial h)^2\right]^{-1}\ \ln(h/\varepsilon) \tag{6.8c}$$

It can be seen that the substitution of the expression of ϕ in (6.8a) and (6.8b) provides the qualitative dependence as $N_{\mathrm{m}} \sim h^{-4}$ and $\lambda_{\mathrm{m}} \sim h^2$. These scaling relations have been found to be valid in many experiments [19, 26].

The above analysis on film instability including the scaling relations is only valid for spinodally unstable films or for metastable film having thickness $h < h_{\mathrm{C}}$. For stable films and for metastable films above $h > h_{\mathrm{C}}$ this mechanism will not cause any instability, as the interfacial interaction is too weak (for thicker films) or there is predominant long-range repulsion between the two interfaces. However, many experiments show dewetting of stable films as well as thick films and surprisingly in many such cases the morphological evolution sequence (discussed later) as well as the final patterns are found similar to spinodal instability. In such films rupture and dewetting can only result due to nucleation around substrate heterogeneities [19, 27–31, 34, 35]. Under real experimental conditions, such heterogeneities can be dust, trapped micro-cavities, chemical contamination, variation in oxide layer thickness on silicon, variable polymer chain adsorption at different locations, etc. [19, 31, 35]. The substrate heterogeneity in many cases result in large amplitude initial surface defects in the film during the coating process itself, eventually resulting in a local film thickness which is thinner than the critical thickness below which the film is unstable. In such a case, the film ruptures over these areas and the retraction of the contact line leads to further dewetting. This type of nucleation is generally termed as "true" or "heterogeneous" nucleation [19, 35]. The other mechanism of nucleation has been identified as "thermal" or "homogeneous" nucleation. Such type of nucleation is possible only in a metastable film having thickness close to the spinodal boundary ($h \sim h_C$) [19, 28, 31]. For these films, the thermally excited capillary waves may spontaneously amplify, sometimes resulting in a local reversal of sign of $\partial\phi/\partial h$. However, as no specific energy barrier is overcome in this case, it is debatable if this regime undergoes a true nucleation. Experts like Sharma prefer to term this mechanism as "defect-sensitive spinodal regime" (DSSR) [19]. It is also important to note that an unstable film may also undergo nucleation in the presence of substrate heterogeneity or defects. This scenario is an example of "nucleation within spinodal regime" and is different from DSSR [19]. In most real settings, for a spinodally unstable film, both modes of instability are co-operative; the morphological evolution is governed by the mode with a faster timescale and the final morphology often bears the signature of both the forms of instability [19]. On a heterogeneous surface, ϕ shows spatial variation [$\phi(x, y)$], which results in a non-zero value for the in-plane potential gradient ($(\partial\varphi/\partial x)|_h$) in (6.4). This term represents a micro-scale wettability contrast on the substrate and leads in a flow of liquid from a less wettable area to a more wettable region. This flow mechanism is in many ways analogous to the well-known Marangoni flow, except that here the flow originates due to the gradient of free energy at the solid–film interface rather than at the free liquid surface, as in classical Marangoni flow [19, 35].

In addition to the above two mechanisms, several mechanisms have been proposed which may lead to rupture and dewetting of a thin polymer film. These include localized density variation within the film [36], additional force field arising out of confinement of thermally excited acoustic waves [24], influence of molecular recoiling [37], surface melting and associated lowering of glass transition temperature [38], and substrate preparation and cleaning procedures [39]. Unfortunately, none of them have been able to seamlessly address all the experimentally observed complex dewetting behaviors across different settings. Most of them are yet to be theoretically established in a convincing fashion.

A recent work by Reiter and coworkers attributes the residual stresses accumulated within a film during the coating process itself to be responsible for dewetting

[40]. The stresses accumulate due to rapid evaporation of solvent during spin coating. The existence of these accumulated stresses significantly influences the probability of rupture. The stresses are attributed to a non-equilibrium conformation of the entangled long-chain polymer molecules. With time, these molecules attain conformations closer to thermodynamic equilibrium, resulting in a reduction in the level of accumulated stresses [31, 40]. The stress release occurs even at temperatures below T_G. However, at room temperature the rate of release of these stresses is very slow. Consequently, the level of stress remnant in a film becomes a function of the time of aging. Thus, films aged for different durations have different levels of residual stresses and therefore have varied probability of rupture, when the films are heated above T_G. Thus, a film aged for a lower period will have higher probability of rupture and hole formation as compared to a film which has been aged for a longer period of time. The magnitudes of the force field arising from the release of the stresses are much higher as compared to forces arising due to interfacial interactions and therefore tend to dominate the dewetting scenario [40]. However, the theoretical understanding of the role of residual stress on dewetting is far from complete as of now.

3 Experimental Studies on Dewetting of Thin Polymer Films

The first experimental evidence of dewetting of a thin polymer film was reported by Reiter in his landmark paper in 1992 [20]. The system they investigated was the thermal dewetting of a PS film coated on a hydrophobized silicon substrate [20–22]. Ever since this work, many groups have investigated various aspects of thin film dewetting experimentally [23–34]. Ultra-thin films of simple homopolymers like PS (polystyrene) and PMMA (polymethyl methacrylate) coated on cleaned silicon wafer, quartz, or glass substrates have traditionally been preferred model systems for studies on thin film instability. In most cases, spin coating is adopted as the preferred coating technique, as other methods like dip coating and drop casting fail to produce films thinner than $\sim$50 nm, which is essential for successfully studying spontaneous dewetting. However, dewetting is not limited to homopolymer thin films alone as the phenomena is rather material non-specific. Dewetting has been observed in various settings involving thin films of a variety of complex functional polymers like liquid crystals [41], polyelectrolytes, conjugated polymers [42] blends, and block co-polymers [43]. In most experiments, the films are generally in a glassy solid state at room temperature and are liquefied, either by heating beyond T_G or by exposing to solvent vapor, to engender dewetting. Heating above T_G results in a rapid drop in the film viscosity (about four to five orders), thereby allowing the evolution to occur at realistic timescales. A higher molecular weight polymer film will have higher viscosity and will therefore exhibit slower dynamics. Exposing the film to solvent vapor reduces the effective glass transition temperature of the polymer as the solvent molecules penetrate into the polymer matrix, reducing the intermolecular cohesion. It is in the liquid state where the rupture and the growth of instability as well as all the subsequent morphological evolution takes place. In the solid state, the polymer film is amorphous and behaves like a supercooled liquid with extremely high viscosity. Thus, even a thermodynamically unstable film at room temperature does not exhibit any instability, which can be attributed to extremely high viscosity of the polymer film at that temperature. Thus, an intact thin polymer film at room temperature can be kinematically stabilized even if it is thermodynamically unstable. Precise signature of spinodal instability alone is rare, as under true experimental conditions heterogeneities always dominate. One of the first such examples of pure spinodal instability was reported by Xie et al. in an ultra-thin PS film (<5 nm thickness), where it was seen that the amplification of the capillary wave spectrum resulted in the formation of bi-continuous patterns [23]. The structures, though random typically possess a characteristic length scale (λ), which can be obtained from fast Fourier transform (FFT) of the image. With time, the amplitude of the undulations grow, finally resulting in the rupture and disintegration of the film into small droplets. These droplets are often not stable and get transformed to larger droplets due to a late stage coarsening.

For thicker films (20–50 nm thickness), combined signatures of nucleation and spinodal dewetting are observed in most experiments. For such films, the onset of instability is in the form of appearance of nearly equal sized, random holes (Fig. 6.3a). With time, more holes appear and the already existing holes grow in size, with the retraction of the three-phase contact line. Hole growth in most cases is also associated with the formation of a distinct rim along the periphery of the hole. The rims appear due to mismatch in the rates at which polymer is dislodged from the substrate and at which it gets redistributed to the other intact parts of the film, which is slower. This results in a localized accumulation of polymer just ahead of the moving contact line, which gets manifested in the form of a rim. In typical optical micro-graphs, the rim appears as a bright circular border around each hole, as can be seen in Fig. 6.3b. Interestingly, the shape of the rim carries a distinct signature of the rheology (viscoelasticity) of the film as well as indicates the possible presence of interfacial slippage [31]. While the cross section of a typical hole rim is circular for a purely viscous film, it shows a highly anisotropic geometry in viscoelastic thin films. A detailed discussion on this theme is, however, beyond the scope of this chapter.

Subsequent growth of holes results in coalescence of adjacent rims thereby forming a network of polymer ribbons. This morphology is often referred to as the cellular pattern (Fig. 6.3c). With time these polymer ribbons break down into isolated droplets due to Rayleigh instability. This particular sequence of thin film dewetting has been widely reported by various groups. In some experiments, the holes did not grow cleanly as the rims themselves became unstable in the form of undulations and undulatory structures (Fig. 6.3a). In cases where a rim instability is observed, the cellular pattern formation stage is suppressed and a completely random array of droplets result (Fig. 6.3d). The shapes of the final dewetted structures are governed by the equilibrium contact angle (θ_E) of the polymer on the substrate material, as can be seen from the AFM image in Fig. 6.3e. The droplet size as well as the periodicity of the features becomes larger with increase in film

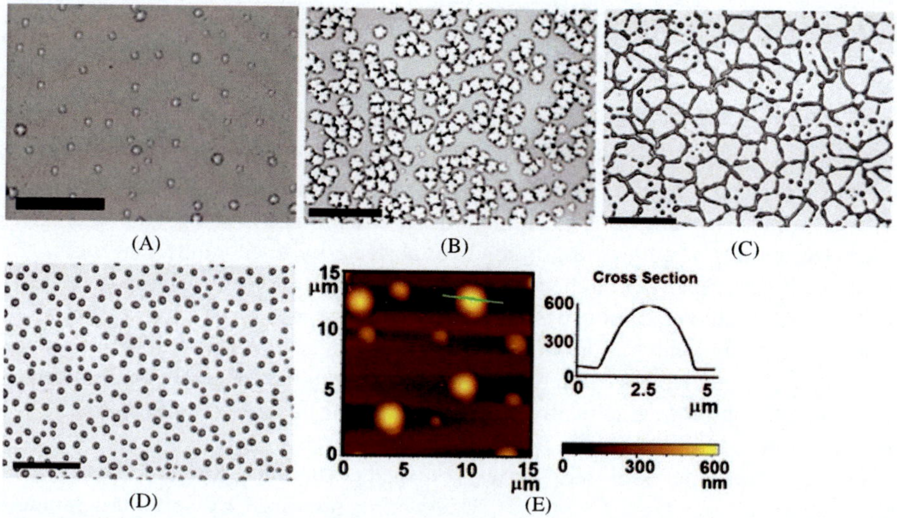

Fig. 6.3 Evolution of morphological patterns at different stages of solvent (toluene) vapor-assisted dewetting of a 24 nm thick PS film on a flat, smooth cross-linked PDMS surface. Frames show different stages of dewetting after exposure to toluene vapor for (**a**) 85 s (initiation of random holes), (**b**) 3 min (growth of the holes), (**c**) 7 min (coalescence of growing holes leading to the formation of polymer ribbons) (scale bar 30 µm), and (**d**) 16 min (formation of isolated dewetted droplets). (**e**) AFM scan of the dewetted structures, showing the cross section of an individual droplet. The *scale bars* correspond to 50 µm in all images except (**c**). Reproduced with permission from [51], © Royal Society of Chemistry 2008

thickness, though the overall morphology of the patterns remains nearly identical. This provides a clue for controlling the feature size by simply adjusting the initial conditions like film thickness. This particular aspect has made instability-mediated pattern formation a versatile surface patterning technique, which might eventually become a viable alternative to soft lithography. Further, the morphological evolution can be arrested at any intermediate stage by simply quenching the film below glass transition temperature, thereby creating structures on demand. Thus, from the spontaneous instability and dewetting of an ultra-thin polymer film, it might be possible to create either an array of holes or an array of droplets, by allowing the dewetting sequence to progress up to a desired level. However, lack of long-range order of the structures severely affects the utility of instability-mediated morphological evolution being used as a viable surface patterning technique. Significant recent attention is therefore on working out strategies by which long-range order can be imposed to the dewetting films. Typical approaches include dewetting on a chemically or a topographically patterned substrate. This approach, which is often referred as template-guided self-organization, is an example of pattern formation based on a combination of top-down and bottom-up approaches. Some examples on substrate pattern-guided dewetting is presented in the next section.

4 Ordered Structure Formation by Dewetting on a Patterned Substrate

That the isotropic and randomly oriented dewetted patterns can be aligned was first demonstrated by Higgins and Jones, as they simply rubbed the substrate onto which the film was coated with a lens tissue before coating the film [44]. The dewetted patterns were highly anisotropic, aligned in the direction of rubbing. Reproducible ordered structures are, however, difficult to fabricate by a technique like rubbing. Using a topographically or a chemically patterned substrate can result in dewetted patterns with high degree of order spanning over large areas. Here we present one such exciting case study on dewetting of a thin polymer film on a topographically patterned substrate. However, there are several other published papers on dewetting on chemically and topographically patterned substrates and an interested reader is encouraged to consult them [44–56].

The substrate under consideration was a striped cross-linked Sylgard 184 (a two-part elastomer from Dow Chemicals, USA) film, produced by soft lithographic techniques [57]. A key question to raise at this point is if a film is directly spin coated on a topographically patterned substrate, would it form a film of uniform thickness on the patterned substrate. In most cases, it is seen that direct spin coating on a topographically patterned substrate results in thickness variation [58]. In such a case, dewetting invariably engenders from the location where the film thickness is minimum. In this particular investigation, to avoid any thickness variation, a novel sample preparation strategy was adopted. The film of uniform thickness was first spin coated on a cleaned silicon wafer and was subsequently transformed onto the patterned substrate by floating on water (Fig. 6.4) [50, 51]. It is seen that depending on the condition of the capture of the film during transfer, two completely different initial film morphologies with respect to the underlying substrate are possible. A near-horizontal position of the patterned substrate while capturing the freely floating PS film combined with a rapid pull off resulted in the so-called focal adhesion,

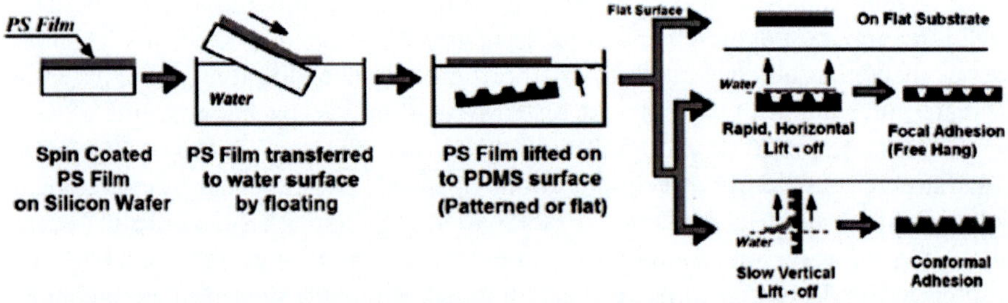

Fig. 6.4 Schematic of the protocol used for floating of a thin polystyrene (PS) film from water bath and its subsequent transfer onto a patterned cross-linked polydimethylsiloxane (PDMS) surface. Depending on the capture condition, focal or conformal adhesion of the polymer film on the patterned substrate is possible. Reproduced with permission from [51], © Royal Society of Chemistry 2008

where the top surface was nearly flat, as the film was only in contact with the raised protrusions of the substrate pattern [50]. In contrast, when the patterned substrate was pulled off at a slower pace and was held vertically, the transferred polymer film was found to be rather closely adhering to the contours of the substrate pattern resulting in the so-called conformal adhesion [50, 51]. While in both the cases the dewetted droplets get aligned due to the substrate pattern, the positioning of the drops is strongly influenced by the nature of initial conformation of the film, as will be clear from the subsequent discussion.

In this particular work, solvent vapor-assisted dewetting of a 38.3 nm thick PS film transferred onto a stripe-patterned substrate (stripe periodicity 1.5 μm, stripe width ~700 nm, stripe height ~120 nm) has been investigated. Figure 6.5a1, b1 shows the initial morphology of the transferred PS films on the topographically patterned substrates. It can be clearly seen from the insets that Fig. 6.5a1 shows an initial *focal adhesion* of the film on the substrate. In contrast, Fig. 6.5b1 shows a scenario where the film is in *conformal adhesion* with the substrate pattern. When these films are exposed to solvent vapor both the films rupture. However, the location of the rupture varies depending on the nature of adhesion. In *focal adhesion*,

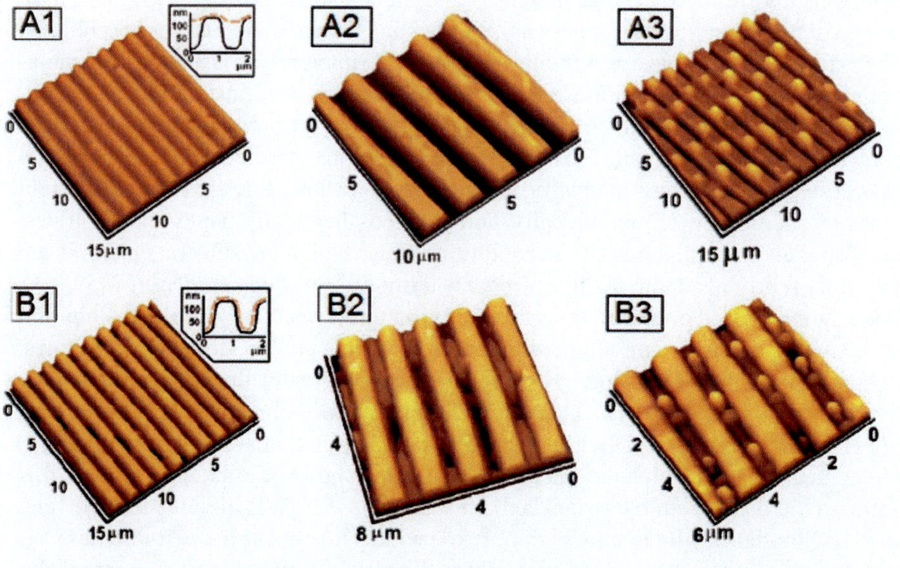

Fig. 6.5 Morphological evolution of a 38.3 nm thick PS film on a stripe-patterned Sylgard 184 substrate ($\lambda_P = 1.5$ μm, stripe width 700 nm, height 130 nm), for two distinct initial morphologies: (**a**) focal adhesion and (**b**) conformal adhesion. **a1** and **b1** show the initial morphology of the transferred PS film over the patterned substrate, before exposure to solvent vapor. **a2** and **b2** show the intermediate morphology in both cases, respectively, after 5 min of solvent vapor exposure. In both cases the morphology comprises of aligned long polymer strips, but in **a2** (focal adhesion) they are on top of the stripes and in **b2** (conformal adhesion) they are within the channels. **a3** and **b3** show the final morphology after 20 min of solvent vapor exposure comprising aligned droplets, on top of the stripes (**a3**) and along the stripes (**b3**), respectively. Reproduced with permission from [50], © American Scientific Publishers 2007

the rupture of the film takes place over the areas where it hangs freely between two adjacent stripes. The higher effective Hamaker constant and zero viscous resistance over the freely hanging zones as compared to areas of the film in contact with the substrate lead to a much faster rupture. Subsequently, the ruptured film snaps back onto the top of the stripes, eventually resulting in aligned polymer ribbons on top of each stripe (Fig. 6.5a2) [50]. With prolonged solvent vapor, exposure the long strips of polymer eventually disintegrate into aligned droplets on top of the stripes (Fig. 6.5a3) due to Rayleigh instability [50]. However, in conformal adhesion (Fig. 6.5b3), though the droplets are of nearly equal size as compared to that seen for focal adhesion (Fig. 6.5a3), they are aligned inside the channels [50]. In the latter case, film rupture initiates along the stripe edges and the ruptured film results in the formation polymer ribbons confined within the channels (Fig. 6.5b2). Eventually, these ribbons also break down into droplets due to Rayleigh instability which remain aligned within the channels [50].

There are several other examples of substrate pattern-directed dewetting and ordering of thin polymer films. For example, on a 2-D-patterned substrate, it is further seen that a perfectly ordered and perfectly filled dewetted droplet array forms only for a narrow range of film thickness and depends greatly on the relative commensuration between film thickness and the substrate geometry [51]. A slight change in the initial film thickness drastically changes the final pattern morphology. Such drastic transitions are difficult to track experimentally and therefore simulations constitute an important aspect of research on thin film instability in general, and on patterned substrates in particular. Kargupta and Sharma have significantly contributed in this area with 3-D nonlinear simulations, predicting the morphology as well as the dewetting pathways for films with various thicknesses on different types of chemically, topographically, and physico-chemically patterned substrates, providing an understanding of the conditions under which the substrate patterns are faithfully reproduced into the film, which is termed as "perfect templating" [52, 53]. For example, based on simulations the condition for perfect templating of a thin film on a substrate comprising alternating less and more wettable stripes is as follows: (1) the periodicity of substrate pattern (λ_S) must be greater than the characteristic length scale of instability (λ_D) corresponding to the film thickness, but less than an upper cutoff, which is $\sim 2\lambda_D$; (2) film rupture is initiated near the stripe edges; (3) the contact line rests close to the stripe boundary; and (4) the liquid cylinders that form on the more wettable stripes remain stable [52–54]. It is obvious that the relative magnitudes of film thickness (h_0), stripe width (l_P), and stripe periodicities (λ_S) and their commensuration with each other crucially influence the final morphology. A variety of parametric studies have been done using simulation tools, thereby predicting the likely final morphologies with excellent accuracy. Many of the predicted morphologies on 1-D- and 2-D-patterned substrates by the Sharma group have been verified experimentally by other researchers, establishing the robustness of the simulations. The availability of robust-simulated phase diagrams is extremely useful in substantially reducing experimental efforts as they can be used for selecting the appropriate initial conditions (i.e., film thickness, pattern size, geometry, etc.) to experimentally produce the desired final structures [52–54].

5 Suppression of Dewetting

So far we have discussed about the spontaneous instability and associated pattern formation in an ultra-thin film [12–58]. However, from the standpoint of a coating, spontaneous disintegration of a film is completely undesirable. Thus, one of the key motivations of the study of dewetting is to understand the science that leads to thin film instability and based on that knowledge work out strategies to suppress or inhibit the film degradation. Several methods and techniques with limited success have been proposed all of which aim at inhibiting dewetting. Some examples of such approaches include grafting of compatibility-enhancing polymer layers at the film–substrate interface [61], incorporation of polymer brushes having end groups with specific affinity toward the substrate material [62], metal complexation or sulfonation of the polymer [63], phase segregation-induced stabilization of polymer thin films (PS) by adding another polymer (PMMA) in low amounts [65], in situ photochemical crosslinking using photoactive molecules [66] etc. Addition of nanoparticles (NP) or nanofillers to the polymer film in extremely low quantities [67–74] has been shown to be successful in suppressing dewetting. We discuss another recent case study involving this approach and show that the NP concentration is a critical parameter that influences the stability of the film significantly [75]. This approach was first reported by Barnes et al. with ultra-thin polystyrene and polybutadiene thin films (20–50 nm) to which low quantities (0.1–5% w/w) of fullerene nanoparticles were added [67].

In a recent paper, involving solvent vapor-induced dewetting of thin polystyrene (PS) films having thickness in the range of 12–42 nm containing *uncapped* gold nanoparticles, several new findings have been observed [75]. The first major observation involves a transition from complete dewetting for low NP concentration to complete inhibition of dewetting for higher NP concentration. The existence of three distinct stability regimes was identified based on NP concentration: *Regime 1*, complete dewetting all the way up to the stage of droplet formation for NP concentration up to 2% (w/w); *Regime 2*: where dewetting occurs only partially, leading to some intermediate stage like the hole formation or ribbon formation only, for intermediate NP concentrations (3–6%); and *Regime 3*: where dewetting is completely inhibited at high NP concentrations (>7.0%). This transition is clearly observed in Fig. 6.6, which shows the final morphology of thin films having identical thickness (~17 nm) with different NP concentrations and exposed to toluene vapor for 60 h, an adequate time for complete dewetting of a particle-free 17 nm thick PS film. Figure 6.6b shows the final morphology of a fully dewetted particle-free film, which is nearly identical to that observed for a regime 1 film with NP concentration of 1.0% (Fig. 6.6a). Frames of Fig. 6.6c, d capture the morphology of regime 2 films with NP concentrations of 3.2 and 6.4%, respectively, where dewetting progressed only upto certain intermediate stage in both the cases. It is further observed that with increase in NP concentration, dewetting progresses to a lesser extent [75]. Figure 6.6e shows a completely intact film with high (8%) NP concentration, which does not show any signature of instability or dewetting. This is further validated by the AFM image showing a smooth film surface in Fig. 6.6f. Further, it has been observed that the

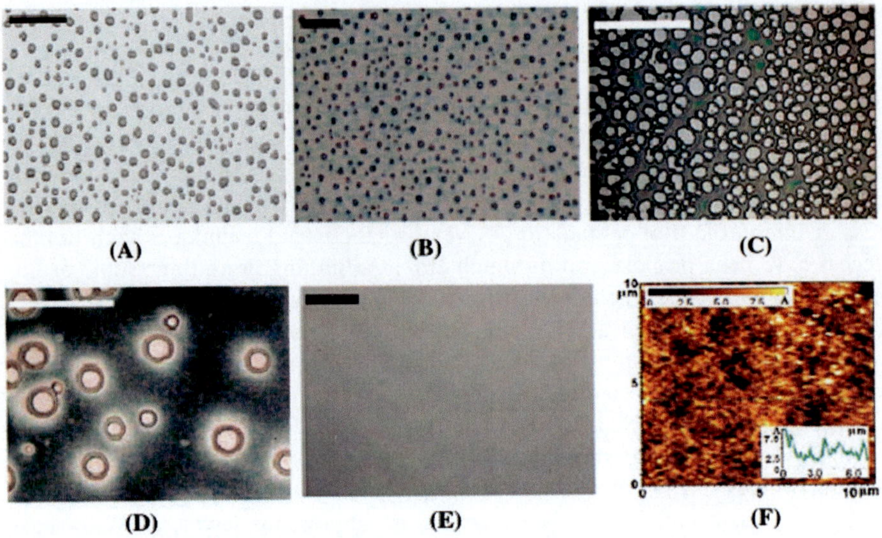

Fig. 6.6 Influence of NP concentration on the final morphology of polystyrene films exposed to toluene vapor. Completely dewetted structures are seen for regime 1 (low NP concentration) films: (**a**) 1.0% NP concentration, exposure time ~3 h; (**b**) dewetted structure resulting from a particle-free film, exposed for 12 min. Partial dewetting for regime 2 films, after 60 h of exposure: dewetting up to the level of (**c**) coalescence of rims for a film with NP concentration of 3.2% and (**d**) up to formation of holes for a film with 6.4% NP concentration. (**e**) Completely stable film after 60 h of solvent vapor exposure when the NP concentration was 8.0% (regime 3). (**f**) AFM image of a stable (regime 3) film, showing a smooth film surface (rms roughness ~1 nm as verified by AFM) with no surface undulations even after 60 h of solvent vapor exposure. Film thickness ~17 nm in all cases and scale bar 30 μm in all micro-graphs except (**f**). Reproduced with permission from [75], © American Chemical Society 2010

length scale of the structures remains unaltered for regime 1 films to that observed in particle-free films. However, the structure length scale gradually increases with enhanced NP concentrations for regime 2 films. Further the dynamics of film evolution becomes progressively sluggish with increase in NP concentration in all regimes [75]. At low NP concentrations, particularly for regime 1 films, it is also observed that there are multiple stages of dewetting velocity with clear crossover, a rapid initial stage followed by a sluggish and prolonged late stage. However, this distinction progressively vanishes with enhanced NP concentration [75].

The slower dynamics of dewetting with enhanced NP concentration is manifested in two ways: (a) longer solvent vapor exposure time for the appearance of the first hole as well as to reach the stage where the holes start coalescing and also (b) slower dewetting velocity. Critical examination of the possible mechanisms responsible for the change in instability length and timescales and their observed dependence on particle concentration throw further insight to the complexities associated with the whole process. It has been argued that no single factor is responsible for the observations affecting the stability and the length scales alone, but a combination of several factors controls the scenario [75]. For example, addition of nanoparticles can influence (a) slippage of the film on its substrate or (b) rheology (viscoelasticity) of the film or (c) can even introduce additional sites for heterogeneous nucleation-based mechanism for rupture. However, slippage alone would ideally enhance the dewetting velocity and would reduce the length scale exponent. But the experimental observations show exactly the opposite trend. Increase in viscoelasticity with higher NP concentration would indeed make the dynamics sluggish, but would have again lead to reduction of length scale of the features due to enhanced resistance against dewetting and instability brought in by the higher level of elasticity. Particle-induced nucleation sites would have also resulted in reduction of instability length scale by providing more nucleation sites. Thus it is obvious that none of the reasons above in isolation are responsible for the observed behaviors, indicating the complex nature of the system [75], which indeed requires further investigation.

6 Conclusion

In this chapter we have presented glimpses of how spontaneous instability can result in an ultra-thin polymer film and can lead to flow of liquid polymer from one location to the other by dewetting. We have discussed how van der Waals force-driven interfacial attraction can lead to such instabilities [10–42]. We have also discussed in very simple terms why such instability is not possible in a thicker film. Further classification of a thin film into a stable, a metastable, or an unstable one has been done based on the sign and expression of the potential of the film (ϕ) [19, 28]. Based on the linear stability analysis we have presented the necessary conditions for film instability in an ultra-thin film in terms of ϕ [26]. Further, we have shown the existence of a dominant length scale in spontaneous instability, which depends on initial film thickness. We have also correlated the origin of capillary waves to the molecular level fluctuations in very simple terms.

We have also discussed how under real experimental condition such instability is manifested. The morphological evolution during dewetting provides a clue for creating patterns on demand, by simply controlling the extent of dewetting, which can be stopped at any point of time by simply quenching the sample below room temperature. Instability structures also show scaling with initial film properties like film thickness and surface tension in terms of feature size and structure length scale. This can further be exploited to control feature size by adjusting the initial conditions. However, the practical utility of the instability-mediated structures is strongly limited due to their lack of long-range order [43–56]. In order to achieve ordering, dewetting on a topographically or a chemically patterned substrate can be adopted. In this chapter we have discussed in detail one example on dewetting of a polymer thin film on a topographically patterned substrate and have shown how the initial configuration of the film with respect to the underlying substrate influences the positioning of the droplets [50].

The final issue that has been discussed in this chapter, again with reference to a recent work, was on working out strategies that can be adopted for stabilizing a thermodynamically unstable film [61–75]. Addition of nanoparticles or nanofillers in very less quantity is found to be a valid approach. We have discussed how the concentration of the particles plays a key role in determining the stability of a film vis-a-vis the extent of dewetting. Depending on the particle concentration, a film may completely or partly dewet or can even become completely stable. This approach might be of extreme use in stabilizing ultra-thin coatings [75].

Finally, I would like to point out that this chapter is in no way a comprehensive review on dewetting of thin polymer films. An interested reader is encouraged to consult many relevant publications as well as some excellent and exhaustive review papers [60, 75, 81–85] that are already available.

Acknowledgment The author wishes to thank Ashutosh Sharma, Department of Chemical Engineering, IIT, Kanpur, for valuable discussion and useful advices. Funding from a DST Nano Mission project is gratefully acknowledged.

References

1. Lloyd CR, Gorman CB (2002) The genesis of molecular electronics. Angew. Chem. Int. Ed. 41:4378–4400
2. Kim DH, Rogers JA (2008) Stretchable electronics: Materials strategies and devices. Adv. Mater. 20:4887–4892
3. Scott BJ, Wirnsberger G, Stucky GD (2001) Mesoporous and mesostructured materials for optical applications. Chem. Mater. 13:3140–3150
4. Sato C, Kubo S, Gu ZZ (2009) Structural color films with lotus effects, superhydrophilicity, and tunable stop-bands. Acc. Chem. Res. 42:1–10
5. Geim AK, Dubonos SV, Grigorieva IV, Novoselov KS, Zhukov AA, Shapoval SY (2003) Microfabricated adhesive mimicking gecko foot-hair. Nat. Mater. 2:461–463
6. Sun M, Luo C, Xu L, Ji H, Ouyang Q, Yu D, Chen Y (2005) Artificial lotus leaf by nanocasting. Langmuir 21:8978–8981
7. Chin VI, Taupin P, Sanga S, Scheel J, Gage FH, Bhatia SN (2004) Microfabricated platform for studying stem cell fates. Biotechnol. Bioeng. 88:399–415
8. Sharma A, Ruckenstein E (1986) The role of lipid abnormalities, aqueous and mucus deficiencies in the tear film breakup, and implications for tear substitutes and contact lens tolerance. J. Colloid Interface Sci. 111:8–34
9. Ruckenstein E, Jain RK (1974) Spontaneous rupture of thin liquid films. J. Chem. Soc., Faraday Trans. II 70:132
10. Derjaguin BV (1989) Theory of Stability of Thin Films and Colloids. New York, NY: Plenum
11. Israelachvilli JN (1985) Intermolecular and Surface Forces. New York, NY: Academic
12. Williams MB, Davis SH (1982) Nonlinear theory of film rupture. J. Colloid Interface Sci. 90:220–228
13. Sharma A, Ruckenstein E (1986) An analytical nonlinear theory of thin film rupture and its application to wetting Films. J. Colloid Interface Sci. 113:456–479
14. Sharma A (1993) Relationship of thin film stability and macroscopic parameters of wetting in the systems. Langmuir 9:861–869
15. Sharma A, Khanna R (1998) Pattern formation in unstable thin liquid films. Phys. Rev. Lett. 81:3463–3466
16. Sharma A, Khanna R (1999) Pattern formation in unstable thin liquid films under the influence of antagonistic short- and long-range forces. J. Chem. Phys. 110:4929–4936
17. Ghatak A, Khanna R, Sharma A (1999) Dynamics and morphology of holes in dewetting of thin films. J. Colloid Interface Sci. 212:483–494

18. Sharma A, Reiter G (2002) Morphological phase transitions in spontaneous dewetting of thin films on homogeneous and heterogeneous surfaces. Phase Trans. 75:377–399
19. Sharma A (2003) Many paths to dewetting of thin films: Anatomy and physiology of surface instability. Euro. Phys. J. E 12:397–408
20. Reiter G (1992) Dewetting of thin polymer films. Phys. Rev. Lett. 68:75–78
21. Reiter G (1993) Unstable thin polymer films: Rupture and dewetting processes. Langmuir 9:1344–1351
22. Reiter G (1994) Dewetting as a probe of polymer mobility in thin films. Macromolecules 27:3046–3052
23. Xie R, Karim A, Douglas JF, Han CC, Weiss RA (1998) Spinodal dewetting of thin polymer films. Phys. Rev. Lett. 81:1251–1254
24. Morariu MD, Schäffer E, Steiner U (2004) Molecular forces caused by the confinement of thermal noise. Phys. Rev. Lett. 92:156102 (4 pages)
25. Reiter G, Khanna R, Sharma A (2000) Enhanced instability in thin liquid films by improved compatibility. Phys. Rev. Lett. 85:1432–1435
26. Reiter G, Sharma A (1996) Instability of thin polymer films on coated substrates: Rupture, dewetting, and drop formation. J. Colloid Interface Sci. 178:383–399
27. Seemann R, Herminghaus S, Jacobs K (2001) Shape of a liquid front upon dewetting. Phys. Rev. Lett. 87:196101 (4 pages)
28. Seemann R, Herminghaus S, Jacobs K (2001) Dewetting patterns and molecular forces: A reconciliation. Phys. Rev. Lett. 86:5534–5537
29. Seemann R, Herminghaus S, Jacobs K (2001) Gaining control of pattern formation of dewetting liquid films. J. Phys.: Condens. Matter 13:4915–4938
30. Becker J, Grun G, Seeman R, Mantz H, Jacobs K, Mecke KR, Blossey R (2003) Complex dewetting scenarios captured by thin-film models. Nat. Mater. 2:59–64
31. Al Akhrass S, Ostaci RV, Grohens Y, Drockenmuller E, Reiter G (2008) Influence of progressive cross-linking on dewetting of polystyrene thin films. Langmuir 24:1884–1890
32. Müller M, MacDowell LG, Müller-Buschbaum P, Wunnicke O, Stamm M (2001) Nano-dewetting: Interplay between van der Waals- and short-ranged interactions. J. Chem. Phys. 115:9960–9969
33. Ashley KM, Meredith JC, Amis E, Raghavan D, Karim A (2003) Combinatorial investigation of dewetting: Polystyrene thin films on gradient hydrophilic surfaces. Polymer 44: 769–772
34. Du B, Xie F, Wang Y, Yang Z, Tsui OKC (2002) Dewetting of polymer films with built-in topographical defects. Langmuir 18:8510–8517
35. Konnur R, Kargupta K, Sharma A (2000) Instability and morphology of thin liquid films on chemically heterogeneous substrates. Phys. Rev. Lett. 84:931–934
36. Mittal J, Sharma A (2002) Instability of thin liquid films by density variations: A new mechanism that mimics spinodal dewetting. Phys. Rev. Lett. 89:186101 (4 pages)
37. Yang MH, Hou SY, Chang YL, Yang ACM (2006) Molecular recoiling in polymer thin film dewetting. Phys. Rev. Lett. 96:066105 (4 pages)
38. Seeman R, Jacobs K, Landfester K, Herminghaus S (2006) Freezing of polymer thin films and surfaces: The small molecular weight puzzle. J. Polym. Sci.: Part B: Polym. Phys. 44: 2968–2979
39. Muller-Buschabaum P (2003) Influence of surface cleaning on dewetting of thin polystyrene films. Eur. Phys. J. E 12:443–448
40. Reiter G, Hamieh M, Damman P, Sclavons S, Gabriele S, Vilmin T, Raphaël E (2005) Residual stresses in thin polymer films cause rupture and dominate early stages of dewetting. Nat. Mater. 4:754–758
41. Bardon S, Valignat MP, Cazabat AM, Stocker W, Rabe JP (2004) Study of liquid crystal prewetting films by atomic force microscopy in tapping mode. Langmuir 14:2916–2924
42. Fukuzawa K, Shimuta T, Yoshida T, Mitsuya Y, Zhang H (2006) Direct visualization of dewetting of molecularly thin liquid films on solid surfaces. Langmuir 22:6951–6955

43. Huang E, Rockford L, Russell TP, Hawker CJ (1998) Nanodomain control in copolymer thin films. Nature 395:757–756
44. Higgins M, Jones RAL (2000) Anisotropic spinodal dewetting as a route to self-assembly of patterned surfaces. Nature 404:476–478
45. Mayer E, Braun HG (2000) Controlled dewetting processes on microstructured surfaces – a new procedure for thin film microstructuring. Macromol. Mater. Eng. 276/277:44–50
46. Sehgal A, Ferreiro V, Douglas JF, Amis EJ, Karim A (2002) Pattern-directed dewetting of ultrathin polymer films. Langmuir 18:7041–7048
47. Zhang Z, Wang Z, Xing R, Han Y (2003) How to form regular polymer microstructures by surface-pattern-directed dewetting. Surf. Sci. 539:129–136
48. Julthongpiput D, Zhang W, Douglas JF, Karim A, Fasolka MJ (2007) Pattern-directed to isotropic dewetting transition in polymer films on micropatterned surfaces with differential surface energy contrast. Soft Matter 3:613–618
49. Geoghegan M, Wang C, Rhese N, Magerle R, Krausch G (2005) Thin polymer films on chemically patterned, corrugated substrates. J. Phys. Condens. Matter 17:S389–S402
50. Mukherjee R, Gonuguntla M, Sharma A (2007) Meso-patterning of thin polymer films by controlled dewetting: From nano-droplet arrays to membranes. J. Nanosci. Nanotechnol.7: 2069–2075
51. Mukherjee R, Bandyopadhyay D, Sharma A (2008) Control of morphology in pattern directed dewetting of thin polymer films. Soft Matter 4:2086–2097
52. Kargupta K, Sharma A (2001) Templating of thin films induced by dewetting on patterned surfaces. Phys. Rev. Lett. 86:4536–4539
53. Kargupta K, Sharma A (2003) Mesopatterning of thin liquid films by templating on chemically patterned complex substrates. Langmuir 19:5153–5163
54. Kargupta K, Sharma A (2002) Morphological self-organization by dewetting in thin films on chemically patterned substrates. J. Chem. Phys. 116:3042–3051
55. Kargupta K, Sharma A (2002) Creation of ordered patterns by dewetting of thin films on homogeneous and heterogeneous substrates. J. Colloid Interface Sci. 245:99–115
56. Kargupta K, Sharma A (2002) Dewetting of thin films on periodic physically and chemically patterned surfaces. Langmuir 18:1893–1903
57. Mukherjee R, Sharma A, Gonuguntla M, Patil GK (2008) Adhesive force assisted imprinting of soft solid polymer films by flexible foils. J. Nanosci. Nanotechnol. 8:3406–3415
58. Mukherjee R, Pangule RC, Sharma A, Banerjee I (2007) Contact instability of thin elastic films on patterned substrates. J. Chem. Phys. 127:064703 (6 pages)
59. Reynolds O (1886) On the theory and of lubrication and its application to Mr. Beauchamp Tower's experiments, including an experimental determination of viscosity of olive oil. Philos. Trans. R. Soc. Lond. 177:157–234
60. Craster RV, Matar OK (2009) Dynamics and stability of thin liquid films. Rev. Mod. Phys. 81:1139–1198
61. Yerushalmi-Rozen R, Klein J, Fetters L (1994) Suppression of rupture in thin, nonwetting liquid films. Science 263:793–795
62. Henn G, Bucknall DG, Stamm M, Vanhoorne P, Jerome R (1996) Chain end effects and dewetting in thin polymer films. Macromolecules 29:4305–4313
63. Feng Y, Karim A, Weiss RA, Douglas JF, Han CC (1998) Control of polystyrene film dewetting through sulfonation and metal complexation. Macromolecules 31:484–493
64. Li X, Han Y, An L (2003) Inhibition of thin polystyrene film dewetting via phase separation. Polymer 44:5833–5841
65. Kropka JM, Green PF (2006) Control of interfacial instabilities in thin polymer films with the addition of a miscible component. Macromolecules 39:8758–8762
66. Carroll GT, Sojka ME, Lei X, Turro NJ, Koberstein JT (2006) Photoactive additives for cross-linking polymer films: Inhibition of dewetting in thin polymer films. Langmuir 22:7748–7754
67. Barnes KA, Karim A, Douglas JF, Nakatani AI, Gruell H, Amis EJ (2000) Suppression of dewetting in nanoparticle-filled polymer films. Macromolecules 33:4177–4185

68. Mackay ME, Hong Y, Jeong M, Hong S, Russell TP, Hawker CJ, Vestberg R, Douglas JF (2002) Influence of dendrimer additives on the dewetting of thin polystyrene films. Langmuir 18:1877–1882
69. Hosaka N, Tanaka K, Otsuka H, Otsuka H, Takahara A (2004) Influence of the addition of silsesquioxane on the dewetting behavior of polystyrene thin film. Compos. Interfaces 11:297–306
70. Kropka, JM, Garcia Sakai V, Green PF (2008) Local polymer dynamics in polymer-C60 mixtures. Nano Lett. 8:1061–1065
71. Sharma S, Rafailovich MH, Peiffer D, Sokolov J (2001) Control of dewetting dynamics by adding nanoparticle fillers. Nano Lett. 1:511–514
72. Xavier JH, Sharma S, Seo YS, Isseroff R, Koga T, White H, Ulman A, Shin K, Satija SK, Sokolov J, Rafailovich MH (2006) Effect of nanoscopic fillers on dewetting dynamics. Macromolecules 39:2972–2980
73. Luo H, Gersappe D (2004) Dewetting dynamics of nanofilled polymer thin films. Macromolecules 37:5792–5799
74. Hosaka N, Tanaka K, Otsuka H, Otsuka H, Takahara A (2007) Structure and dewetting behavior of polyhedral oligomeric silsesquioxane-filled polystyrene thin films. Langmuir 23:902–907
75. Mukherjee R, Das S, Das A, Sharma SK, Raychaudhuri AK, Sharma A (2010) Stability and dewetting of metal nanoparticle filled thin polymer films: Control of instability length scale and dynamics. ACS Nano 4:3709–3724
76. de Gennes PG (1985) Wetting: Statics and dynamics. Rev. Mod. Phys. 57:827–863
77. Wenzel RN (1936) Resistance of solid surfaces to wetting by water. Ind. Eng. Chem. 28: 988–994
78. Cassie ABD, Baxter S (1944) Wettability of porous surfaces. Trans. Faraday Soc. 40:546–551
79. Nosonovsky M, Bhushan B (2008) Biologically inspired surfaces: Broadening the scope of roughness. Adv. Funct. Mater. 18:843–855
80. Marmur A (1994) Thermodynamic aspects of contact angle hysteresis. Adv. Colloid Interface Sci. 50:121–141
81. Krausch G (1995) Surface induced self assembly in thin polymer films. Mater. Sci. Eng. R14:1–95
82. Blossey R (2001) Dimple-assisted dewetting: Heterogeneous nucleation in undercooled wetting films. Ann. Phys. (Leipzig) 10:733–775
83. Muller-Buschbaum P (2003) Dewetting and pattern formation in thin polymer films as investigated in real and reciprocal space. J. Phys.: Condens. Matter 15: R1549–R1582
84. Bucknall DG (2004) Influence of interfaces on thin polymer film behaviour. Prog. Mater. Sci. 49:713–786
85. Baumchen O, Jacobs K (2010) Slip effects in polymer thin films. J. Phys.: Condens. Matter 22:033102 (21 pages)

Chapter 7
Electrodics in Electrochemical Energy Conversion Systems: A Mesoscopic Formalism

Partha P. Mukherjee and Qinjun Kang

Abstract In recent years, amidst a global thrust toward clean energy research, electrochemical energy conversion systems, i.e., fuel cells, have received significant attention. Among the different types of fuel cells, the polymer electrolyte fuel cell (PEFC) is being considered as the primary candidate for variety of applications. In the PEFC, the electrode is a key component where the electrochemical reaction occurs. The electrode is the host to multi-scale, coupled physicochemical interactions including charge, liquid/vapor/gas transport inside a porous microstructure, which affect the overall cell performance. In this chapter, a mesoscopic modeling framework is presented in order to elucidate the intricate microstructure–transport–performance interplay inherent in the PEFC electrode.

Keywords Electrochemical energy conversion · Fuel cell · Electrode · Transport phenomena · Microstructure · Electrochemical reaction · Species · Charge · Two-phase · Mesoscopic modeling · Stochastic microstructure reconstruction · Direct numerical simulation · Lattice Boltzmann model · Microstructure optimization

1 Introduction

Recent years have witnessed an explosion of interest in the development of electrochemical energy systems as the potential clean energy provider for automotive, stationary, and portable power. Winter and Brodd [1] have recently provided an excellent overview of different electrochemical energy systems. Fuel cells, owing to their high energy efficiency, environmental friendliness, and low noise, are widely considered as the 21st century energy conversion devices. Unlike the conventional Carnot cycle-based energy conversion devices with intermediate heat and mechanical energy generation, fuel cells convert the chemical energy of a fuel directly into electricity. Among the different types of fuel cells, the polymer electrolyte fuel cell

P.P. Mukherjee (✉)
Computer Science and Mathematics Division, Oak Ridge National Laboratory, Oak Ridge, TN 37831, USA
e-mail: mukherjeepp@ornl.gov

S. Chakraborty (ed.), *Mechanics Over Micro and Nano Scales*,
DOI 10.1007/978-1-4419-9601-5_7,

(PEFC) has emerged as the front runner for a wide range of applications. In this chapter, we focus on the PEFC electrode.

A typical PEFC, schematically shown in Fig. 7.1, consists of seven subregions: the anode gas channel, anode gas diffusion layer (GDL), anode catalyst layer (CL), ionomeric membrane, cathode CL, cathode GDL, and cathode gas channel. The proton exchange membrane electrolyte is a distinctive feature of the PEFC. Usually, the two thin catalyst layers are coated on both sides of the membrane, forming a membrane electrode assembly (MEA). The anode feed, generally, consists of hydrogen, water vapor, and nitrogen or hydrogen/water binary gas, whereas humidified air is fed into the cathode. Hydrogen and oxygen combine electrochemically within the electrode to produce electricity, water, and waste heat. The catalyst layer of thickness around 10 μm is, therefore, a critical component of a PEFC and requires extensive treatment. Gottesfeld and Zawodzinski [2] provided a good overview of the catalyst layer structure and functions. The hydrogen oxidation reaction (HOR) occurs at the anode side catalyst layer and protons are generated according to the following reaction:

$$H_2 \rightarrow 2H^+ + 2e^- \tag{7.1}$$

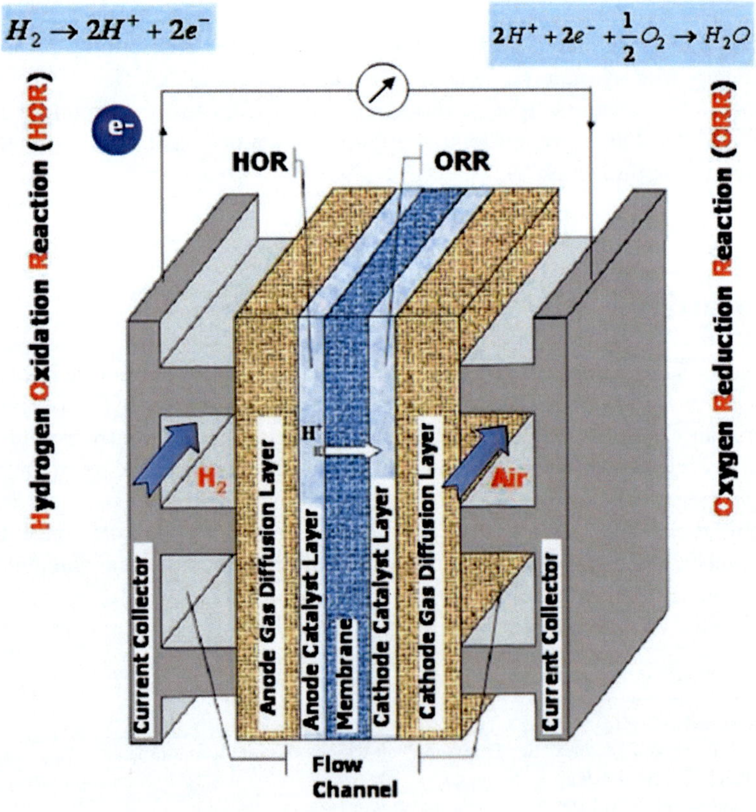

Fig. 7.1 Schematic diagram of a polymer electrolyte fuel cell

The oxygen reduction reaction (ORR) takes place at the cathode catalyst layer and produces water:

$$O_2 + 4H^+ + 4e^- \rightarrow 2H_2O \tag{7.2}$$

Thus, the overall cell reaction is

$$2H_2 + O_2 \rightarrow 2H_2O \tag{7.3}$$

HOR has orders of magnitude higher reaction rate than ORR, which leaves ORR as a potential source of large voltage loss in PEFCs and hence the cathode catalyst layer is the electrode of primary importance in a PEFC. Due to the acid nature of the polymer membrane and low-temperature operation, Pt or Pt alloys are the best-known catalysts for PEFCs.

The performance of a PEFC is characterized by the *polarization curve* giving the relation between cell voltage and current density. Figure 7.2 shows a typical polarization curve for a PEFC with three distinct voltage loss regimes [3]. At low current density operation, the voltage loss is primarily due to the sluggish ORR at the cathode catalyst layer and is referred to as the "kinetic loss" or "activation loss." At intermediate current densities, the voltage loss characterized by resistance to ion transport in the polymer electrolyte membrane and the catalyst layers dominates and is known as "ohmic loss." At high current density operation, "mass transport limitations" come into play due to the excessive liquid water build-up mainly in the cathode side. Liquid water blocks the porous pathways in the CL and GDL thus

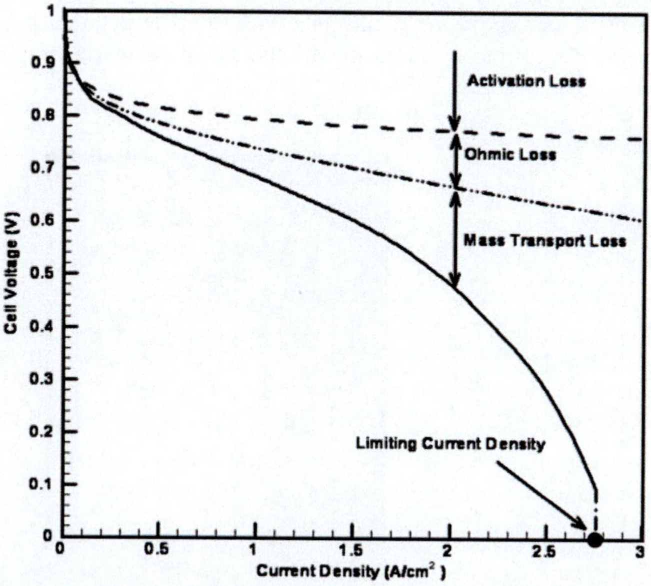

Fig. 7.2 Typical polarization curve of a PEFC

causing hindered oxygen transport to the reaction sites as well as covers the electrochemically active sites in the CL thereby increasing surface overpotential. This phenomenon is known as "flooding" and is perceived as the primary mechanism leading to the limiting current behavior in the cell performance.

For the electrochemical reaction to occur in the catalyst layer, it must provide access for oxygen molecules, protons, and electrons. The state-of-the-art CL in a PEFC is thus a three-phase composite consisting of (1) ionomers, i.e., the ionic phase which is typically Nafion® to provide a passage for protons to be transported in or out; (2) metal (Pt) catalysts supported on carbon, i.e., the electronic phase for electron conduction; and (3) pores for the oxygen gas to be transferred in and product water out. Figure 7.3 shows a high-resolution transmission electron microscope (TEM) image of a typical PEFC electrode [4] with the three-phase interface, where the electrochemical reaction takes place. The salient phenomena occurring in the complex catalyst layer, therefore, include interfacial reaction at the electrochemically active sites, proton transport in the electrolyte, electron conduction in the electronic phase (i.e., Pt/C), oxygen diffusion, liquid water transport through the porous network, and phase change. *Electrodics*, herein, refers to the interplay among the diverse and competing electrochemistry-coupled multiphase and multicomponent reactive transport phenomena in the porous electrode microstructure. Throughout the chapter, the "electrode" and "catalyst layer" terms are used interchangeably in order to refer to the electrocatalytically active porous structure in the PEFC sandwich, which is responsible for the electrochemical reaction to occur.

Different approaches have been undertaken in the literature to model the catalyst layer of a PEFC. In most of the macroscopic models reported in the literature, the active catalyst layer was treated either as an infinitely thin interface or a macrohomogeneous porous layer. A few electrode-specific detailed models were developed for PEFCs primarily based on the theory of volume averaging, which

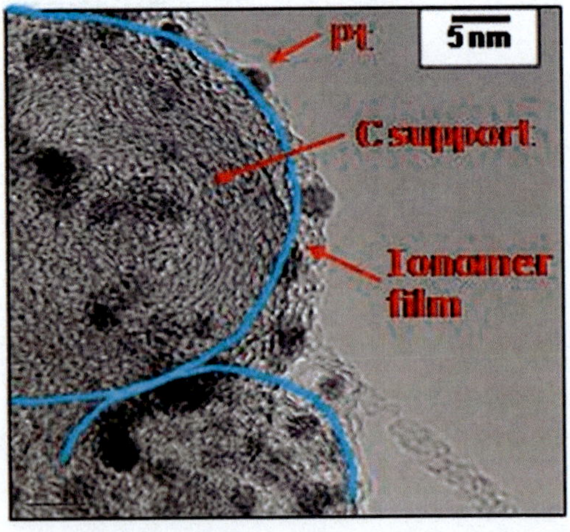

Fig. 7.3 High-resolution TEM image of a PEFC catalyst layer microstructure (reproduced from [4] with permission from K. L. More)

can be further distinguished as film model, homogeneous model, and agglomerate model. Several analytical and numerical solutions for the cathode catalyst layer under various conditions were provided by Springer and Gottesfeld [5], Perry et al. [6], and Eikerling and Kornyshev [7]. Comprehensive overviews of the various electrode models were furnished in the recent reviews by Wang [8] and Weber and Newman [9].

Since the catalyst layer consists of both the ionomer and void phase, water management becomes an important issue especially for the cathode catalyst layer. The ionomer requires water for good proton conductivity. On the other hand, water is produced due to ORR and also migrates from the anode side through the polymer membrane due to electro-osmotic drag, thus causing flooding of the cathode catalyst layer leading to hindered oxygen transport to the reaction sites. Hence, investigating water transport in the cathode catalyst layer is of paramount importance. Several groups have modeled water transport in PEFCs at various levels of complexity. Among the various water transport models for the catalyst layer, developed within the general framework of computational fuel cell dynamics (CFCD), notable works include Wang and coworkers [10, 11], Dutta et al. [12, 13], Berning et al. [14], and Mazumder and Cole [15].

However, none of the above-mentioned models resolve the underlying structural influence on the catalyst layer performance and water transport in it. Due to the macroscopic nature of the current models, they employ constitutive closure relations based on the effective medium approximation (EMA), originally developed by Bruggeman (1935) and Landauer (1952) [16], thereby replacing the disordered medium with an equivalent uniform system with certain effective transport properties, which mimic the actual medium. In the two-phase regime, liquid water covers the reaction sites rendering reduced catalytic activity as well as blocks the porous pathways thus causing hindered oxygen transport to the active reaction sites, thereby leading to CL flooding and severe performance degradation. This leads to an extra level of complexity in macroscopic two-phase fuel cell models in terms of employing appropriate two-phase closure relations, namely capillary pressure and relative permeability correlations as function of liquid water saturation. Unfortunately, no such correlations for the catalyst layer are available in the literature, primarily because a viable experimental approach in obtaining such constitutive relations in the 10 μm thick structure is exceedingly difficult and probably impossible in the foreseeable future. To this extent, although substantial research, both theoretical and experimental, has been conducted as far as the overall PEFC is concerned, there is serious paucity of fundamental understanding about the overall structure–performance relation as well as the underlying two-phase dynamics in the catalyst layer, especially, the following outstanding questions pertaining to the PEFC electrode arise:

- What is a reasonable estimate of the resistance to oxidant diffusion and ion transport owing to the underlying complex morphology of the CL structure?
- What could be the optimum composition of the catalyst layer in terms of the relative volume fractions of the constituent phases?

- Can we predict two-phase correlations, namely capillary pressure and relative permeability as functions of liquid water saturation, which can be employed as closure relations in two-phase computational fuel cell dynamics (CFCD) models?
- What could be a methodology toward quantitative estimation of the catalytic site coverage and the pore blockage effects due to liquid water in the catalyst layer, which can be further used as inputs into macroscopic two-phase models? How does the CL flooding affect the cell performance?

Mesoscopic modeling offers great promise in unveiling the underlying microstructure–transport interplay in the PEFC electrode owing to the information available at the microscopic scale. Recent advancements in the microstructure generation techniques coupled with high-performance computation have truly set the stage for predictive mesoscopic modeling and macroscopic upscaling in multi-physics transport in microporous media.

This chapter provides a systematic overview of the development of a comprehensive, predictive mesoscopic modeling framework, which attempts to answer the aforementioned questions and garners fundamental understanding of the underlying multi-physics transport mechanisms in the complex electrode of a PEFC.

2 Catalyst Layer Microfabrication and Microstructure

The thin-film technique, originally proposed by Wilson in his pioneering work in 1993 at Los Alamos National Laboratory [17], is the present state of the art in fabricating PEFC catalyst layers. In this method, Wilson [17] used perfluorosulfonate ionomer, typically Nafion®, as the binding agent leading to enhanced proton conductivity and larger electrochemically active area in the thin-film electrodes. The resulting thin-film CL is a three-phase composite consisting of Pt catalyst on carbon support (C/Pt), ionomer, and pore, in order to facilitate the electrochemical reaction to occur, as elucidated in Fig. 7.3.

The procedure for fabricating a thin-film CL, based on the work by Wilson [17], is as follows:

1. Combine 5% solution of solubilized perfluorosulfonate ionomer (such as Nafion®) and 20 wt% Pt/C support catalyst in a ratio of 1:3 Nafion®/catalyst.
2. Add water and glycerol to weight ratios of 1:5:20 (carbon:water:glycerol) to form the ink.
3. Mix the solution ultrasonically until the catalyst is uniformly mixed and the ink is adequately viscous for coating.
4. Ion exchange the protonated Nafion® membrane to the Na^+ form by soaking in NaOH solution with subsequent rinsing and drying.
5. Apply the C/Pt–water–glycerol ink to one side of the membrane to achieve adequate uniformity and catalyst loading.
6. Dry the membrane in a preheated vacuum chamber at approximately 160°C.

7. Repeat steps 5 and 6 for the other side of the membrane.
8. Ion exchange the membrane electrode assembly to the protonated form by boiling the MEA in 0.1 M H_2SO_4 solution and rinsing in de-ionized water.

Carbon paper and/or cloth GDL can then be placed against the MEA to form the MEA-GDL sandwich. The catalyst layer can also be formed by using a transfer printing method in which the catalyst layer is cast to a PTFE (Teflon) blank and then decaled onto the membrane [18]. Furthermore, direct coating methods, e.g., catalyst-coated membrane (CCM) and catalyzed diffusion medium (CDM) with the catalyst layer being cast directly onto the membrane or diffusion layer, respectively, have been developed for improved electrode performance [19]. Litster and McLean [20] have presented an extensive overview of different electrode fabrication methods for PEFCs.

3 Catalyst Layer Microstructure Generation

Detailed description of a porous microstructure can be obtained in the form of 3-D volume data, either by experimental imaging or by stochastic reconstruction method. Several experimental techniques can be deployed to image the pore structure in 3-D. Earlier attempts include destructive serial sectioning [21, 22] of pore casts to reconstruct the complex pore space. Recently, non-invasive techniques such as X-ray microtomography [23], magnetic resonance imaging [24], and confocal microscopy [25] are the preferred choices over the earlier destructive methods. Additionally, 3-D porous structure can be generated using stochastic simulation technique, originally developed by Joshi [26] in 2-D and later extended to 3-D by Quiblier [27]. The stochastic reconstruction method creates 3-D replicas of the microstructure based on specified statistical information (e.g., porosity, two-point correlation function) of a porous sample. Adler and Thovert [28] and Torquato [29] provided comprehensive overviews of the stochastic reconstruction techniques along with statistical description of porous microstructures. In the absence of adequate 3-D volume data, the low cost and high speed of data generation, as well as the ability to overcome present resolution constraints of computed microtomography (ca. 1–5 μm per voxel) have established the base for the wide acceptance of the stochastic generation method as a viable alternative to experimental acquisition of 3-D volume data, especially for the thin ($\sim$10 μm) PEFC electrode. In the following section, the stochastic reconstruction method in the context of the PEFC electrode microstructure generation is discussed.

As described earlier, the state-of-the-art catalyst layer of a PEFC is a three-phase composite consisting of C/Pt, ionomer, and pore, for the electrochemical reaction to occur. In this study, the catalyst layer is delineated as a two-phase (pore/solid) structure consisting of the gas phase (i.e., the void space) and a mixed electrolyte/electronic phase (i.e., the solid matrix). The assumption of the mixed electrolyte/electronic phase is well justified from the perspective of ion transport in the electrolyte phase as the limiting mechanism as compared to the electron conduction via the electronic (C/Pt) phase within the CL and henceforth is referred to as the "electrolyte" phase [30–32]. Figure 7.3 further illustrates a pseudo-dual phase (pore/solid) scenario with the ionomer film impregnated C/Pt as an effective solid phase and the adjacent pore space leading toward an electrochemically active interface.

The stochastic reconstruction technique is based on the idea that an arbitrarily complex pore structure can be described by the values of a phase function, $Z(\vec{r})$, at each point, $\vec{r}$, within the porous medium. The phase function takes the values of zero or unity depending on whether the point corresponds to void or solid, respectively, and can be defined as [33]

$$Z(\vec{r}) = \begin{cases} 0 & \text{if } \vec{r} \text{ is in the pore space} \\ 1 & \text{otherwise} \end{cases} \tag{7.4}$$

The intrinsic randomness of the phase function can be adequately qualified by the low-order statistical moments, namely porosity and two-point autocorrelation function [33]. The porosity is the probability that a voxel is in the pore space. The two-point autocorrelation function is the probability that two voxels at a specific distance are both in the pore space [33]. Details about the CL microstructure reconstruction along with the underlying assumptions are elaborated in our recent work [30, 32], which is based on the stochastic generation method originally reported by Adler et al. [34] and Bentz and Martys [35]. In brief, the stochastic reconstruction technique starts with a Gaussian distribution which is filtered with the two-point autocorrelation function and finally thresholded with the porosity, which creates the 3-D realization of the CL structure. The autocorrelation function is computed from a 2-D TEM image of an actual CL [36]. The porosity can be calculated by converting the mass loading data of the constituent components available from the CL fabrication process. For example, the porosity of the CL can be evaluated using the following relation [31, 32]:

$$\varepsilon_{\mathrm{CL}} = 1 - \left[\frac{1}{\rho_{\mathrm{Pt}}} + \frac{R_{\mathrm{C/Pt}}}{\rho_{\mathrm{C}}} + \frac{R_{\mathrm{C/Pt}} \cdot R_{\mathrm{I/C}}}{\rho_{\mathrm{Nafion}}}\right] \frac{L_{\mathrm{Pt}}}{\Delta X_{\mathrm{CL}}} \tag{7.5}$$

where ρ_{Nafion}, ρ_{Pt}, and ρ_{C} are the density of Nafion®, Pt, and carbon, respectively. $R_{\mathrm{I/C}}$ is the weight ratio of ionomer to carbon, $R_{\mathrm{C/Pt}}$ is the weight ratio of carbon to Pt in 40% Pt/C catalyst, L_{Pt} is the Pt loading, and ΔX_{CL} is the CL thickness. Now, using (7.5), with $\Delta X_{\mathrm{CL}} = 10\,\mu\mathrm{m}$, $L_{\mathrm{Pt}} = 0.35$ mg Pt/cm^2, $R_{\mathrm{C/Pt}} = 1.5$, $R_{\mathrm{I/C}} = 0.42$, $\rho_{\mathrm{Nafion}} = \rho_{\mathrm{C}} = 2$ g/cc, and $\rho_{\mathrm{Pt}} = 21.5$ g/cc, the nominal porosity of the CL microstructure can be evaluated as $\varepsilon_{\mathrm{CL}} \sim 60\%$.

The pore/solid phase is further distinguished as "transport" and "dead" phase. The basic idea is that a pore phase unit cell surrounded by solid phase-only cells does not take part in species transport and hence in the electrochemical reaction and can, therefore, be treated as a "dead" pore and similarly for the electrolyte phase [30].

The interface between the "transport" pore and the "transport" electrolyte phases is referred to as the electrochemically active area (ECA), and the ratio of ECA and the nominal CL cross-sectional area provides the "ECA ratio." It is noted that in this chapter, ECA is normalized with the apparent electrode area and therefore differs from the definition in terms of the electrochemically active area per Pt loading reported elsewhere in the literature.

In this study, two types of catalyst layer microstructures pertaining to the typical fabrication methods, namely catalyst-coated membrane (CCM) and catalyzed diffusion medium (CDM), are reconstructed. Figure 7.4 shows the reconstructed microstructure of a typical CCM CL with nominal porosity of 60% and thickness of 10 μm along with the input TEM image and the evaluated cross-section averaged

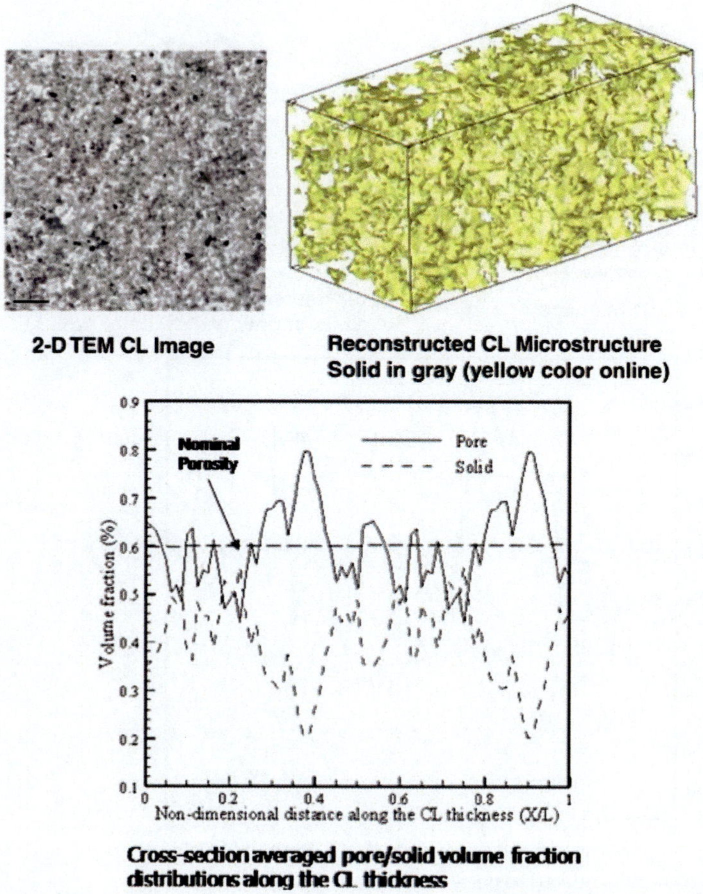

Fig. 7.4 Reconstructed CCM catalyst layer microstructure along with pore and electrolyte phase volume fractions distribution (reproduced from Ref. [30] with permission from ECS – The Electrochemical Society)

pore and electrolyte phase volume fraction distributions across the CL thickness. Figure 7.5 exhibits the reconstructed CL microstructure pertaining to the CDM fabrication method with nominal porosity of 60% and thickness of 10 μm along with the input TEM image and the evaluated cross-section averaged pore and electrolyte phase volume fraction distributions across the CL thickness. The cross-section averaged pore/electrolyte volume fraction distributions illustrate the local tortuosity variation along the electrode thickness. The concept of tortuosity is often introduced in the context of defining the closure relations for solving transport in porous media and is derived from the fact that the actual path followed by the transported material, such as species and ion in the case of the catalyst layer, is microscopically complex, or "tortuous" [37, 38]. Several 3-D structures are generated with varying

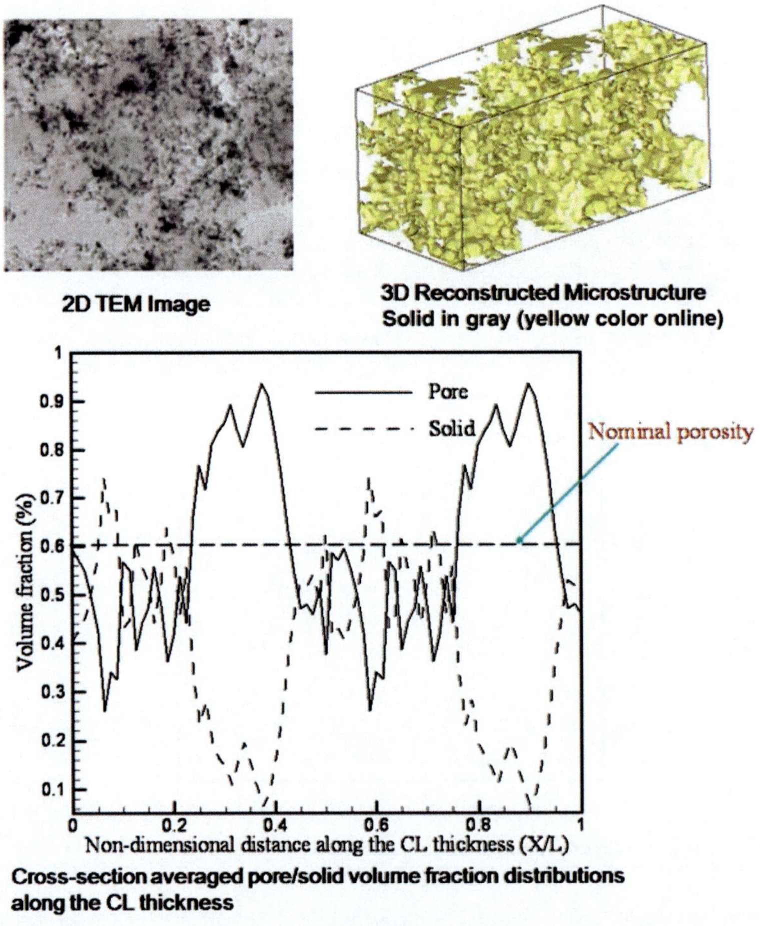

Fig. 7.5 Reconstructed CDM catalyst layer microstructure along with pore and electrolyte phase volume fractions distribution

Table 7.1 ECA ratio values for the CCM and CDM CLs

CCM reconstructed	CCM experimental (CV)	CDM reconstructed	CDM experimental (CV)
46	45	30	39

number of unit cells. However, the reconstructed microstructure with the electrochemically active interfacial area ratio complying closely with the experimentally measured value using cyclic voltammetry (CV) is selected for the subsequent simulations. The 3-D reconstructed microstructures with 100 elements in the thickness direction and 50 elements each in the spanwise are selected for this study. The ECA ratio values obtained from the reconstructed microstructures are listed in Table 7.1 along with the ECA ratio data measured experimentally using cyclic voltammetry (CV) for both the CCM and CDM CLs. The ECA ratio value for the CCM CL matches quite well with the experimental data; however, the reconstructed CDM CL exhibits slightly lower value than the experimental data. The discrepancy in the ECA ratio value for the CDM microstructure could be due to the lack of phase resolution in the input 2-D TEM image resulting in a moderately resolved pore/solid binary micrograph, shown in Fig. 7.5, which is subsequently used as input for the two-point autocorrelation function calculation.

4 Electrochemistry-Coupled Direct Numerical Simulation

Traditionally, porous electrodes in fuel cells are modeled using the macrohomogeneous method, where the properties and variables of each phase are volume averaged over a representative elementary volume containing a sufficient number of particles. In such an approach, microscopic details of the pore structure are smeared and the electrode is described using the porosity, interfacial area per unit volume, effective conductivity, diffusivity, etc., through a homogenized porous medium. With these volume-averaged variables, macroscopic governing equations are derived from their microscopic counterparts by assuming the uniformity of the microscopic properties within the representative elementary volume, which implies existence of phase equilibrium. In these volume-averaged equations, empirical correlations are used to describe the effective properties as a function of porosity and tortuosity, which are characteristics of the porous structure. Therefore, in the macrohomogeneous model, both the structure and the variables are homogenized microscopically. The effects of the microstructural morphology are ignored and also empirical transport properties are introduced in the macroscopic models, which do not address localized phenomena at the pore level. In order to reveal the interplay between the catalyst layer structure and the underlying species and charge transport, as well as to predict reliable microstructural parameters in terms of Bruggeman factors, a direct numerical simulation (DNS) model is developed. The DNS model solves transport equations for charge, oxygen, and water vapor directly at the pore level on a realistic, statistically rigorous 3-D description of the catalyst layer microstructure generated using the stochastic reconstruction technique, detailed earlier. The model is also ideally suited for studying the structure–transport–performance relationships for different microstructures toward performance optimization of the catalyst layer based on the underlying composition and structure.

4.1 DNS Model

Once the microstructure is generated and the constituent phases (i.e., "transport" and "dead" pore and electrolyte phases) are identified, the DNS model solves point-wise accurate transport equations for charge and species conservation directly on the reconstructed catalyst layer microstructure.

4.1.1 Physico-electro-chemical Processes

The key processes considered in the current DNS model, which describe several interlinked electrochemical and transport phenomena occurring within the catalyst layer, are the following:

- The oxygen reduction reaction (ORR) at the electrochemically active surface represented by the interface between a "transport" pore and a "transport" electrolyte cell
- Diffusion of oxygen and water vapor through the "transport" pore phase
- Charge transport through the "transport" electrolyte phase

4.1.2 Assumptions

The primary assumptions employed in the model are as follows:

- isothermal and steady-state operation;
- existence of thermodynamic equilibrium at the reaction interface between the oxygen concentration in the gas phase and that dissolved in the electrolyte phase, i.e., negligible oxygen diffusion in the electrolyte phase due to the small ionomer film thickness ($\sim$2 nm), as approximately estimated by Kocha [39], covering the C/Pt surface;
- water is in the gas phase even if water vapor concentration slightly exceeds the saturation value corresponding to the cell operation temperature (i.e., slight oversaturation is allowed);
- water in the electrolyte phase is in equilibrium with the water vapor, thus water transport through the gas phase is only considered and the electro-osmotic drag due to water through the electrolyte phase in the CL is neglected;
- uniform electronic phase potential since the electrode is very thin and its electronic conductivity is very high and hence the electron transport is not considered. The ionic conductivity, κ, of the mixed phase is thus normalized with regard to the electrolyte phase volume fraction as follows:

$$\kappa = \kappa_0 \cdot \left(\frac{\varepsilon_e}{\varepsilon_e + \varepsilon_s}\right)^{1.5} = \kappa_0 \cdot \left(\frac{\varepsilon_e}{1 - \varepsilon_g}\right)^{1.5} \tag{7.6}$$

where κ_0 is the intrinsic conductivity of the electrolyte phase; ε_e, ε_s, and ε_g are the electrolyte phase, electronic phase, and gas pore volume fractions, respectively. It should be noted that the aforementioned Bruggeman-like correction factor in the effective, intrinsic ionic conductivity is strictly due to the mixed electrolyte/electronic phase assumption in the two-phase reconstruction model. However, similar to the pore path tortuosity owing to the underlying structure, the local variation of the electrolyte phase volume fraction and its impact on the proton path tortuosity are still enforced via the reconstruction model as evidenced by the local variations of the solid phase volume fractions in Figs. 7.4 and 7.5.

4.1.3 Governing Equations

A single set of differential equations valid for all the phases is developed, which obviates the specification of internal boundary conditions at the phase interfaces. Due to slow kinetics of the ORR, the electrochemical reaction is described by the Tafel kinetics as follows:

$$j = -i_0 \left(\frac{c_{O_2}}{c_{O_2,\text{ref}}}\right) \exp\left(\frac{-\alpha_c F}{RT}\eta\right) \tag{7.7}$$

where i_0 is the exchange current density, c_{O_2} and $c_{O_2,\text{ref}}$ refer to local oxygen concentration and reference oxygen concentrations, respectively, α_c is the cathode transfer coefficient for ORR, F is Faraday's constant, R is the universal gas constant, and T is the cell operating temperature. A value of 50 nA/cm^2 for the exchange current density is used in this work. The overpotential, η, is defined as

$$\eta = \phi_s - \phi_e - U_0 \tag{7.8}$$

where ϕ_s and ϕ_e stand for the electronic and electrolyte phase potentials at the reaction sites, respectively. U_0 is the reference open-circuit potential of the cathode under the cell operation temperature.

The conservation equations for the transport of proton, O_2, and water vapor, respectively, can be expressed as follows:

$$\left.\begin{aligned}
&\nabla \cdot (\kappa_e \nabla \phi_e) + a \int_\Gamma j\delta(x - x_{\text{interface}})ds = 0 \\
&\nabla \cdot \left(D^g_{O_2} \nabla c_{O_2}\right) + a \int_\Gamma \frac{j}{4F}\delta(x - x_{\text{interface}})ds = 0 \\
&\nabla \cdot \left(D^g_{H_2O} \nabla c_{H_2O}\right) + a \int_\Gamma \frac{j}{2F}\delta(x - x_{\text{interface}})ds = 0
\end{aligned}\right\} \tag{7.9}$$

where a represents the specific interfacial area and is defined as the interfacial surface area where the reaction occurs per unit volume of the catalyst layer, s is the non-dimensional interface, Γ represents the interfacial surface over which the surface integral is taken, $\delta(x - x_{\text{interface}})$ is a delta function which is zero everywhere but unity at the interface where the reaction occurs. The second term in the above equations, therefore, represents a source/sink term at the catalyzed interface where the electrochemical reaction takes place. It is important to note that the transfer current density, j, is negative for the electrolyte phase. With the CL thickness of 10 μm, the reconstructed microstructures for the CCM and CDM CLs give rise to the specific interfacial area, a, of 45×10^5 m^{-1} and 30×10^5 m^{-1}, respectively, which is an important input to the macrohomogeneous model detailed later.

The above-governing equations are extended to be valid for the entire computational domain by introducing a discrete phase function f. The phase function, f, at each elementary cell center (i,j,k) is defined as follows:

$$f(i,j,k) = \begin{cases} 0 & \text{“transport” pores} \\ 1 & \text{“transport” electrolytes} \\ 2 & \text{“dead” pores} \\ 3 & \text{“dead” electrolytes} \end{cases} \tag{7.10}$$

The details about the boundary conditions, geometry, physico-chemical parameters, and solution procedure are explained in our recent works [30–32].

4.2 DNS Predictions: Single-Layer CL

Figure 7.6 shows the DNS-predicted polarization curve with air as the oxidant at 100% inlet humidity with inlet pressure of 200 kPa and cell temperature of 70°C along with the experimental measurement. It should be noted that the term “polarization curve” refers to the cathode overpotential vs. current density curve in this case. As a general trend, the predicted cathode polarization curve depicts a fast drop in the small current density region controlled by the ORR kinetics followed by a linear voltage drop in the mixed control regime and finally at higher current densities (~1 A/cm^2), the mass transport limitation appears with a fast voltage drop resulting from oxygen depletion. The experimentally obtained cell voltage (V_{cell}) vs. current density (I) data was processed to extract the variation of cathode overpotential (η_c) with current density according to the following relation:

$$\eta_c = V_{\text{cell}} + I \times \text{HFR} - U_0 \tag{7.11a}$$

where HFR refers to the high-frequency resistance measured experimentally and U_0 is the thermodynamic equilibrium potential corresponding to the fuel cell operating temperature. It is important to note that in the above equation, the anodic overpotential for hydrogen oxidation and protonic resistance in the anode catalyst layer

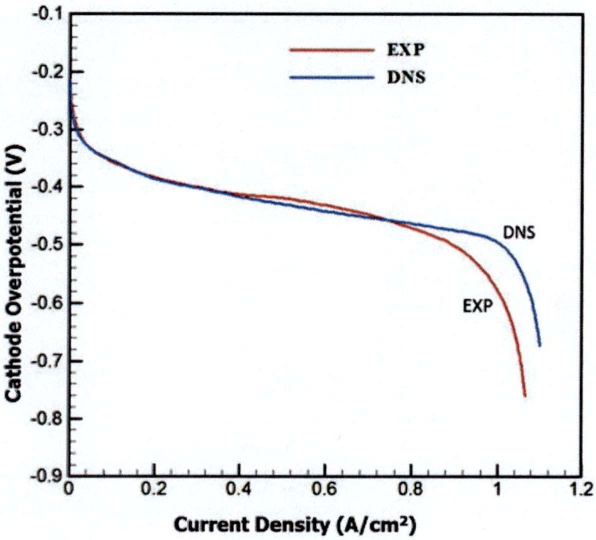

Fig. 7.6 Comparison of the cathode polarization curves between DNS predictions and experimental observations for 100% RH air as the oxidant for the CCM CL (reproduced from [30] with permission from ECS – The Electrochemical Society)

are assumed to be negligible. However, the cathodic overpotential defined in (7.11) contains the protonic resistance or ohmic loss in the cathode catalyst layer. From Fig. 7.6, it is clear that there are reasonable agreements between the DNS predictions and experimental observations in the kinetic control and ohmic control regimes. However, the ohmic control regime seems to be slightly extended in the DNS calculations. Since water transport has been modeled only in the gas phase, it does not include any water condensation effect and fails to capture the mass transport resistance due to liquid water flooding. It is important to note that the mass transport resistance through the GDL was adjusted in the simulations by properly tuning the structural properties, namely tortuosity, in order to achieve a reasonably realistic limiting current density compared to the experimental data. Figure 7.6, however, demonstrates that the DNS model is not only able to capture the general trend of the fuel cell performance curve on a realistic CL microstructure but also exhibits sufficient agreement with experimental results.

Figure 7.7 shows the effect of the inlet humidity on the local cross-section averaged reaction current and overpotential distributions along the catalyst layer thickness at an average current density of 0.6 A/cm^2 with air as the oxidant. It is clear that the reaction zone shifts toward the membrane–CL interface with lower inlet humidity. Apparently, this is due to the poorer proton conductivity or higher ionic resistance in the electrolyte phase and results in a much lower surface overpotential near the CL–GDL interface. In order to compensate for the lower reaction current produced near the CL–GDL interface, the vicinity of the membrane–CL interface shows higher reaction current as the average current density is fixed. This leads to a higher surface overpotential needed at the membrane–CL and hence leads to higher total voltage loss in the cathode for the low humidity operation.

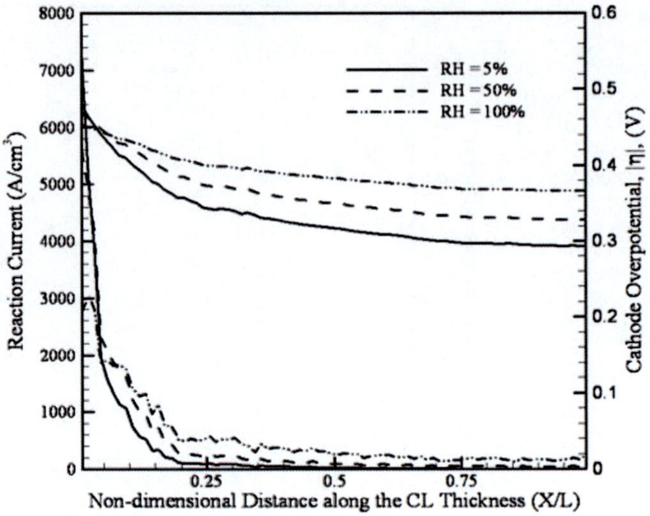

Fig. 7.7 Local overpotential and reaction current density distributions across the thickness of the CCM CL for different inlet humidities with air as the oxidant (reproduced from [30] with permission from ECS – The Electrochemical Society)

One major implication of the DNS calculation is that the Bruggeman correlation, required for the macrohomogeneous models, can be evaluated using the DNS data. Details about the macrohomogeneous model can be found in the original work by Springer and coworkers [5, 40]. In brief, the governing equations for charge (proton) and species (oxygen and water vapor) transport are solved in the CL domain, which does not contain any microstructural information as in the DNS model, however, with resistance due to the porous medium structure taken into consideration through effective transport properties. Bruggeman correction factor, ξ, is commonly applied to determine the effective transport property as follows:

$$\Gamma_k^{\text{eff}} = \Gamma_k \cdot \varepsilon_k{}^{\xi} \tag{7.11b}$$

In the 1-D macrohomogeneous model, the same specific surface area, a (cm^2/cm^3), as that in the constructed 3-D catalyst layer microstructure, is used in the Tafel equation to represent the volumetric reaction current and is expressed by

$$j = -a \cdot i_0 \left(\frac{c_{O_2}}{c_{O_2,\text{ref}}}\right) \exp\left(-\frac{\alpha_c F}{RT}\eta\right) \tag{7.12}$$

Other input parameters for the macrohomogeneous model are maintained same as in the DNS model for comparison.

Figure 7.8 shows the comparison of the cross-section averaged reaction current distribution at an average current density of 0.6 A/cm^2 with air as the oxidant across

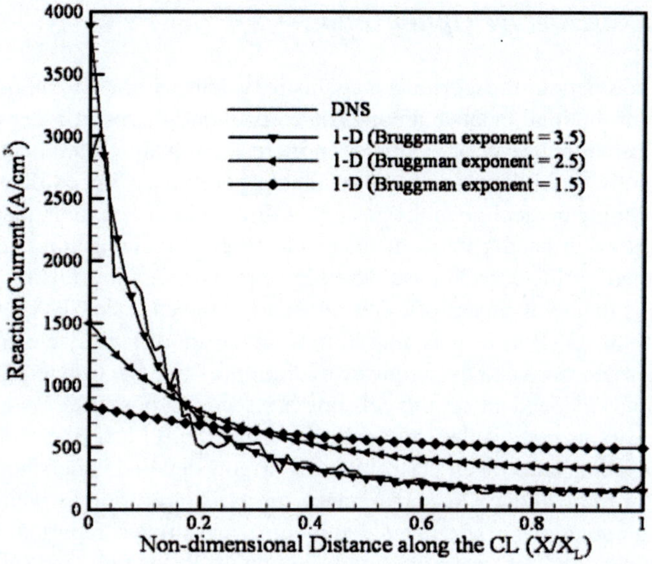

Fig. 7.8 Comparison between the cross-section averaged reaction current distributions across the thickness of the CCM CL from the DNS and 1-D macrohomogeneous models (reproduced from [30] with permission from ECS – The Electrochemical Society)

the CL thickness between the DNS and 1-D macrohomogeneous model predictions. Different Bruggeman factors have been attempted. The DNS result exhibits good agreement with the 1-D macrohomogeneous model prediction with the Bruggeman factor of 3.5. It is also important to note that the high reaction current in the 15–20% of the catalyst layer thickness in the vicinity of the membrane can be attributed to the limited ionomer conductivity resulting from the low electrolyte phase volume fraction (~11%) throughout the CL. Similar to the Bruggeman factor, another measure of the transport resistance often encountered in porous media literature is through the tortuosity, τ, and can be expressed as [41, 42]

$$\Gamma_k^{\mathrm{eff}} = \Gamma_k \cdot \frac{\varepsilon_k}{\tau_k} \tag{7.13}$$

Physically, the tortuosity is defined as the ratio of the actual distance traversed by the species between two points to the shortest distance between those two points. By combining (7.12) and (7.14), the tortuosity can be evaluated. With the Bruggeman factor of 3.5, the pore path tortuosity representative of the resistance to oxygen diffusion is calculated around 4 and the proton path tortuosity indicative of the ion transport resistance through the electrolyte phase around 18. The high proton path tortuosity could be attributed to the significantly low volume fraction (e.g., $<20\%$) of the electrolyte phase typically considered in the standard PEFC catalyst layers.

4.3 CL Microstructure Optimization

Diverse and competing transport mechanisms prevalent within the cathode catalyst layer require an optimal balance among the constituent phases in order to achieve the best cell performances. For example, an increase in Nafion® content improves proton conductivity while reduces the available pore space for oxidant transport resulting in significant decline in the gas phase diffusivity. Increase in platinum (Pt) loading, on the other hand, causes enhanced electrochemical reaction rate, which is however limited by increase in cost. Another important factor affecting the PEFC performance is the involvement of water transport in the cathode CL, via water production due to the ORR as well as migration from the anode side by electro-osmotic drag. Good proton conductivity requires hydration of the electrolyte phase which is again strongly dependent on the cell operating conditions in terms of the inlet relative humidity as well as the cell operating temperature. It is, therefore, evident that achieving an enhanced cell performance warrants detailed understanding of the dependence of the cathode CL performance on its composition as well as on the cell operating conditions. Experimental and numerical studies, reported in the literature, investigated the influence of Nafion® content, Pt loading, and thickness for optimum composition on the CL performance for both single-layer and functionally graded CLs [43–47]. The computational models deployed in the literature are mainly based on the macroscopic theory of volume averaging and fail to resolve the pore-scale heterogeneity on the reaction zone penetration and spatial variation of the electrochemical activity for CL composition optimization toward enhanced performance.

In this section, the DNS approach detailed earlier is advanced further for oxygen, water, and proton transport through a 3-D, bilayer catalyst layer microstructure. The statistical description of the 3-D CL microstructure is realized using the stochastic reconstruction technique developed for the single-layer CL as elaborated earlier. The pore-level description of the underlying transport phenomena through different realizations of the bilayer cathode CL is presented. Finally, the salient predictions from the present DNS model are furnished elucidating the tri-fold influence of the CL structure, composition, and the cell operating conditions on the performance.

4.3.1 Bilayer CL Microstructure

In the bilayer catalyst layer, two catalyst-coated membrane (CCM) layers, **A** and **B**, of thickness around 5 μm each are physically juxtaposed to develop the catalyst layer. Catalyst layers **A** and **B** have the same Pt loadings, but different I/C (ionomer to carbon) weight ratios of 0.417 and 0.667, respectively. By placing **A** and **B** either close to the membrane or close to the GDL, two different composite CLs with stair-step structures, in terms of nominal composition volume fractions, can be prepared. These two composite structures are hereafter referred to as **A/B** CL or **B/A** CL as illustrated in Fig. 7.9. Using (7.5), with $\Delta X_{CL} = 5\ \mu m$, $L_{Pt} = 0.199\ mg\ Pt/cm^2$,

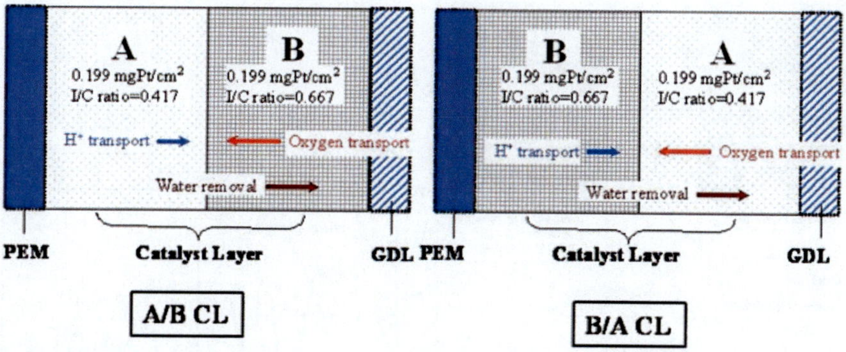

Fig. 7.9 Schematic representation of the bilayer CLs (reproduced from [31] with permission from ECS – The Electrochemical Society)

$R_{C/Pt} = 1.5$, $\rho_{Nafion} = \rho_C = 2$ g/cc, and $\rho_{Pt} = 21.5$ g/cc, the electrolyte and pore volume fractions for CL-**A** and CL-**B** can be calculated:

$$\varepsilon_{CL/\mathbf{A}} = 0.56, \quad \varepsilon_{Nafion/\mathbf{A}} = 0.12$$
$$\varepsilon_{CL/\mathbf{B}} = 0.48, \quad \varepsilon_{Nafion/\mathbf{B}} = 0.2$$

With the evaluated volume fractions of pore, electrolyte, and electronic phases, the microstructures of **A/B** CL and **B/A** CL are reconstructed using the stochastic generation method detailed earlier. It is to be noted that for simplicity, the two-point autocorrelation function evaluated from a 2-D TEM image of a CCM CL-**A** is used for the CCM CL-**B** as well as for the reconstruction of the **A/B** and **B/A** CLs. This simplification could be well justified since both the **A/B** and **B/A** CLs exhibit very similar electrochemically active interfacial area ratio (~29) measured using cyclic voltammetry, which in turn indicates that a very similar pore-space correlation would suffice for a reasonable two-phase reconstruction. Figure 7.10 shows the reconstructed microstructure of the **A/B** CL with the four-phase description, namely the "transport" and "dead" pore and electrolyte phases, the corresponding 2-D TEM image used for evaluation of the autocorrelation function, and the cross-section averaged pore and electrolyte volume fractions distributions across the CL thickness for the **A/B** CL. Figure 7.11 shows the reconstructed **B/A** CL and the corresponding cross-section averaged pore/electrolyte volume fraction distributions across the CL thickness. Similar to the single-layer CL microstructures, several 3-D structures are generated with varying number of unit cells. However, the reconstructed microstructure with the electrochemically active interfacial area ratio complying closely with the experimentally measured value using cyclic voltammetry is selected for the subsequent DNS calculations. The reconstructed, 3-D microstructure with 100 elements in the thickness direction and 50 elements each in the span-wise directions produces an active interfacial area ratio of around 30, which matches reasonably well with the measured value (~29) and thereby reproduces the most important structural parameter responsible for electrochemical activity of the CL. Once the

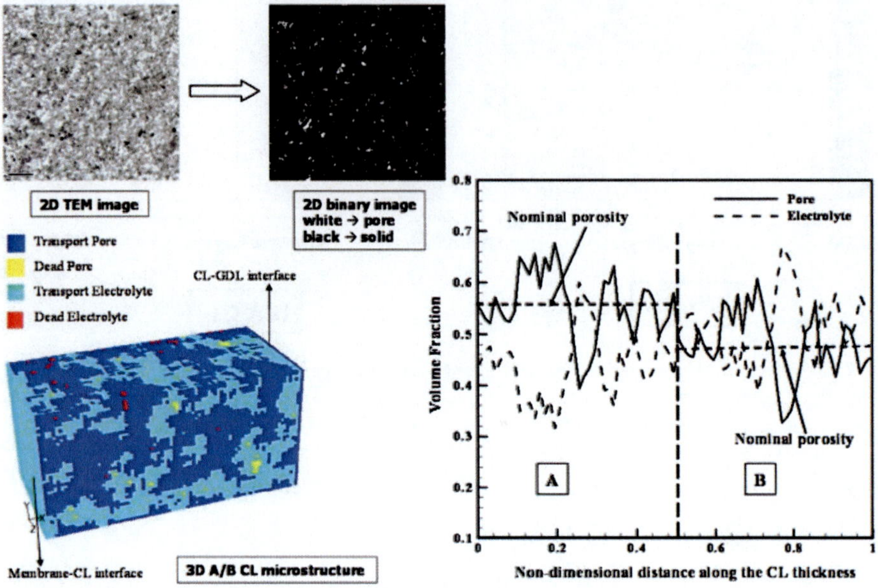

Fig. 7.10 Reconstructed bilayer *A*/*B* CL microstructure (reproduced from [31] with permission from ECS – The Electrochemical Society)

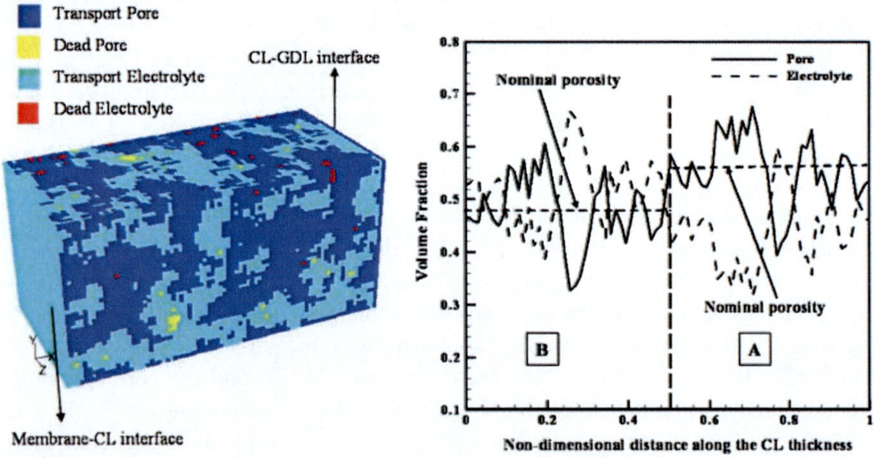

Fig. 7.11 Reconstructed bilayer *B*/*A* CL microstructure (reproduced from [31] with permission from ECS – The Electrochemical Society)

bilayer CL microstructures are generated, the governing equations for species and charge transport are solved directly on the underlying structures. Detailed description of the primary transport processes, governing equations, model assumptions, boundary conditions, model input parameters, and solution strategy are furnished in our recent works [31, 32].

4.3.2 DNS Predictions – Bilayer CL

Figure 7.12 shows the cross-section averaged reaction current and cathode overpotential distributions along the CL thickness for the **A/B** and **B/A** CLs with 100% relative humidity (RH) air, at 70°C cell operation temperature, 200 kPa inlet pressure, and current density of 0.4 A/cm^2. The catalyst layer performance is influenced by two competing factors, namely protonic resistance and oxygen transport. Higher ionomer volume fraction reduces protonic resistance while higher pore volume fraction favors oxygen transport. In the **B/A** CL, near the membrane–CL interface, higher ionomer content reduces protonic resistance, which in turn aids in the reduction of cathode overpotential and results in an extended reaction zone as is evident from the reaction current distribution. In the **A/B** CL, due to increased protonic resistance resulting from lower ionomer content in the vicinity of the membrane–CL interface and in order to overcome the corresponding higher overpotential, the reaction zone is concentrated to a relatively smaller region as compared to the **B/A** CL. While the benefit of higher ionomer volume fraction near the membrane–CL interface is clearly pronounced via reduced overpotential and extended reaction zone from Fig. 7.12 at low current density (i.e., 0.4 A/cm^2) where transport limitation is primarily due to hindered proton transport, the resulting reaction current distribution due to the varying ionomer content could also be harnessed to enhance cell performance at higher current density where transport limitation owing to hindered oxygen transport will dominate. For the **B/A** CL, even at higher current density, there might still be some active reaction sites due to an extended reaction zone resulting from a favorable distribution of the ionomer content along the CL thickness, and thus will exhibit a higher limiting current density as compared to the **A/B** CL. Furthermore, a higher pore volume fraction near the CL–GDL interface in the **B/A** CL will aid in enhanced oxygen transport. On the contrary, in the **A/B** CL, the concentration of the active reaction zone toward the vicinity of the membrane–CL interface as well

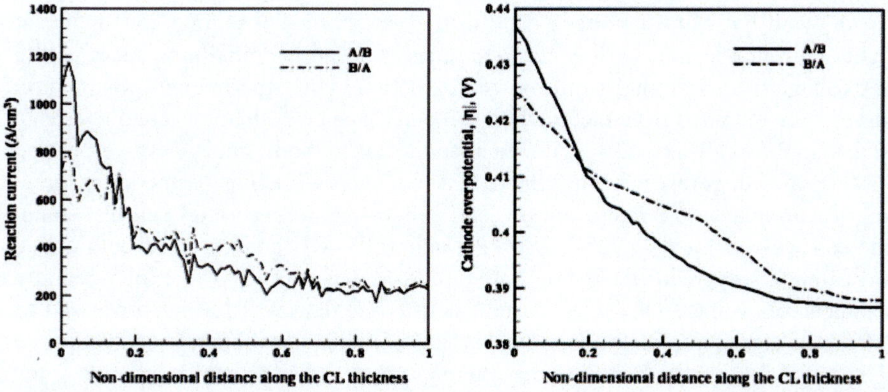

Fig. 7.12 Cross-section averaged reaction current and overpotential distributions along the thickness of the bilayer *A/B* and *B/A* CLs (reproduced from [31] with permission from ECS – The Electrochemical Society)

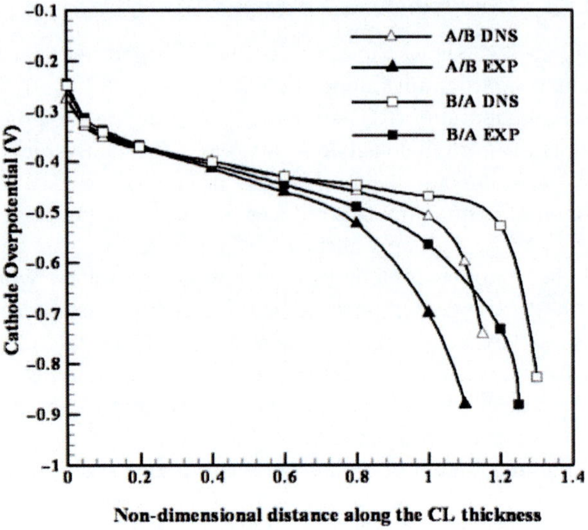

Fig. 7.13 Polarization curves for the bilayer *A*/*B* and *B*/*A* CLs (reproduced from [31] with permission from ECS – The Electrochemical Society)

as relatively lower void phase fraction near the CL–GDL interface will negatively impact the transport of oxygen to the reaction sites at higher current density and subsequently will exhibit a lower limiting current and hence a poorer performance as compared to the **B/A** CL.

The simulated polarization curves with air as the oxidant under the aforementioned operating condition are shown in Fig. 7.13, along with the corresponding experimental data. The "polarization curve" refers to the cathode overpotential vs. current density curve in this study similar to that reported in the single-layer CL. As a general trend, the predicted cathode polarization curves by the DNS model depict a fast drop in the small current density region controlled by the ORR kinetics followed by a linear voltage drop in the mixed control regime and finally at higher current densities ($\sim$1 A/cm^2), the mass transport limitation appears with a fast voltage drop resulting from oxygen depletion. The experimental polarization curves were obtained from electrochemical performance evaluations conducted in a 5 cm^2 graphite cell fixture with identical anode and cathode single-pass, serpentine flow fields with computer-controllable test parameters, such as temperature, pressure, fuel/oxidant flow rates, current, and cell voltage using a fuel cell test stand. The cell was operated at 70°C, 200 kPa, and 100% RH conditions at both anode and cathode sides with fixed flow rates of hydrogen and air. The experimentally obtained cell voltage (V_{cell}) vs. current density (I) data was further processed to extract the variation of cathode overpotential (η_c) with current density according to (7.11). It is clear from Fig. 7.13 that there are reasonable agreements between the DNS predictions and experimental observations in the kinetic-control and ohmic control regimes. However, for both **A/B** and **B/A** CLs, the ohmic control regime seems to be slightly extended in the DNS calculations. Since water transport has been modeled only in the gas phase and the effect of liquid water transport in the CL is not considered, the DNS model overpredicts the ohmic control regime. Finally, from the figure, it is evident that the DNS calculations show better performance for the **B/A** CL as compared to the **A/B** CL in accordance with the experimental observations and conforms to the explanations provided earlier with respect to the relative distributions of ionomer and pore volume fractions in the CL and the resulting reaction current distributions. Finally, it can be inferred that a higher ionomer content near the membrane–CL interface along with a higher void fraction near the CL–GDL interface prove to be beneficial for a bilayer CL performance.

Figure 7.14 shows the cross-section averaged reaction current and cathode overpotential distributions across the CL thickness with air as the oxidant at 5 and 100% RH, at 70°C cell operation temperature, 200 kPa inlet pressure, and current density of 0.4 A/cm^2. From the cathode overpotential distribution, it is evident that low humidity operation results in enhanced voltage drop. This is primarily due to the poorer proton conductivity associated with the partially hydrated electrolyte phase, leading to enhanced protonic resistance. At 100% RH, the electrolyte phase remains fully humidified with saturated water vapor while the water vapor concentration is significantly reduced at 5% RH causing increased ohmic resistance. The enhanced protonic resistance at low humidity further exacerbates the performance of the **A/B** CL where the reaction zone further shifts toward the vicinity of the membrane–CL interface as compared to that for the 100% RH operation. In the **A/B** CL, at 5% RH operation, the lower electrolyte phase volume fraction near the membrane–CL interface along with the reduced ionic conductivity owing to low humidity result in significantly higher protonic resistance, which in turn increases the overpotential. In order to overcome the localized elevated overpotential, the reaction zone shrinks further and concentrates primarily in the 10–15% of the CL thickness near

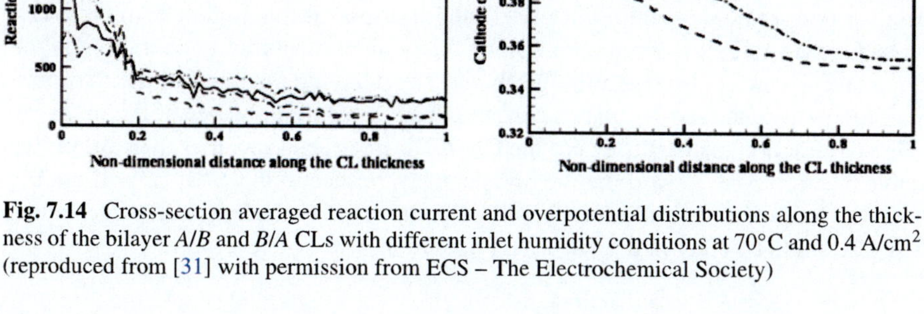

Fig. 7.14 Cross-section averaged reaction current and overpotential distributions along the thickness of the bilayer *A/B* and *B/A* CLs with different inlet humidity conditions at 70°C and 0.4 A/cm^2 (reproduced from [31] with permission from ECS – The Electrochemical Society)

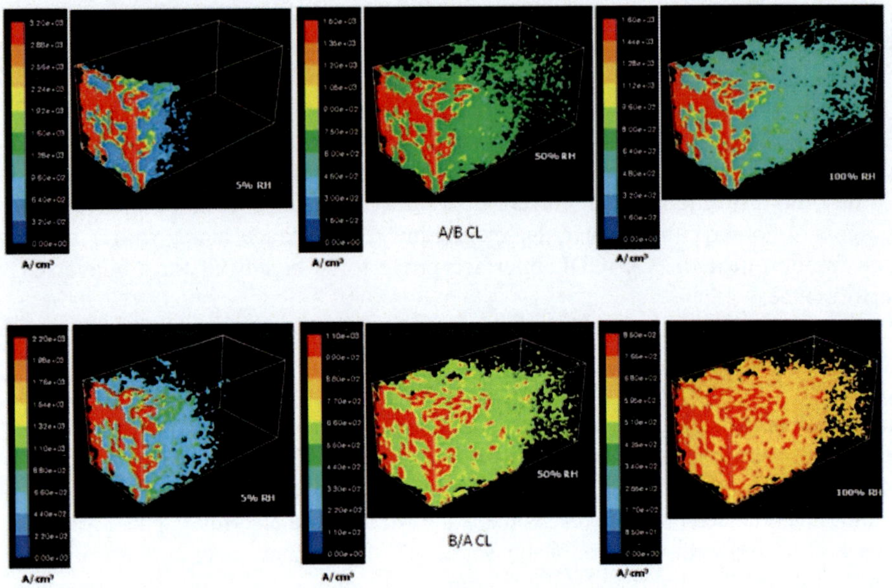

Fig. 7.15 Three-dimensional reaction current distribution contours for different inlet humidity conditions at 70°C and 0.4 A/cm^2 for the *A/B* and *B/A* CLs (reproduced from [31] with permission from ECS – The Electrochemical Society)

the membrane–CL interface in the **A/B** CL. Overall, it can be conjectured that low humidity operation leads to a steeper overpotential gradient as well as an apparent reduction in the active reaction zone extent as compared to that in the fully humidified operation.

Figure 7.15 shows the 3-D reaction current contours at different inlet relative humidity for the **A/B** and **B/A** CL microstructures. It can be observed that low humidity operation indeed causes reaction current snap-off and renders a significant portion of the CL virtually inactive, which is however more pronounced for the **A/B** CL as explained earlier. At fully humidified operation, the reaction current contours show that high value of reaction current, represented by the red color map in the contours, prevails over significant portion of the **B/A** CL as compared to the **A/B** CL and further emphasizes the importance of the current DNS model in elucidating detailed pore-scale description of underlying transport through the CL microstructures. Furthermore, it is worth mentioning that while low humidity operation might extend the ohmic control regime slightly thereby delaying the onset of the transport limitation characterized by higher current density operation, however, the unfavorable void fraction distribution near the CL–GDL interface as well as the shift of the active reaction zone toward the membrane–CL interface in the **A/B** CL will inhibit oxygen transport and would finally exhibit poorer performance as compared to the **B/A** CL, similar to that in the 100% RH operation.

5 Two-Phase Transport in the PEFC Electrode

Owing to the complex microstructure characterized by significantly small pore size, e.g., around 0.05–0.1 μm, the two-phase transport in the PEFC electrode exhibits the dominance of surface forces as compared to the gravity, viscous, and inertia forces. This can be further illustrated by the following non-dimensional numbers representative of the PEFC CL.

$$\text{Reynolds number: Re} = \frac{\rho_2 U_2 D}{\mu_2} \sim 10^{-4}$$

$$\text{Capillary number: Ca} = \frac{\mu_2 U_2}{\sigma} \sim 10^{-6}$$

$$\text{Bond number: Bo} = \frac{g(\rho_2 - \rho_2)D^2}{\sigma} \sim 10^{-10}$$

U_2 and μ_2 are the non-wetting phase velocity and dynamic viscosity, respectively, σ is the surface tension, and g is the gravitational acceleration. It should be noted that for a slightly hydrophobic PEFC CL, water is the non-wetting phase (NWP) and air the wetting phase (WP). In this chapter, we identify the non-wetting phase with subscript 2 and the wetting phase with 1. The bond number, defined as the ratio of gravitational force to the surface tension force, shows that the effect of gravity force is negligible with respect to the surface tension force, thereby indicating strong capillary force dominance. The Reynolds number, representing the ratio of inertia force to viscous force, further demonstrates that the inertial effect, with significantly small velocity (U_2) in the PEFC CL, is negligible as compared to the viscous force. The capillary number, Ca, which represents the ratio of viscous force to the surface tension force, illustrates that the effect of viscous force is also negligible as compared to the surface tension force. Combining the implications of low Ca, Re, and Bo numbers, it can be inferred that the surface tension forces predominate and the density ratio, which is ~1000 for air–water two-phase flow for the PEFC operation, should have significantly small influence on the overall transport in the CL. It should be noted that the viscosity ratio (M) for the non-wetting and wetting phases in a fuel cell operating at 80°C is estimated to be $M = \mu_2/\mu_1 \sim 18$. Based on the calculated representative viscosity ratio and capillary number values, the typical operating two-phase regime for the PEFC electrode belongs to the capillary fingering zone on the "phase diagram" proposed by Lenormand et al. [48] and is shown in Fig. 7.16. The notion of the "phase diagram," by Lenormand et al. [48], is based on their experiment, involving immiscible displacement of a wetting phase by a non-wetting phase, in a flat and horizontal porous medium where gravity forces were neglected. This phase diagram further bolsters the fact that for air–water two-phase transport in the PEFC CL, the principal driving force is owing to the action of capillarity and the two-phase regime should lie in the capillary fingering region. Typical fluid displacement patterns pertaining to the three flow regimes are also shown in Fig. 7.16 along with the phase diagram and are adapted from the work by Ewing and

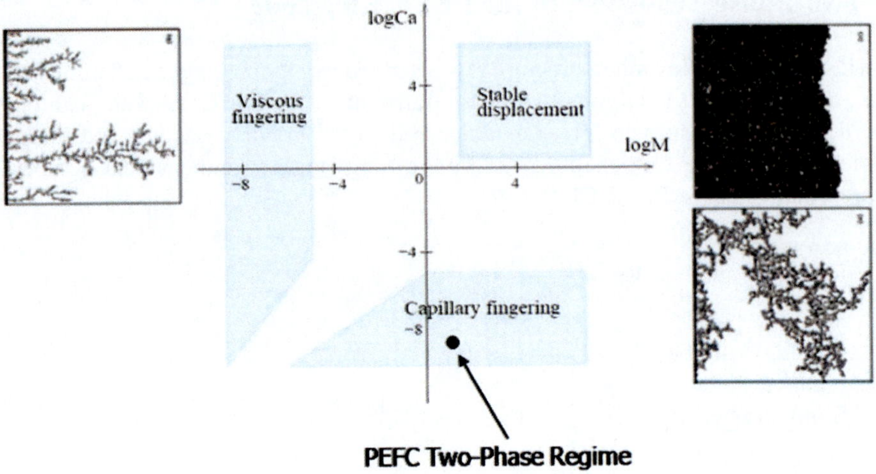

Fig. 7.16 Phase diagram along with fluid displacement patterns (reproduced from [65] with permission from Elsevier)

Berkowitz [49]. Furthermore, the effect of gravity force in the overall PEFC system, let alone in the electrode, has been shown to be truly insignificant [50]. Using this analysis, it can be safely adjudged that for modeling air–water two-phase transport in the PEFC CL the effects of high density ratio (∼1000) and viscosity ratio (∼18) variation can be assumed to be negligible represented by the significantly low capillary number. In the following section, the development of the lattice Boltzmann model for simulating two-phase transport in the PEFC electrode is discussed.

5.1 Two-Phase Lattice Boltzmann Model

In recent years, the lattice Boltzmann (LB) method, owing to its excellent numerical stability and constitutive versatility, has developed into a powerful technique for simulating fluid flows and is particularly successful in applications involving interfacial dynamics and complex geometries [51]. The LB method is a first-principle-based numerical approach. Unlike the conventional Navier–Stokes solvers based on the discretization of the macroscopic continuum equations, lattice Boltzmann methods consider flows to be composed of a collection of pseudo-particles residing on the nodes of an underlying lattice structure which interact according to a velocity distribution function. The lattice Boltzmann method is also an ideal scale-bridging numerical scheme which incorporates simplified kinetic models to capture microscopic or mesoscopic flow physics and yet the macroscopic averaged quantities satisfy the desired macroscopic equations, e.g., Navier–Stokes equation [51]. Due to its underlying kinetic nature, the LB method has been found to be particularly useful in applications involving interfacial dynamics and complex boundaries, e.g., multiphase or multicomponent flows, and flow in porous medium. As opposed to the front-tracking and front-capturing multiphase models in traditional CFD (computational fluid dynamics) approach, due to its kinetic nature, the LB model incorporates phase segregation and surface tension in multiphase flow through interparticle force/interactions, which are difficult to implement in traditional methods. While the LB modeling approach better represents the pore morphology in terms of a realistic digital realization of the actual porous medium and incorporates rigorous physical description of the flow processes, it is computationally very demanding. However, because of the inherently parallel nature of the LB algorithm, it is also very efficient running on massively parallel computers [51].

Several LB models have been presented in the literature to study multiphase/multicomponent flows. Gunstensen et al. [52] developed a multicomponent LB model based on a two-component lattice gas model. Shan and Chen [53, 54] proposed an LB model with interparticle potential for multiphase and multicomponent fluid flows. Swift et al. [55] developed an LB multiphase and multicomponent model by using the free-energy approach. He et al. [56] proposed an LB multiphase model using the kinetic equation for multiphase flow. Among the aforementioned multiphase LB models, the interaction potential-based approach is widely used due to its simplicity in implementing boundary conditions in complex porous structures and remarkable versatility in terms of handling fluid phases with different densities, viscosities, and wettabilities, as well as the capability of incorporating different equations of state. In this work, we have developed the interaction potential-based two-phase LB model to study the structure–wettability influence on the underlying two-phase dynamics in the CL of a PEFC.

In brief, the S-C model [53, 54] introduces k distribution functions for a fluid mixture comprising of k components. Each distribution function represents a fluid component and satisfies the evolution equation. The non-local interaction between particles at neighboring lattice sites is included in the kinetics through a set of potentials. The evolution equation for the kth component can be written as

$$f_i^k(\mathbf{x} + \mathbf{e}_i\delta_t, t + \delta_t) - f_i^k(\mathbf{x}, t) = -\frac{f_i^k(\mathbf{x}, t) - f_i^{k(\mathrm{eq})}(\mathbf{x}, t)}{\tau_k} \tag{7.14}$$

$f_i^k(\mathbf{x}, t)$ is the number density distribution function for the kth component in the ith velocity direction at position $\mathbf{x}$ and time t and δ_t is the time increment. In the term on the right-hand side, τ_k is the relaxation time of the kth component in lattice unit and $f_i^{k(\mathrm{eq})}(\mathbf{x}, t)$ is the corresponding equilibrium distribution function. The right-hand side of (7.14) represents the collision term based on the BGK (Bhatnagar–Gross–Krook) or the single-time relaxation approximation [57]. The phase separation between different fluid phases, the wettability of a particular fluid phase to the solid, and the body force are taken into account by modifying the velocity used to calculate the equilibrium distribution function. An extra component-specific velocity due to interparticle interaction is added on top of a common velocity for each component. Interparticle interaction is realized through the total force, $\mathbf{F}_k$, acting on the *k*th component, including fluid/fluid interaction, fluid/solid interaction, and external force. More details can be found in [58, 59].

The continuity and momentum equations can be obtained for the fluid mixture as a single fluid using Chapman–Enskog expansion procedure in the nearly incompressible limit:

$$\left.\begin{aligned} &\frac{\partial \rho}{\partial t} + \nabla \cdot (\rho \mathbf{u}) = 0 \\ \rho &\left[\frac{\partial \mathbf{u}}{\partial t} + (\mathbf{u} \cdot \nabla)\mathbf{u}\right] = -\nabla p + \nabla \cdot [\rho \nu (\nabla \mathbf{u} + \mathbf{u}\nabla)] + \rho \mathbf{g} \end{aligned}\right\} \tag{7.15}$$

where the total density and velocity of the fluid mixture are given, respectively, by

$$\left.\begin{aligned} \rho &= \sum\nolimits_k \rho_k \\ \rho \mathbf{u} &= \sum\nolimits_k \rho_k \mathbf{u}_k + \frac{1}{2}\sum\nolimits_k \mathbf{F}_k \end{aligned}\right\} \tag{7.16}$$

with a non-ideal gas equation of state [60].

The primary physical parameters, such as the fluid/fluid and fluid/solid interaction parameters, need a priori evaluation through model calibration using numerical experiments. The fluid/fluid interaction gives rise to the surface tension force and the fluid/solid interaction manifests in the wall adhesion force. The fluid/fluid and fluid/solid interaction parameters are evaluated by designing two numerical experiments, *bubble test* in the absence of solid phase and *static droplet test* in the presence of solid wall, respectively. The details of these numerical experiments are detailed elsewhere [32, 58].

5.2 Two-Phase Numerical Experiments and Setup

Two numerical experiments are designed specifically for investigating liquid water transport and two-phase dynamics in the reconstructed CL microstructure in an ex situ setup. Additionally, isothermal condition is assumed as a first and reasonable approximation in the subsequent two-phase numerical simulations.

The first numerical setup is designed to simulate a *quasi-static displacement experiment*, typically devised in the petroleum/reservoir engineering applications and detailed elsewhere in the literature [37, 38, 61], for simulating immiscible, two-phase transport in the CL microstructures. Figure 7.17 schematically shows the computational domain and setup of the displacement, i.e., *primary drainage* experiment. An NWP reservoir is added to the porous structure at the front end and a WP reservoir is added at the back end [32, 62]. These two end reservoirs added to the CL domain in the through-plane (i.e., thickness) direction are composed of void space. It should be noted that for the primary drainage (PD) simulation in the slightly hydrophobic CL, liquid water is the NWP and air is the WP.

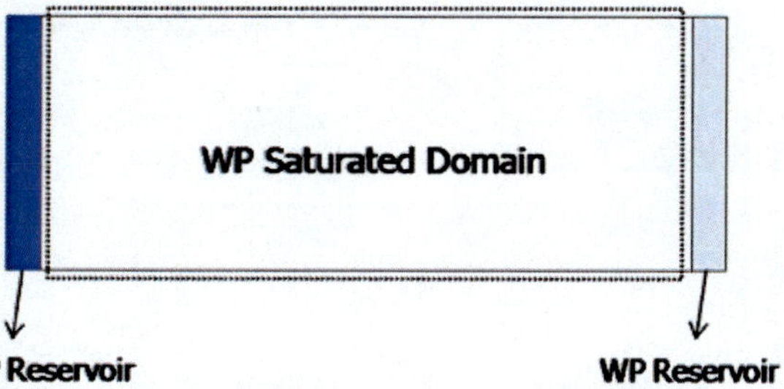

Fig. 7.17 Schematic diagram of the capillary pressure experiment domain

The primary drainage process is simulated starting with zero capillary pressure, by fixing the NWP and WP reservoir pressures to be equal. Then the capillary pressure is increased incrementally by decreasing the WP reservoir pressure while maintaining the NWP reservoir pressure at the fixed initial value. The pressure gradient drives liquid water into the initially air-saturated CL by displacing it. The primary objective of the quasi-static displacement simulation is to study liquid water behavior through the CL structures and the concurrent response to capillarity as a direct manifestation of the underlying pore morphology.

The second numerical experiment is designed based on the *steady-state flow experiment*, typically devised in the petroleum/reservoir engineering applications and detailed elsewhere in the literature [37, 38, 61], in order to garner further insight into the underlying two-phase transport and interfacial dynamics in the PEFC electrode. In the steady-state flow experiment, two immiscible fluids are allowed to flow simultaneously until equilibrium is attained and the corresponding saturations, fluid flow rates, and pressure gradients can be directly measured and correlated using Darcy's law. In the *steady-state flow simulation*, initially both the NWP and WP are randomly distributed throughout the CL microstructure such that the desired NWP saturation is achieved [32, 63]. The initial random distribution of the liquid water phase (i.e., NWP) in the otherwise air (i.e., WP) occupied CL closely represents the physically perceived scenario of liquid water generation due to the electrochemical reaction at different catalytically active sites within the CL structure and subsequent transport by the action of capillarity. Details about the numerical setup, boundary conditions, solution procedure regarding the two-phase LB modeling in the PEFC electrode are furnished in our recent works [32, 64, 65].

5.3 Two-Phase Transport and Flooding Behavior

The aforementioned numerical experiments, namely *quasi-static drainage* and *steady-state flow simulations*, are specifically designed to study the influence of microstructure and wetting characteristics on the underlying two-phase behavior and flooding dynamics in the PEFC electrode.

Figure 7.18 displays the steady state invading liquid water fronts corresponding to increasing capillary pressures from the *primary drainage simulation* in the

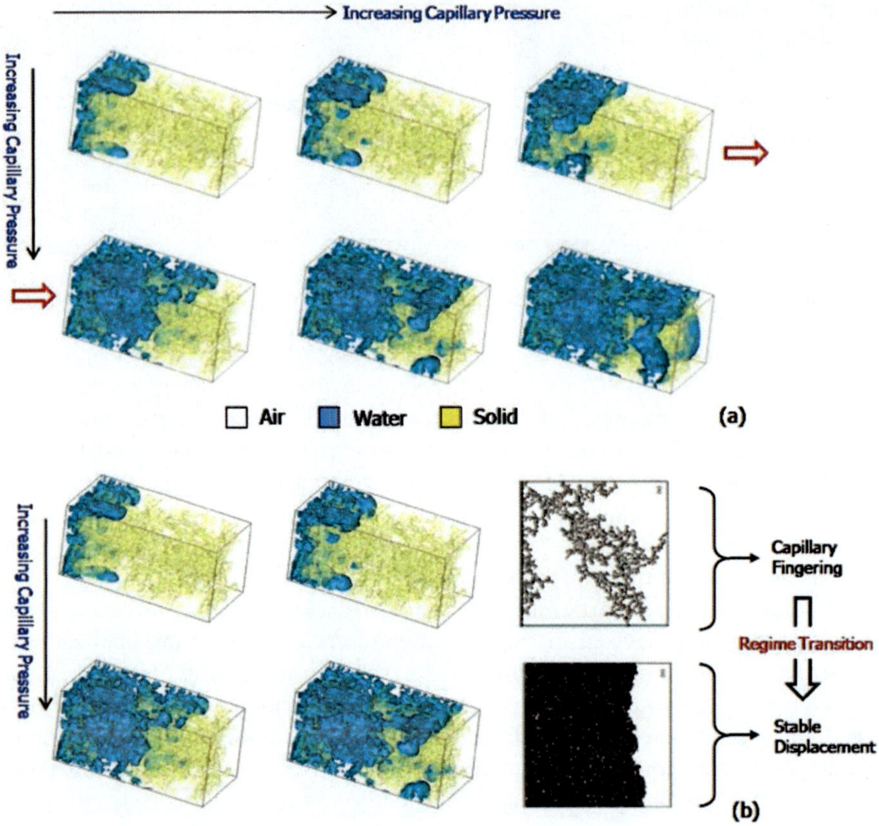

Fig. 7.18 Advancing liquid water front with increasing capillary pressure through the initially air-saturated reconstructed CCM CL microstructure from the primary drainage simulation (reproduced from [65] with permission from Elsevier)

reconstructed CCM CL microstructure characterized by slightly hydrophobic wetting characteristics with a static contact angle of 100°. At lower capillary pressures, the liquid water saturation front exhibits finger-like pattern, similar to the displacement pattern observed typically in the capillary fingering regime. The displacing liquid water phase penetrates into the body of the resident wetting phase (i.e., air) in the shape of fingers owing to the surface tension-driven capillary force. However, at high saturation levels, the invading non-wetting phase tends to exhibit a somewhat flat advancing front. This observation, highlighted in Fig. 7.18, indicates that with increasing capillary pressure, even at very low capillary number (Ca), several penetrating saturation fronts tend to merge and form a stable front. The invasion pattern transitions from the capillary fingering regime to the stable displacement regime and potentially lies in the transition zone in between.

Figure 7.19 shows the liquid water invasion pattern from the *primary drainage simulation* in the reconstructed CDM CL microstructure characterized by slightly

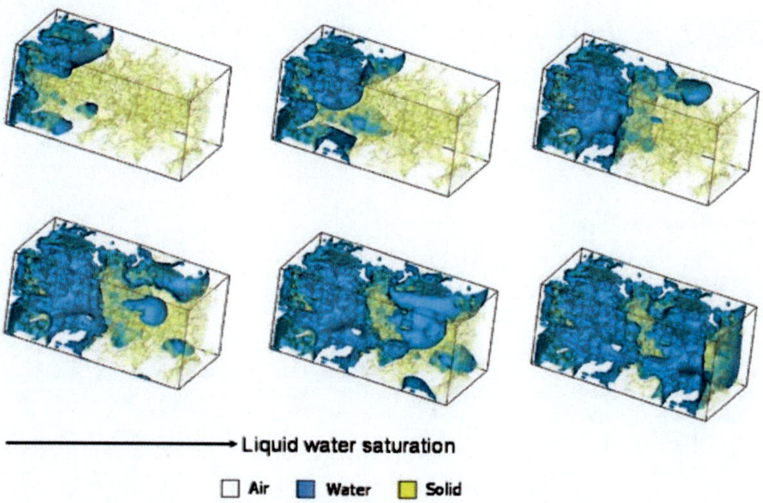

Fig. 7.19 Advancing liquid water front with increasing capillary pressure through the initially air-saturated reconstructed CDM CL microstructure from the primary drainage simulation

hydrophobic wetting characteristics with a static contact angle of 100°. Overall, the CDM CL microstructure exhibits similar two-phase behavior leading to transition from capillary finger to stable displacement at higher capillary pressures. Taking a closer look at the saturation fronts between the CCM and CDM CLs reveals that the CCM CL exhibits a transition from the capillary fingering regime to the stable displacement regime at lower saturation level as compared to the CDM CL which further indicates the influence of the underlying pore morphology on the liquid water transport. In an operating fuel cell, the resulting liquid water displacement pattern pertaining to the underlying pore morphology and wetting characteristics would play a vital role in the transport of the liquid water and hence the overall flooding behavior.

Figure 7.20 exhibits the 3-D liquid water distributions corresponding to several low saturation levels (below 15%) for the CCM and CDM CL microstructures from the *steady-state flow simulation* at equilibrium. It can be observed that below 10% saturation level there is hardly any connected pathway for the liquid water phase to transport through the CL structure and hence the relative mobility of the liquid water phase with respect to the incumbent air phase is negligible. As the saturation level increases, the initially random liquid water phase redistributes owing to the action of capillarity and finds a connected pathway for transport through the CL structure. Figure 7.21 exhibits the 2-D liquid water saturation maps corresponding to a representative saturation level of 20% on several cross sections along the thickness of the reconstructed CCM and CDM CL microstructures. The implication of the 2-D saturation maps is that the reduction in electrochemically active interfacial area (ECA) owing to liquid water coverage in the CL can be estimated. Furthermore, it is imperative to emphasize that this pore-scale two-phase model delineates the impact

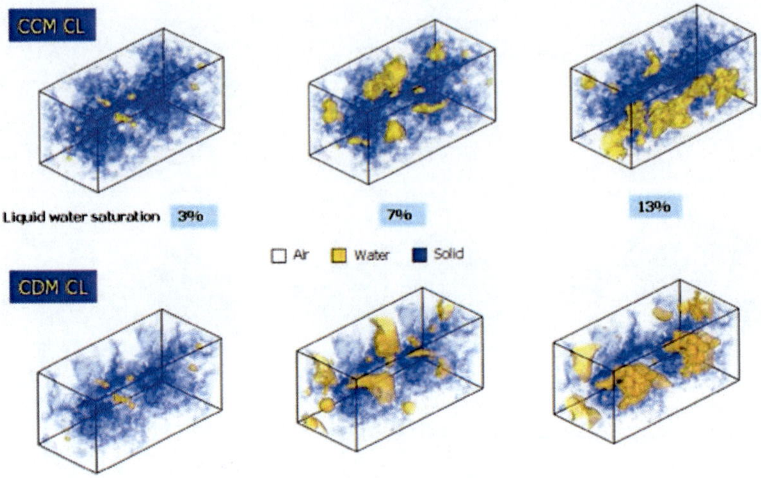

Fig. 7.20 Three-dimensional liquid water distributions in the CCM and CDM CLs

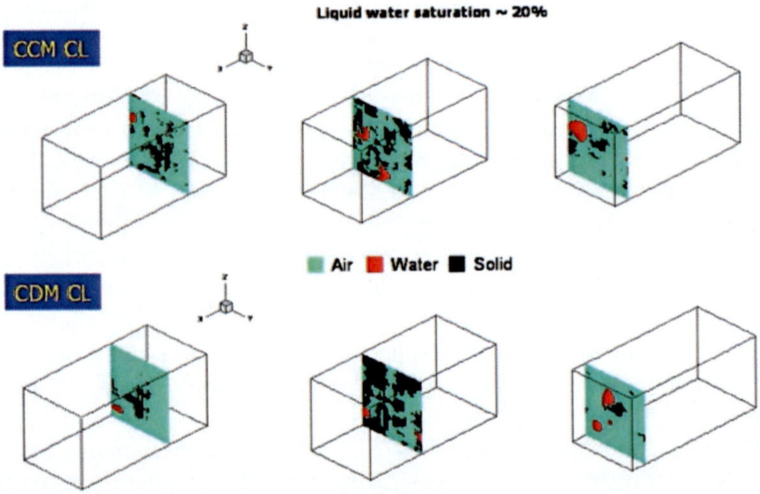

Fig. 7.21 Two-dimensional maps of liquid water distributions in the CCM and CDM CLs

of the underlying structure and wetting characteristics of the CL on liquid water distribution and dynamics which would subsequently influence the relative phase mobility and transport.

5.3.1 Capillary Pressure–Saturation Relation

The *capillary pressure* response can be estimated as a direct manifestation of the underlying pore morphology of the CL microstructures from the aforementioned two-phase primary drainage simulations. Figure 7.22 shows the capillary pressure relation as a function of liquid water saturation for the CCM and CDM CL

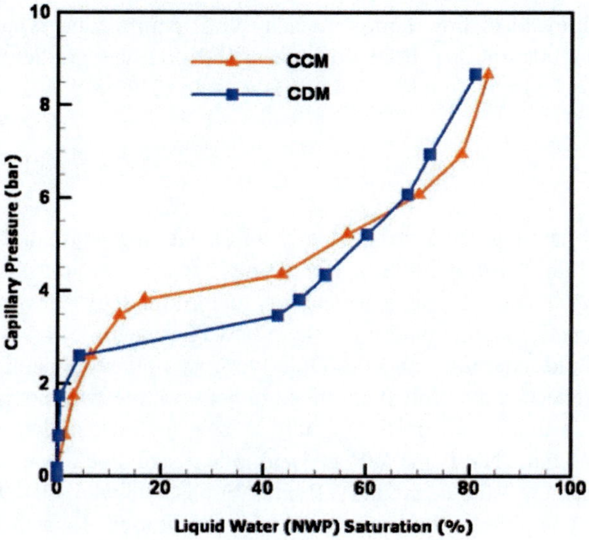

Fig. 7.22 Capillary pressure vs. liquid water saturation for the CCM and CDM CLs

microstructures [32]. The overall shape of the capillary pressure curve agrees well with those reported in the literature for synthetic porous medium [62]. The capillary pressure–saturation curve exhibits a non-zero entry pressure for the NWP to initiate invading into the WP-saturated CL structure characterized by slightly hydrophobic wetting characteristics. From the plot, it is evident that the initial entry pressure for the NWP to initiate invading into the WP-saturated domain is relatively higher for the CDM CL as compared to the CCM CL. This could be due to the higher open pore volume fraction in the vicinity of the NWP reservoir for the CCM CL with respect to the CDM CL as can be observed from the respective pore volume fraction distributions. Another point to be noted is that for the CCM CL, the NWP saturation increases gradually with increase in capillary pressure till around 20%, after which NWP saturation jumps to around 45% with negligible increase in capillary pressure. The CDM CL, however, exhibits a rapid jump in NWP saturation from around 5 to 45% with negligible increase in capillary pressure. This behavior is again closely linked to the underlying structures of the respective CLs and this observation is in accordance with the higher fraction of available pore space for the invading NWP in the CDM CL. Beyond 45% saturation, the CDM CL exhibits a steeper capillary pressure-saturation gradient as compared to the CCM CL. However, both the CLs exhibit quite similar residual WP saturation of around 18%.

5.3.2 Relative Permeability–Saturation Relation

The steady-state flow experiment in a porous medium is ideally suited for evaluating the *relative permeability* relations for the two phases involved in the underlying transport. Briefly, in the steady-state flow experiment two immiscible fluids are allowed to flow simultaneously until equilibrium is attained and the corresponding saturations, fluid flow rates, and pressure gradients can be directly measured and correlated using Darcy's law, defined below:

$$v_i = -\frac{\kappa \kappa_{ri}}{\mu_i}(\nabla p_i - \rho_i g) \tag{7.17}$$

where v_i is the Darcy velocity for the wetting phase and non-wetting phase, p_i is the fluid phase pressure, $\rho_i g$ is the body force, μ_i is the dynamic viscosity of the fluid, κ is the intrinsic permeability determined by the pore structure of the porous medium alone, and κ_{ri} is the relative permeability of each phase that depends upon fluid saturation and the underlying two-phase dynamics. The term "steady state" reflects a dynamical equilibrium between the two moving and macroscopically stable fluids [66]. In the numerical experiment, detailed elsewhere [32, 63], initially, both the NWP and WP are randomly distributed throughout the CL porous structure such that the desired NWP saturation is achieved. Counter-current flow is simulated by applying a body force for both the phases along the flow direction which mimics the redistribution of the initial phases under the capillary force corresponding to the typical capillary number (Ca $\sim 10^{-6}$). Once steady state is achieved, the flux of each phase is calculated. The corresponding absolute flux is calculated by modifying the two-phase LB model, where a body force is applied to one phase and the density of the other phase is rendered zero at all locations. Finally, the ratio of flux of each phase from the two-phase calculation to the one obtained from the single-phase calculation gives the relative permeability related to the saturation level.

The phase distributions from such two-phase simulations is already shown in Fig. 7.20 for the reconstructed CCM and CDM CL microstructures. Figure 7.23 shows the NWP and WP relative permeability as functions of liquid water (NWP) saturation for the CCM and CDM CL microstructures [32]. It can be observed that the liquid water relative permeability values below 10% saturation is negligibly small due to the non-existence of a connected pathway for transport apropos of the 3-D liquid water distributions in Fig. 7.20. In general it is observed that at this low Ca the relative permeability curves exhibit non-linear relationships for both phases between the flow rate and the driving pressure gradient and this observation is in agreement with the results reported by Li et al. [63] for geologic porous samples. This observation further emphasizes that for the capillary number of interest for fuel cell operation, i.e., Ca $\sim 10^{-6}$, the relative permeability relations for both phases will exhibit strong non-linearity. The relative permeability–saturation relations can be further deployed in the two-phase computational fuel cell dynamic models for reliable transport predictions.

5.4 Effect of Liquid Water on Electrode Performance

The effect of liquid water on the electrochemical performance manifests in terms of coverage of the electrochemically active area in the CL leading to reduced catalytic activity and blockage of the porous pathways in the CL rendering hindered

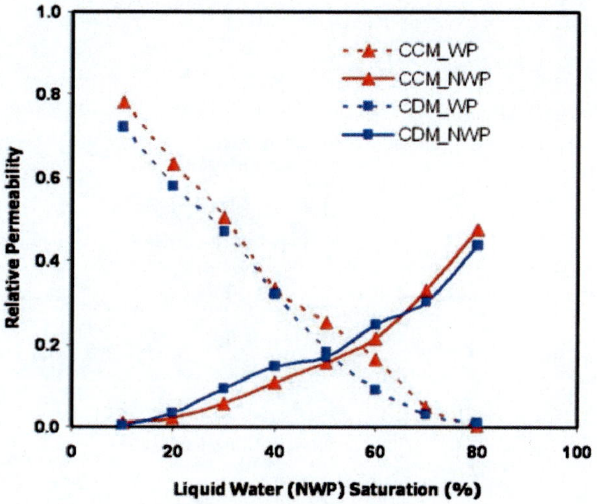

Fig. 7.23 Relative permeability vs. liquid water saturation relations for the CCM and CDM CLs

oxygen transport to the active reaction sites. In the macroscopic fuel cell models, where the CL is treated as a macrohomogeneous porous layer, the site coverage and pore blockage effects owing to liquid water are taken into account through an electrochemical area reduction relation and the Bruggeman-type correction for the effective oxygen diffusivity, respectively. These two empirical correlations cannot be separately discerned through experimental techniques.

A computational approach coupling the two-phase LB model for liquid water transport and the DNS model for species and charge transport can be employed to quantify the site coverage and pore blockage effects in the PEFC electrode. At steady state, 3-D liquid water distributions can be obtained in the CL microstructure, representatively shown earlier in Fig. 7.20, from the two-phase LB simulations. From the liquid water distributions within the CL structure, the information about the catalytic site coverage effect can be extracted directly. The DNS model can be deployed subsequently on the liquid water-blocked CL structure pertaining to a saturation level for the evaluation of the hindered oxygen transport. Figure 7.24 schematically shows the coupled modeling framework.

The reduction in electrochemically active interfacial area (ECA) owing to liquid water coverage can be estimated from the 2-D saturation maps, resulting from the liquid water distribution available from the two-phase LB simulation corresponding to a saturation level, and subsequently a correlation between the effective ECA and the liquid water saturation can be established as the following [32, 67]:

$$\mathrm{ECA}^{\mathrm{eff}} = \mathrm{ECA}(1 - S_{\mathrm{r}})^{c} \tag{7.18}$$

where S_{r} is the liquid water saturation and c the site coverage factor.

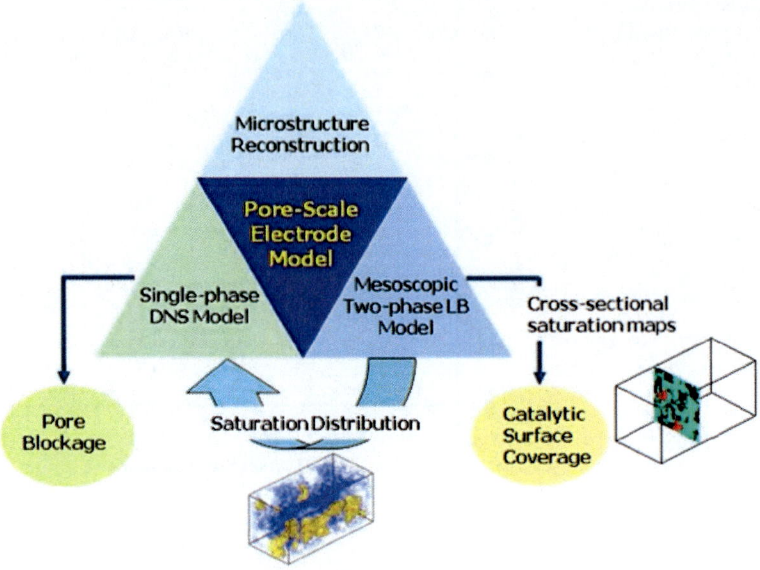

Fig. 7.24 DNS and LBM coupling schematic diagram

Table 7.2 Catalytic site coverage factor

	CCM CL	CDM CL
Surface coverage factor (c)	1.05	1.2

Using the 2-D saturation maps from the two-phase LB simulation for the CL microstructures, representatively illustrated in Fig. 7.21, the relation for the effective ECA as a function of liquid water saturation can be evaluated according to (7.18). Based on several liquid water saturation levels, the catalytic surface coverage factors for the CCM and CDM microstructures are estimated and furnished in Table 7.2, which can be used as valuable inputs to macroscopic two-phase fuel cell models [32, 67]. Figure 7.25 shows the variation of the effective ECA with liquid water saturation from the evaluated correlation along with the typical correlations with ad hoc fitting of the coverage parameter otherwise used in the macroscopic fuel cell models [8]. It is to be noted that the effect of liquid water is manifested via a reduction of the active area available for electrochemical reaction through the coverage parameter. In the mass transport control regime, the hindered oxygen transport owing to liquid water blockage, described in the following section, becomes the limiting factor.

In order to evaluate the effect of pore volume blockage in the presence of liquid water causing hindered oxygen transport to the active reaction sites, the DNS model [30–32] is deployed. From the 3-D liquid water signature obtained from the two-phase LB simulations in the CL, representatively shown in Fig. 7.20, the

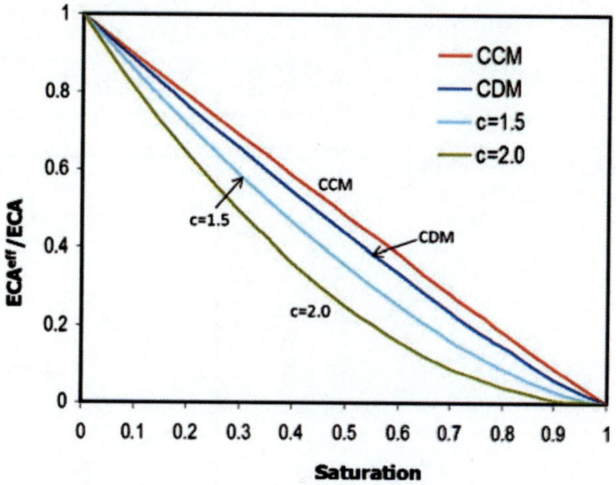

Fig. 7.25 Catalytic site coverage relation as a function of liquid water saturation

pores occupied by liquid water are identified corresponding to a particular saturation level and these pores are rendered as "dead" pores. The liquid water-blocked pores in the modified CL microstructure do not take part in the oxygen reduction reaction (ORR) as well as produce extra resistance by reducing the effective porosity of the structure. With this virtual microstructure of the liquid water-blocked electrode, pointwise accurate species and charge conservation equations are solved within the DNS modeling framework. The pore blockage effect is finally evaluated from the oxygen concentration field and can subsequently be correlated in terms of the effective oxygen diffusion coefficient based on the oxygen flux as the following [32, 67]:

$$D_{O_2}^{\text{eff}} = D_{O_2,0} f(\varepsilon_{CL})\, g(S_r) = D_{O_2,0}(\varepsilon_{CL})^m (1 - S_r)^b \tag{7.19}$$

ε_{CL} is the CL porosity, S_r the liquid water saturation, m the Bruggeman factor for the oxygen transport through the unblocked CL microstructure, and b the volume blockage factor representing the extra resistance to oxygen transport in the presence of liquid water in the CL. The effect of the resistance due to the tortuous pore pathways to oxygen transport in the absence of liquid water is evaluated using the DNS model in terms of a Bruggeman factor, m, and is detailed in the DNS modeling section. The pore blockage factors for the CCM and CDM catalyst layers are evaluated and furnished in Table 7.3. It should be noted that for the CDM CL microstructure the same value of the Bruggeman factor of $m = 3.5$ is assumed in this study. These estimates could prove to be valuable inputs for more accurate representation of the pore blockage effect in the macroscopic two-phase fuel cell models. Figure 7.26 shows the variation of the effective oxygen diffusivity with liquid water saturation from the correlation along with the typical macrohomogeneous correlation with $m = 1.5$ and $b = 1.5$ otherwise used arbitrarily in the macroscopic fuel cell modeling literature.

Table 7.3 Pore blockage factor

	CCM CL	CDM CL
Pore blockage factor (b)	1.97	2.12

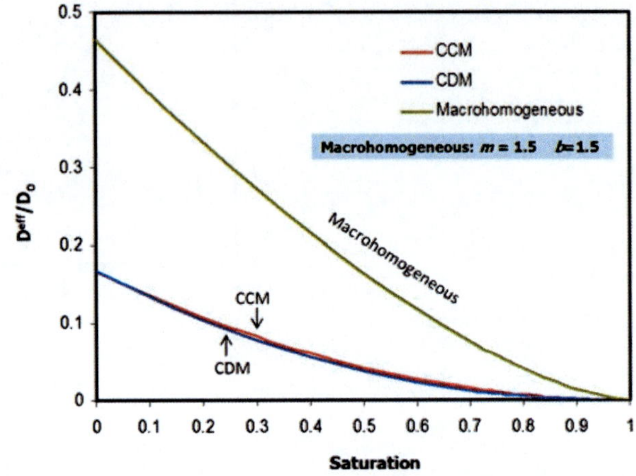

Fig. 7.26 Pore blockage relation as a function of liquid water saturation

The derogatory influence of liquid water on the electrode voltage loss comes from the impeded oxygen transport and reduced electrochemically active area as explained above which can be described by the electrochemical kinetics in terms of the reaction current density, j, through the Tafel equation, given by (7.12). The Tafel equation instead includes, a^{eff}, which represents the effective ECA due to the catalytic site coverage effect.

The pore blockage effect comes into play through the oxygen concentration, c_{O_2}, distribution given by the following equation:

$$\nabla \cdot \left(D_{O_2}^{\text{eff}} \nabla c_{O_2} \right) = \frac{j}{4F} \tag{7.20}$$

With the evaluated site coverage and pore blockage correlations for the effective ECA and oxygen diffusivity, respectively, and the intrinsic active area available from the reconstructed CL microstructure, the electrochemistry-coupled species and charge transport equations can be solved with different liquid water saturation levels within the 1-D macrohomogeneous modeling framework [30, 32], and the cathode overpotential, η, can be estimated.

Figure 7.27 exhibits the polarization curves in terms of the cathode overpotential variation with current density for the CCM CL obtained from the 3-D DNS model prediction, the experimental observation, and the liquid water transport-corrected 1D macrohomogeneous model. In Fig. 7.27, "DNS" refers to the polarization curve predicted by the single phase, electrochemistry-coupled DNS model for the reconstructed CCM CL microstructure, whereas "coverage/blockage" refers to the performance curve predicted by the 1-D macrohomogeneous model with correction for liquid water transport taken into account via the correlations for pore blockage

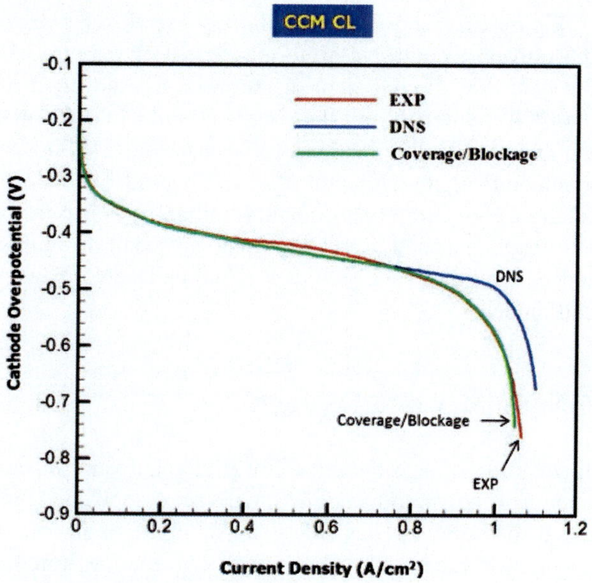

Fig. 7.27 Performance curves for the CCM CL

and surface coverage effects. It can be observed that the estimated catalytic site coverage and pore blockage parameters for the CCM CL from the combined two-phase LB model and the DNS model can indeed capture the transport-limiting regime and agrees well with the experimental data.

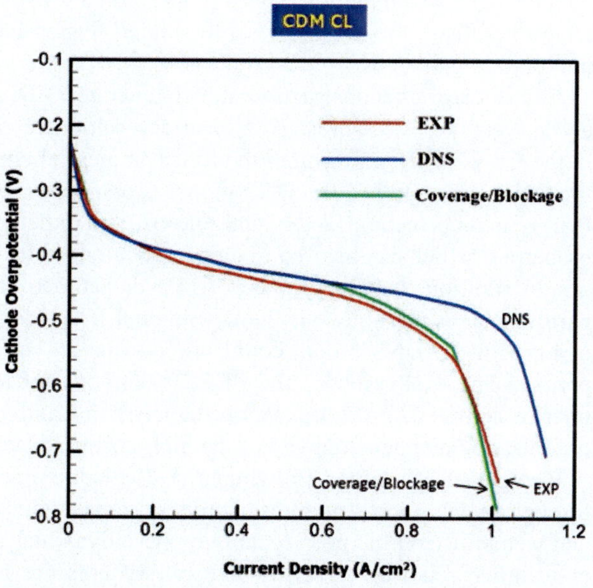

Fig. 7.28 Performance curves for the CDM CL

Figure 7.28 shows the similar polarization curves for the CDM CL. Although the transport control regime is adequately captured in the CDM CL overpotential behavior, the apparent discrepancies could be attributed to the lower value of the ECA, namely 30, in the reconstructed CDM CL microstructure as compared to the experimentally observed value from the CV data (e.g., 39) and also due to the assumed Bruggeman factor of $m = 3.5$ similar to the CCM CL in the pore blockage relation. However, the principal highlight of this pore-scale investigation is that the systematic estimation of the effective transport parameters for the porous CL can indeed quantitatively predict the fuel cell performance from the macroscopic fuel cell models.

6 Summary and Outlook

In the midst of a global effort toward a secure and sustainable energy future, electrochemical energy conversion systems are perceived to play a key role and the polymer electrolyte fuel cell has emerged as a promising power source. Despite tremendous progress in recent years, a pivotal performance limitation in the PEFC comes from the underlying competing transport mechanisms in the catalyst layer due to the sluggish oxygen reduction reaction as well as transport limitation in the presence of liquid water and flooding phenomena. Computational modeling has been extensively employed at different levels of complexities to study fuel cell transport and performance. However, the macroscopic fuel cell models cannot address the effects of the underlying complex pore morphology of the CL. In this chapter, we discuss the development of a comprehensive mesoscopic modeling framework, comprising a stochastic microstructure reconstruction model, an electrochemistry-coupled direct numerical simulation (DNS) model, and a two-phase lattice Boltzmann modeling formalism in order to reveal the underlying structure–transport–performance interplay.

The stochastic reconstruction model generates 3-D, statistically meaningful catalyst layer microstructures. Pore-level description of charge and species transport in the CL is achieved through the direct numerical simulation (DNS) model. The predictive capability of the DNS model is demonstrated with the evaluation of the Bruggeman correction factors and tortuosity parameters for the effective transport properties, which can be used as closure relations in the macroscopic fuel cell models. In addition, the importance of bilayer cathode catalyst layers on the PEFC performance is presented via the DNS model. The stochastic reconstruction method along with the DNS model could prove to be an effective screening tool for the performance evaluation of the PEFC electrodes reflecting the interaction of the microstructure with the transport characteristics and hence would aid toward the optimization and development of the high-performance electrodes.

The two-phase lattice Boltzmann (LB) model simulates liquid water transport through the CL microstructures in order to gain insight into the influence of structure and wettability on the pore-scale two-phase dynamics and evaluates the two-phase constitutive relations in terms of capillary pressure and relative permeability as functions of liquid water saturation. A quantitative estimate of the detrimental consequence of liquid water transport in the CL on the cell performance in terms of the pore blockage and catalytic site coverage effects is predicted by combining LB and DNS models. In the dearth of two-phase correlations for the PEFC electrode, the two-phase transport parameters evaluated from such pore-scale study could be adapted into two-phase computational fuel cell dynamic models for more reliable performance predictions.

Finally, this chapter aims to emphasize the need for the development of a predictive mesoscopic modeling framework, which will not only foster enhanced understanding of the complex structure–transport interlinks but also enable virtual microstructure design for high-performance PEFC electrodes.

Acknowledgments Financial support from Los Alamos National Laboratory LDRD Program to PPM for the Director's Fellowship (2008–2009) and UC Lab Fees research project UCD-09-15 is gratefully acknowledged. The authors acknowledge Elsevier and Electrochemical Society for the figures reproduced in this chapter from the referenced publications of their respective journals.

References

1. M. Winter and R. J. Brodd, *Chem. Rev.*, **104**, 4245 (2004).
2. S. Gottesfeld and T. A. Zawodzinski, in *Advances in Electrochemical Science and Engineering*, R. C. Alkire, H. Gerischer, D. M. Kolb, and C. W. Tobias, Editors, Vol. 5, p. 195, Wiley, New York, NY (1997).
3. Private communication with F. Q. Liu and C. Y. Wang, Electrochemical Engine Center (ECEC), Pennsylvania State University.
4. K. L. More and K. S. Reeves, *DOE Hydrogen Program Annual Merit Review Proceedings*, Arlington, VA, USA, May 23–26 (2005).
5. T. E. Springer and S. Gottesfeld, in *Modeling of Batteries and Fuel Cells*, R. E. White, Editor, p. 197, The Electrochemical Society Proceedings, The ECS Publications (1991).
6. M. L. Perry, J. Newman, and E. J. Cairns, *J. Electrochem. Soc.*, **145**, 5 (1998).
7. M. Eikerling and A. A. Kornyshev, *J. Electroanal. Chem.*, **453**, 89 (1998).
8. C. Y. Wang, *Chem. Rev.*, **104**, 4727 (2004).
9. A. Z. Weber and J. Newman, *Chem. Rev.*, **104**, 4679 (2004).
10. S. Um, C. Y. Wang, and K. S. Chen, *J. Electrochem. Soc.*, **147**, 4485 (2000).
11. S. Um and C. Y. Wang, *J. Power Sources*, **125**, 40 (2004).
12. S. Dutta, S. Shimpalee, and J. W. Van Zee, *J. Appl. Electrochem.*, **30**, 135 (2000).
13. S. Dutta, S. Shimpalee, and J. W. Van Zee, *Int. J. Heat Mass Transfer*, **44**, 2029 (2001).
14. T. Berning, D. M. Lu, and N. Djilali, *J. Power Sources*, **106**, 284 (2002).
15. S. Mazumder and J. V. Cole, *J. Electrochem. Soc.*, **150**, A1503 (2003).
16. M. Sahimi, *Application of Percolation Theory*, Taylor & Francis, Bristol, PA (1994).
17. M. S. Wilson, U.S. Pat. No. 5, 234, 777 (1993).
18. M. S. Wilson, J. A. Valerio, and S. Gottesfeld, *Electrochim. Acta*, **40**, 355 (1995).
19. Y.-G. Chun, C.-S. Kim, D.-H. Peck, D.-R. Shin, *J. Power Sources*, **71**, 174 (1998).
20. S. Litster and G. McLean, *J. Power Sources*, **130**, 61 (2004).
21. M. J. Kwiecien, I. F. Macdonald, and F. A. L. Dullien, *J. Microsc.*, **159**, 343 (1990).
22. D. P. Lymberopoulos and A. C. Payatakes, *J. Colloid Interface Sci.*, **150**, 61 (1992).
23. P. Spanne, J. F. Thovert, J. C. Jacquin, W B. Lindquist, K. W. Jones, and P. M. Adler, *Phys. Rev. Lett.*, **73**, 2001 (1994).
24. C. A. Baldwin, A. J. Sederman, M. D. Mantle, P. Alexander, and L. F. Gladden, *J. Colloid Interface Sci.*, **181**, 79 (1996).

25. J. T. Fredrich, B. Menendez, and T. F. Wong, *Science*, **268**, 276 (1995).
26. M. Joshi, Ph.D. Dissertation, University of Kansas, Lawrence, Kansas (1974).
27. J. A. Quiblier, *J. Colloid Interface Sci.*, **98**, 84 (1984).
28. P. M. Adler and J. F. Thovert, *Appl. Mech. Rev.*, **51**, 537 (1998).
29. S. Torquato, *Annu. Rev. Mater. Res.*, **32**, 77 (2002).
30. P. P. Mukherjee and C. Y. Wang, *J. Electrochem. Soc.*, **153**, A840 (2006).
31. P. P. Mukherjee and C. Y. Wang, *J. Electrochem. Soc.*, **154**, B1121 (2007).
32. P. P. Mukherjee, Ph.D. Dissertation, Pennsylvania State University, University Park, PA (2007).
33. P. M. Adler, *Porous Media: Geometry and Transports*, Butterworth-Heinemann, Stoneham (1992).
34. P. M. Adler, C. G. Jacquin, and J. A. Quiblier, *Int. J. Multiphase Flow*, **16**, 691 (1990).
35. D. P. Bentz and N. S. Martys, *Transport in Porous Media*, **17**, 221 (1995).
36. J. G. Berryman, *J. Appl. Phys.*, **57**, 2374 (1985).
37. F. A. L. Dullien, *Porous Media: Fluid Transport and Pore Structure*, Academic, San Diego, CA (1992).
38. J. Bear, *Dynamics of Fluids in Porous Media*, Dover, New York, NY (1972).
39. S. S. Kocha, in *Handbook of Fuel Cells—Fundamentals, Technology and Applications*, W. Vielstich, A. Lamm, and H. A. Gasteiger, Editors, Ch. 43, Vol. 3, p. 538, Wiley, Chichester (2003).
40. T. E. Springer, T. A. Zawodzinski, and S. Gottesfeld, *J. Electrochem. Soc.*, **138**, 2334 (1991).
41. N. Epstein, *Chem. Eng. Sci.*, **44**, 777 (1989).
42. A. Koponen, M. Kataja, and J. Timonen, *Phys. Rev. E*, **54**, 406 (1996).
43. M. Uchida, Y. Fukuoka, Y. Sugawara, H. Obara, and A. Ohta, *J. Electrochem. Soc.*, **145**, 3708 (1998).
44. E. Passalacqua, F. Lufrano, G. Squadrito, A. Patti, and L. Giorgi, *Electrochim. Acta*, **43**, 3665 (1998).
45. Z. Qi and A. Kaufman, *J. Power Sources*, **113**, 37 (2003).
46. G. Sasikumar, J. W. Ihm, and H. Ryu, *J. Power Sources*, **132**, 11 (2004).
47. D. Song, Q. Wang, Z. Liu, T. Navessin, and S. Holdcroft, *Electrochim. Acta*, **50**, 731 (2004).
48. R. Lenormand, E. Touboul, and C. Zarcone, *J. Fluid Mech.*, **189**,165 (1988).
49. R. P. Ewing and B. Berkowitz, *Adv. Water Res.*, **24**, 309 (2001).
50. X. G. Yang, F. Y. Zhang, A. Lubawy, and C. Y. Wang, *Electrochem. Solid-State Lett.*, **7**, A408 (2004).
51. S. Chen and G. Doolen, *Ann. Rev. Fluid Mech.*, **30**, 329 (1998).
52. A. K. Gunstensen, D. H. Rothman, S. Zaleski, and G. Zanetti, *Phys. Rev. A*, **43**, 4320 (1991).
53. X. Shan and H. Chen, *Phys. Rev. E*, **47**, 1815 (1993).
54. X. Shan and H. Chen, *Phys. Rev. E*, **49**, 2941 (1994).
55. M. R. Swift, W. R. Osborn, and J. M. Yeomans, *Phys. Rev. Lett.*, **75**, 830 (1995).
56. X. Y. He, S. Y. Chen, and R. Y. Zhang, *J. Comput. Phys.*, **152**, 642 (1999).
57. P. Bhatnagar, E. Gross, and M. Krook, *Phys. Rev.*, **94,** 511 (1954).
58. Q. Kang, D. Zhang, and S. Chen, *Phys. Fluids*, **14**, 3203 (2002).
59. Q. Kang, D. Zhang, and S. Chen, *J. Fluid Mech.*, **545**, 41 (2005).
60. X. Shan and G. D. Doolen, *Phys. Rev. E*, **54**, 3614 (1996).
61. D. Tiab and E. Donaldson, *Petrophysics: Theory and Practice of Measuring Reservoir Rock and Transport Properties*, Gulf Publishing Company, Houston (1996).
62. C. Pan, M. Hilpert, and C. T. Miller, *Water Resour. Res.*, **40**, 1 (2004).
63. H. Li, C. Pan, and C. T. Miller, *Phys. Rev. E*, **72**, 026705 (2005).
64. P. P. Mukherjee, P. K. Sinha, and C. Y. Wang, *J. Mater. Chem.*, **17**, 3089 (2007).
65. P. P. Mukherjee, C. Y. Wang, and Q. Kang, *Electrochim. Acta*, **54**, 6861 (2009).
66. D. G. Avraam and A. C. Payatakes, *Transp. Porous Media*, **20**, 135 (1995).
67. P. P. Mukherjee, and C. Y. Wang, *ECS Trans.*, **3**, 1085 (2006).

Index

Note: The letters 'f' and 't' following locators refer to figures and tables respectively.

S. Chakraborty (ed.), *Mechanics Over Micro and Nano Scales*,
DOI 10.1007/978-1-4419-9601-5,